IGNORANCE UNMASKED

IGNORANCE UNMASKED

*Essays in
the New Science
of Agnotology*

Edited by Robert N. Proctor
and Londa Schiebinger

Stanford University Press
Stanford, California

Stanford University Press
Stanford, California

We are grateful for support for this work from the Patrick Suppes Center for the History and Philosophy of Science, the Stanford Humanities Center, and the Department of History, Stanford University. A special thanks for all their assistance to Rosemary Rogers and Julia Fine.

Library of Congress Cataloging-in-Publication Data
Names: Proctor, Robert, 1954– editor. | Schiebinger, Londa L., editor.
Title: Ignorance unmasked : essays in the new science of agnotology / edited by
    Robert N. Proctor and Londa Schiebinger.
Description: Stanford, California : Stanford University Press, [2025] | Includes
    bibliographical references and index.
Identifiers: LCCN 2025003036 (print) | LCCN 2025003037 (ebook) |
    ISBN 9781503643406 (cloth) | ISBN 9781503643956 (paperback) |
    ISBN 9781503643963 (ebook)
Subjects: LCSH: Ignorance (Theory of knowledge)
Classification: LCC BD221 .I46 2025 (print) | LCC BD221 (ebook) | DDC
    121—dc23/eng/20250428
LC record available at https://lccn.loc.gov/2025003036
LC ebook record available at https://lccn.loc.gov/2025003037

Cover design and illustration: Jan Šabach

The authorized representative in the EU for product safety and compliance is: Mare Nostrum Group B.V. | Mauritskade 21D | 1091 GC Amsterdam | The Netherlands | Email address: gpsr@mare-nostrum.co.uk | KVK chamber of commerce number: 96249943

Contents

Preface   vii
ROBERT N. PROCTOR

1 Agnotology, Thirty Years in the Making   1
ROBERT N. PROCTOR

2 How the Most Important Fact of Global   23
Warming Has Been Obscured
BENJAMIN FRANTA

3 Preventing Unwanted Births Can Help Mitigate   42
Climate Change—While Enlarging Human Liberties
LONDA SCHIEBINGER AND ROBERT N. PROCTOR

4 AI-Fueled Ignorance, Confusion, and Profit   57
HANY FARID

5 On Data Loss and Disappearance in Digital Societies   71
NANNA BONDE THYLSTRUP

6 Law against Knowledge: Anti-Epistemology   85
PETER GALISON

7 Why We Wrongly Imagine Adam Smith   100
as a Free Market Fundamentalist
NAOMI ORESKES AND ERIK M. CONWAY

8 Gun-Lobby Agnotology: Degrading the Truth about Firearms   114
JOHN J. DONOHUE

9   Gluttony and Sloth?   133
    Personal Responsibility versus the True Cause of Obesity
    ROBERT H. LUSTIG

10 How Big Meat Has Created and Legitimized Ignorance   147
    JENNIFER JACQUET

11 On the Burial of the Palestinian Nakba   160
    ROSEMARY SAYIGH

12 Euphemism in the Architecture and Language of Treblinka   171
    DANIEL AKSELRAD

13 Hiram Powers, Black Agnotology, and Segregated Art History   187
    CAROLINE A. JONES

14 "Civilian" Ignorance, American Militarism,   209
    and the Post-9/11 Wars
    NADIA ABU EL-HAJ

    Contributors   223

    Notes   227

    Index   287

Preface

ROBERT N. PROCTOR

AGNOTOLOGY EXPLORES THE UNSEEN, the hidden, the invisible, the ignored, the neglected, the overlooked—but also the suppressed, the silenced, the trivialized, the euphemized. Ignorance can be power, but it can also be a consumer product and corporate fetish. It can be weaponized and deployed; it can crush and coerce.

When most people think about ignorance, they think about *personal* ignorance, what any one of us may or may not know, for better or for worse. Our focus here is rather on ignorance as a *social* product—how ignorance is *produced*. How has the most important fact about global warming been obscured? How has artificial intelligence fueled a world of confusion and giga-scale profits? What causes obesity, and why has it so often been conceived in terms of gluttony and sloth? We explore how mis- and disinformation get spread, how truths about firearms get bent out of shape, and how laws get passed that block the creation of knowledge. We look at why scholars overlook some of the most obvious solutions to the global climate crisis and how catastrophes like the Holocaust or the Nakba get euphemized or lost to memory. And then there's the missed opportunity of forgetting the link between human reproductive freedom and preservation of biodiversity. Our focus throughout is on theorizing ignorance but also on how many of our looming catastrophes stem from failures of the imagination, including failure to challenge root causes.

Our first volume, *Agnotology: The Making and Unmaking of Ignorance*, appeared in 2008, exploring social pathologies like climate denial and cigarette chicanery but also white ignorance, the balance routine in journalism (false objectivity), and the erasure of the clitoris in medical texts ("the epistemology of orgasm"). We looked at how archaeology can involve the destruction of history and why paleontologists have ignored indigenous fossil knowledge. We looked at how abortive techniques from the New World were lost and forgotten and how a vast realm of scientific knowledge remains hidden behind a veil of military secrecy.

Since publication of that volume, we've seen an explosion of interest in ignorance following the rise of election denialism, vaccine hesitancy, climate pseudo-solutionism, and (especially) the rise of industrial-scale campaigns of mis- and disinformation, made possible by ubiquitous social media and artificial intelligence (AI). AI is allowing entirely new kinds of fabrications—think deepfakes and synthetic speech—that pose novel threats to the trust we've always placed in what we see and hear. Seventeenth-century European philosophers valorized "the faith of the eyes" to overcome religious dogma or hoary tradition, but today our eyes and ears have become less sure guides given our reliance on feeds from digital screens. A century of consumer technology has also led us to surround ourselves with novelties that seemed innocent enough to early adopters but proved deadly over decades—think sugar, cigarettes, lead paint, or even greenhouse gases—all of which have lulled us into debilitating complacency, contravening millions of years of evolution: a kind of bodily or sensory agnotology.

Agnotology is the science of ignorance, but this is not a book about ignorance per se—because what we do not know, or cannot know, is infinite. Agnotology is about the *production* and maintenance of ignorance, along with the many tricks and techniques by which things become buried and forgotten, lost and ignored, hidden or invisible. The point is to explore *the means* by which ignorance is created, packaged, and delivered—and how and why it persists. Ignorance through this lens is not an unstructured void and is more than the not yet known; it has a geography and, very often, agents who profit from its augmentation. But much of ignorance making is also inadvertent in the sense that once created, it can endure as part of ordinary culture or even what gets euphemized as education. Especially in the academy, ignorance is also generated by siloing, by miseducation, by "following the money," and by uncritically using terms that contain and propagate lies. Words are often condensed ideologies, micropoetic deceptions. Agnotology in this sense has a lot of overlap with marketing and political propaganda, insofar as they all may involve persuasion without regard for truth, or consumption without regard for quality.

Readers should think of this volume not as a finished text but rather as an invitation to explore how ignorance is made and maintained in places and ways we've never imagined. Too often we imagine that knowledge is the center of the universe, when it is really just one of many planets orbiting a radiant but unheeding sun within a universe almost all of which will forever remain unknown. Even here on Earth, ignorance is more the norm, and deviance the exception. Hippocrates observed that while life is short, the art (*techne*) is long—but that can be wishful thinking given that knowledge is so easily lost and forgotten. Agnotology is a kind of complement to what philosophers know as epistemology, much as the study of crime can complement the study of law, or disease the study of health. Agnotology is still in its infancy, however, and it's in this spirit of invitation to discovery that we've decided to publish these essays. We look forward to others joining in the adventure!

––––––

PS: Proofs for this book arrived in the early stages of Donald Trump's assault on science and the arts, following the playbook of the Heritage Foundation's Project 2025. And while it remains to be seen how far we will be plunged into this new Age of Endarkenment, we've already witnessed significant harms to American scientific institutions, especially in the areas of health research and climate and environmental science. Algorithms have been used to sweep through grant applications, in a hunt to root out "woke" or "gender ideology." And prestigious universities are having their funding withheld to force changes in hiring, teaching, and permissible campus speech. And to demand a kind of farcical "viewpoint diversity"—as if geologists should balance established science with flat-Earth follies, or biologists should entertain the foolishness of creationism.

It should not have been hard to see this coming, given that our current president lives in the superficial world of casinos, beauty contests, and Ultimate Fighting Championships—with some of his strongest support coming from Christian evangelicals and a cabal of corporate miscreants. It can be dangerous to ignore the political power of ignorance, but it is equally dangerous to ignore how easily scholars can be captured by ideologies or corporate connivance. Good history is about learning to see the unseen, and these essays are intended to help us diagnose and dismantle some of our more malignant distractions—and threats to human liberty and planetary health.

May 1, 2025

ONE

Agnotology, Thirty Years in the Making

ROBERT N. PROCTOR

Ignorance is power.
—George Orwell, *1984*

IT HAS BEEN THIRTY years since I asked the linguist Iain Boal to help me coin a word to designate the social production of ignorance and the broader study of ignorance.[1] I needed it for a book I was writing on the history of cancer causation (*Cancer Wars*) and included our neologism in a chapter titled "Doubt Is Our Product," bouncing off Brown & Williamson's secret memo from 1969, which reveals how cigarette makers thought they could only keep selling cigarettes so long as they also kept manufacturing ignorance.[2] The thrust of that book was that we already know enough to prevent most cancers, but powerful political forces prevent us from acting intelligently (then as now, by the way). I had been worrying about ignorance since coming to Harvard as a graduate student in 1976; with Ruth Hubbard and Richard Lewontin I was teaching Biology and Social Issues, which looked at the militarization of science, the racialist history of science, the promise of a feminist science, the politics of food and agriculture, and other topics, including the cultural causes of cancer (via David Ozonoff). Michael Oppenheimer was our go-to guy on global warming, and we focused a lot on extinction, in light of Stephen Jay Gould's work on punctuated equilibrium, but also the prospect of nuclear annihilation given Ronald Reagan's militarization of space.

I was disturbed to find how obsessed my fellow historians of science were with the likes of Darwin, Newton, and Einstein—or Hobbes and Boyle—paying little attention to popular knowledge and *popular ignorance*. I'd grown up in southern Texas, where many members of my family were creationists; I'd worried a lot about the growing political power of Evangelical Christianity, and I was stunned to find how little interest there was at Harvard in such topics. The

1

presumption seemed to be that no one of any significance could still hold on to such beliefs—and why would anyone want to study a bunch of backwater hicks and hillbillies? I came to regard Harvard as a rather parochial place where some of the most interesting topics were being ignored. As for ignoring ignorance, I remember thinking that epistemology without agnotology is like the study of medicine without the study of disease or the study of law without the study of crime.

THE CIGARETTE SPUR

In the annals of agnotology, 1981 was an important turning point. I was studying in Berlin and noticed that several of my older professors were former Nazis, so there's that. The "Doubt is our product" memo was also leaked that year. The Federal Trade Commission was trying to organize a lawsuit against Big Tobacco and had obtained several of their internal documents; "Doubt" was one of the first to see the light of day. I had been working on the corruption of science by Big Food and Big Pharma—and the Nazis—and Harvard was also getting grief for being on the payroll of Big Tobacco (fig. 1.1). Three of my four grandparents had died from cigarettes, so I was primed.

There are other reasons, though, that cigarettes play a special role in agnotology. Cigarettes remain the world's leading preventable cause of death, but that gargantuan fact is virtually ignored in popular culture, where shark

FIGURE 1.1. **Harvard as a carcinogen in a Harvard medical student newsletter from the 1970s. Radical medical students in the 1970s protested the university taking millions of dollars from cigarette makers, who privately boasted that the provision of such funds yielded a "public relations plus" for the industry.**

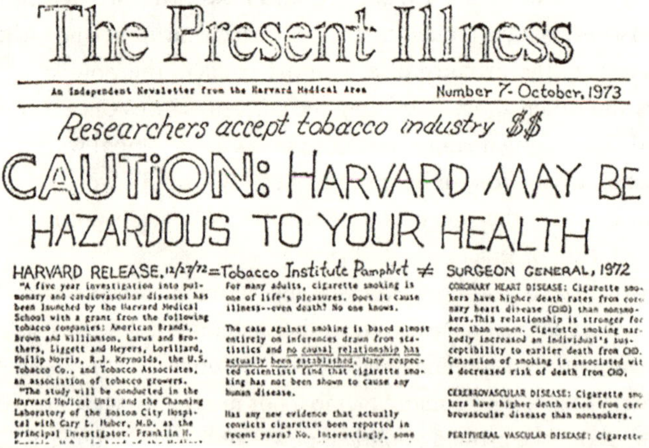
The Present Illness

An Independent Newsletter from the Harvard Medical Area — Number 7- October, 1973

Researchers accept tobacco industry $$

CAUTION: HARVARD MAY BE HAZARDOUS TO YOUR HEALTH

HARVARD RELEASE, 12/27/72 = Tobacco Institute Pamphlet ≠ SURGEON GENERAL, 1972

"A five year investigation into pulmonary and cardiovascular diseases has been launched by the Harvard Medical School with a grant from the following tobacco companies: American Brands, Brown and Williamson, Larus and Brothers, Liggett and Meyers, Lorillard, Philip Morris, R.J. Reynolds, the U.S. Tobacco Co., and Tobacco Associates, an association of tobacco growers.

"The study will be conducted in the Harvard Medical Unit and the Channing Laboratory of the Boston City Hospital with Gary L. Huber, M.D. as the principal investigator. Franklin H.

For many adults, cigarette smoking is one of life's pleasures. Does it cause illness--even death? No one knows.

The case against smoking is based almost entirely on inferences drawn from statistics and no causal relationship has actually been established. Many respected scientists find that cigarette smoking has not been shown to cause any human disease.

Has any new evidence that actually convicts cigarettes been reported in recent years? No. Interestingly, some

CORONARY HEART DISEASE: Cigarette smokers have higher death rates from coronary heart disease (CHD) than nonsmokers. This relationship is stronger for men than women. Cigarette smoking markedly increased an individual's susceptibility to earlier death from CHD. Cessation of smoking is associated wit a decreased risk of death from CHD.

CEREBROVASCULAR DISEASE: Cigarette smokers have higher death rates from cerebrovascular disease than nonsmokers.

PERIPHERAL VASCULAR DISEASE: Cigarette

Source: The Present Illness, Oct. 1973, Truth Tobacco Archive, tid lmnd003.

attacks get more attention. Cigarette makers have effectively invisibilized the epidemic by decomposing it into a set of "personal choices." They've also been very good at using science to produce ignorance, and we'll come back to that, but the cigarette itself is also deceptive in how it causes injury given that (a) cause and effect are displaced in time—the so-called lag or latency effect—and (b) it is the very *mildness* of smoke from cigarettes that makes them far more dangerous than, say, traditional cigars—a kind of sensory or bodily agnotology. That's all in addition to the fact that cigarette chicanery has provided a template or playbook for other kinds of corporate deception.

The year 1981 is also significant because it's the peak of the cigarette epidemic in the United States, with more than 630 billion cigarettes smoked that year. Smoking was still allowed indoors at Harvard at that point, and I'll never forget suffering through seminars in the windowless rooms of the Science Center with nary a protest. This is also a time when cigarette makers were still waging their war against the idea that cigarettes might cause cancer. Helmut Wakeham, head of research at Philip Morris, was the brains behind the company's Science Symposia (1973–1985), which attracted the editor of *Science* (Dan Koshland) and a herd of Nobel laureates, but he had also come up with the idea that smoking cannot cause cancer because some people smoke but don't get cancer, and some people get cancer without having smoked. QED—or rather *quae sunt nugae* ("that's nonsense").

This striking deception was not about *blocking* knowledge, but rather about establishing a metaknowledge regime that could be used to attack the whole idea of causality. Cigarette makers did this by redefining terms but also by funding an army of scientists willing to explore nontobacco causes of disease, things like radon or pesticides or chemicals in our food, air, and water. Cigarette makers funded research into genetics, genomics, and viral causes of cancer as part of an effort to shift biomedical attention from ultimate to proximate causes and (especially) causes *internal to the human body*, including the personal "decision" to smoke. Cigarette interests fogged up big parts of the enterprise of science while turning "cigarettes cause cancer" from a medical fact into a political opinion. This is also why cigarettes were ignored when President Nixon launched his war on cancer in 1971.[3]

That was one early prompt for agnotology—this recognition that *science* can be used as an instrument of deception.[4] Cigarette makers hired the statistician/geneticist R. A. Fisher, for example, to propagate his "itch-in-the-lung" hypothesis—the idea that cancer creates an itch that only cigarettes can scratch, confounding causality. So if epidemiology seems to suggest that smoking causes cancer, the reality is that cancer causes smoking!

Fisher featured big in a lot of cigarette-industry propaganda,[5] but he was part of a much larger army of mercenary scholars. Hans Selye, the "father of stress," was hired to claim that "stress" was the main cause of heart attacks, with the bonus that since smoking *relieves* stress, it could also save lives. Ancel Keys—who put the *K* in K rations—was hired to say that cholesterol was more harmful than cigarettes.[6] And Peter L. Berger, the founder of social constructivism, was hired to claim that "antismoking" was the "new anti-Semitism." Cigarette makers also hired eugenicists, like C. C. Little, inventor of a pure strain of laboratory mice, to amplify his claim that nature was more important than nurture in how we live or die. So if you get cancer, you can blame your genes! Cigarette makers loved genetics—and eugenics—as part of their effort to keep attention focused on causes internal to the human body. For similar reasons they liked the idea that cancer was more common in people born in March—because this would mean that your mom was pregnant during those cold winter months when fresh fruits and vegetables weren't available.[7] So if you come down with cancer, you can blame your mother.

Cigarette makers financed thousands of other scholars, men like Darrell Huff, author of *How to Lie about Statistics*; Hans Eysenck, the racialist psychologist; and Otmar Freiherr von Verschuer, Josef Mengele's thesis advisor with dark connections to Auschwitz. Cigarette makers hired an army of medical historians and at least thirty Nobel laureates to create the impression of a legitimate industry acting responsibly. Cigarette makers even penetrated the National Cancer Institute (through the neutered Tobacco Working Group), which is why cigarettes were ignored when the war on cancer was declared. Billions were spent exploring viral and genetic causes of cancer while ignoring the preventable cause of nearly a third of all cancer deaths.

So it's not entirely correct to think about the cigarette conspiracy as antiscience. Indeed, no one has ever loved science more than Philip Morris. Cigarette makers dumped truckloads of cash into the laps of scholars with the principal goal of distracting from the evidence that cigarettes kill.[8] This allowed cigarette makers, then, to forestall action by claiming that "more research" was needed to resolve a purported controversy—a controversy largely ginned up by the racketeers themselves. Cigarette makers funded science to distract from the truth but also to establish corporate credibility, create alliances, and recruit experts for use in litigation and regulatory hearings—and to create for themselves an impression of prudence and "open-mindedness." Evidence-based deception, you could say.

As for the *scale* of this operation, the Council for Tobacco Research, the industry's principal distraction organ, generated over eight thousand articles in the peer-reviewed medical literature. Humanists have also been corrupted. At least a hundred professional historians have been on Big Tobacco's payroll, and when I came to Stanford in 2004, I was surprised to find two colleagues in my own department working quietly to defend the industry. Philip Morris was also giving millions to our medical school, which is one reason Stanford, as recently as 2011, got an F from Santa Clara County as the most cigarette-friendly university in the Bay Area.

FRAME CAPTURE AND RHETORICAL CAPTURE

Cigarette makers won the war on tobacco largely by making the epidemic invisible and decomposing it into separate and distinct maladies caused by individuals choosing to smoke. Key to that victory has been the reframing of the epidemic from an industrial catastrophe to a disaggregated set of personal choices. This is part of the cartel's broader strategy of *individuation* and *invisibilization*— both of which are deployed as a form of exculpation, which involves truncating causation and redefining what counts as science. The idea has been to push causation "downstream" as far as possible so cancer will be said to begin with, say, a mutation in the genome rather than a decision made in the boardroom of Philip Morris. We see something similar in the reports of the Intergovernmental Panel on Climate Change (IPCC), where corporate connivance is never considered one of the causes of our climate catastrophe—and Chevron and the American Petroleum Institute are treated more as partners than as miscreants.[9] In the popular rendition, fires and floods get blamed on sparks or storms, ignoring the policies that built up the tinder or caused storms to become more violent.

The force of this myopic frame shift (causal truncation) is clear in the tobacco context: Big Nic got cigarettes excluded from the war on cancer by harnessing cigarette-friendly scholars and sympathetic legislators with close friends even on the U.S. Supreme Court.[10] Think about the remarkable fact that the first twenty-odd reports of the U.S. surgeon general on "smoking and health" don't even mention R. J. Reynolds or Philip Morris apart from thanking them for their help preparing those reports. And the cigarette itself is treated as an *uncaused cause, an unmoved mover.*[11] You could call this "frame capture" in the sense of a forced delimitation of the scope of proper inquiry, what can or cannot count as a legitimate cause.[12]

This strategy of individuation and invisibilization is still in full swing. Cigarette makers today will admit that smoking "can" cause cancer, but most eyes are still on the consuming person rather than the product or its maker. And public health scholars often end up carrying water for the industry by adopting its reframes and terminology: Daniel Horn at the Office on Smoking and Health in the 1960s called cigarette use a "personal choice lifestyle behavior" (or PCLB), and the surgeon generals' reports themselves are full of industry-friendly language—starting with their talk of "smoking and health" rather than, say, "cigarettes and death." Tobacco has been very good at getting scholars to use their language and their metrics; they've gotten us to talk about "young adults" rather than youth and "tar and nicotine" instead of cancer and addiction. (Talk of "the smoker" is itself a cigarette-friendly reification, creating an identity out of a condition.) Rhetorical capture and frame capture help us understand the ongoing invisibility of the cigarette epidemic and how important it is to reframe the pandemic as an industrial catastrophe rather than a set of personal choices.[13]

As for targets, it's important to talk about "internal agnotology" in the corporate context, which often takes the form of lexical policing. The canonical example is the censorship of "health information" cigarette-industry employees received from their insurers—part of an effort to maintain morale (= "Vitamin M") in the cigarette workplace.[14] Morale is maintained by high salaries and good benefits combined with incentives not to "spill the beans" or stray from the industry's line on "the main issue."

Internal policing gained an added urgency in 1970 with the establishment of new federal rules of civil procedure allowing legal discovery of confidential corporate records. This gave rise to new fears that the conspiracy could come undone if outsiders were able to access the industry's internal archives. Shook, Hardy & Bacon saw this train coming down the tracks in 1969 and started pressuring cigarette makers to stop or offshore their most sensitive research. The net effect: R. J. Reynolds destroyed its animal-testing facility in March 1970 (in the so-called Mouse House Massacre, where twenty-six scientists were fired and all animals "sacrificed"), and Philip Morris bought a secret research facility in Germany (Institut für biologische Forschung, or INBIFO) to continue the company's most sensitive experiments.[15]

Cigarette makers also ramped up their use of code words and euphemisms, even in their own internal documents. "Compound W" was American Tobacco Company's code for nicotine, and cigarette scribes were instructed to use the term "agrochemicals" instead of "pesticides." This is one reason documents don't always "speak for themselves": cigarette makers in court try to bar all "interpretation" of documents, but the truth is you can't understand a memo talking about "Borstal" causing "zephyr" without realizing that "Borstal" was the industry's code for benzopyrene, and "zephyr," their code for cancer. Thousands of documents are infected with coded rhetoric: some talk about not wanting to target people under eighteen, but this is not innocent speech. I asked Iain Boal to help me coin a term for this kind of rhetoric—which is nominally private but designed to be overheard (think of President Nixon speaking into his tape recorder, "Sure we could break into the Watergate Hotel . . . but it would be wrong!")—and Iain came up with "eavescasting," a portmanteau combining "eavesdropping" and "broadcasting." Cigarette makers talk about the "*New York Times* rule," meaning don't put anything down on paper unless you're willing to see it quoted in the *New York Times*!

Two other points about agnotology in the archives. Cigarette makers saw doubt as their product, but they also went out of their way to *measure* how much ignorance could be produced by a given propaganda campaign. In 1969, for example, Brown & Williamson studied the impact of one of its doubt-mongering editorials, finding that a single viewing diminished the willingness of people to say the harms from smoking were proven from 73 percent down to 60 percent.[16] The Tobacco Institute conducted similar studies of its propaganda films and found even larger agnogenic impacts.[17]

A SHORT TAXONOMY—AND SOME THOUGHTS ON LANGUAGE AND CATARHEUMATICS

Agnotology is close to becoming respectable. Bruno Latour, in a 2014 lecture to Copenhagen's Royal Danish Academy of Sciences, declared agnotology "the most important discipline of the day," and Bonneuil and Fressoz in their 2017 *Shock of the Anthropocene* devoted an entire chapter to "the Agnotocene."[18] Wikipedia now has "agnotology" entries in twenty-eight different languages, and even the *New Yorker* has recognized agnotology as "a subfield of philosophy." Historians and philosophers have started writing books on ignorance,[19] and Routledge has published two important handbooks of ignorance studies.[20]

There are two main reasons for the recent upsurge of interest in agnotology. First is the rise of social media, which has allowed mis- and disinformation to spread like wildfire—which is why we now have "content moderation" and worries over "deepfakes" and the like. Equally important has been the rise to power of authoritarian strongmen and the reversal of progressive policies once regarded as sacrosanct.

All of these have long prehistories, but what is remarkable is how much the world has changed since 2008, when our first agnotology volume was published. Recall that Facebook wasn't even founded until 2004, and Twitter (now X), until 2006. Instagram and TikTok are even more recent (2010 and 2016, respectively). The founder of TikTok was only nine years old when we coined the term "agnotology." This recency is important because the explosion of mis- and disinformation dates from the widespread use of social media, which lowers the cost and vastly increases the reach of communication. "Vaccine hesitancy" wasn't even a thing in 2008, and neither was "filter bubble" or "content moderation." Google search dates only from 1998, and you couldn't even do an Ngram search until 2010.

If agnotology is the study of ignorance, however, we have to think more broadly about the origins and impact of "not knowing," whether from innocence or tradition or deception—or any of the myriad other ways by which something can become or remain unknown. Scholars like to taxonomize, so here I'd like to list some of the ways I've recently been thinking about different kinds of ignorance:

Innocent, or animal, agnotology. Many animals (and even certain plants) have evolved strategies to make themselves seem like something they aren't; this includes visual and auditory biomimicry but also motion mimicry, as when a praying mantis will move like a quaking leaf to evade detection. Orchids famously imitate bees to attract pollinators, but even certain plants have evolved the capacity to blend into their surroundings: *Boquila trifoliolata* is the most contentious, purportedly being able to mimic more than a dozen different leaf types, making it the chameleon of the plant world.[21] Octopi can be placed on a chess board and reproduce its squares, and moths sometimes have large "eyes" on their wings to simulate a nocturnal owl. No one knows what creature first used deception to avoid predation or infection, but it was most likely a microbe, probably a bacterium or a virus, over a billion years ago.[22]

Deliberate, or diabolical, agnotology.[23] This is your quintessential corporate cover-up: creating doubt or throwing sand in the gears of honest science—basically, hiding the truth in order to protect corporate profits or some other higher cause. This often involves a conspiracy and very often fraud—think of the sugar conspiracy,[24] gun-violence denial, lead-poison obfuscation, or climate boomfoggery; we could list a hundred examples. Meat manufacturers hide "how the sausage is made" via ag-gag laws (barring photography inside a slaughterhouse), and poultry trade associations insist that chickens like being cooped up in tiny cages. Many of these are based on the tobacco industry's playbook; I explored many examples in my 1995 *Cancer Wars*, and dozens of others are anatomized in books like David Michaels's *Doubt Is Their Product* and Naomi Oreskes and Erik M. Conway's *Merchants of Doubt*.

Structural agnotology. This is the idea that ignorance often emerges from unwitting myopia or inattention or the selective creation, preservation, or curation of knowledge and skill.[25] Cultural traditions, political or religious ideologies, funding priorities, economic power relations, and disciplinary hierarchies and silos—all of these can work to block the creation or spread of knowledge. In our contribution to this volume, Londa Schiebinger and I explore how enlarging reproductive liberty could help mitigate climate change and why climate modelers have overlooked this. Millions of women give birth as a result of child marriage and bans on abortion, and if it were possible for women (and men) to have only as many babies as they want, this would dramatically lower carbon emissions. Enlarging liberties in this manner is ignored for interesting ideological reasons, leading us to worry more about coercive *contra*ception than coercive *con*ception. This is an important part of agnotology: the study of the interestingly (or pathologically) overlooked.

Bodily agnotology. Many of the most important harms of industrial modernity are initially insensible. Evolution has not prepared us to detect the dangers of radiation or inhaled tobacco, for example, or lead paint or asbestos or refined sugar or ubiquitous plastics. None of these "feel bad" when ingested and typically cause perceptible damage only years or even decades after exposure, obscuring any obvious link between cause and effect, contra millions of years of common sense. Tobacco, sugar, lead, asbestos, and even greenhouse gases— these are all "deepfakes of the body," causing harms that begin insensible but often end up fatal.

Virtuous agnotology (or ignorance). The idea here is that all of us know things we don't want others to know—that's why we have medical confidentiality and statues portraying justice as blind. It's why data get anonymized (and protected); it's why we wear clothes. No one wants to know how a movie ends before it starts—that's why we have spoiler alerts—and no game is fun if everyone knows their opponents' moves in advance. And many things should probably not be known. We don't need to know how to make a deadly virus airborne or the best ways to go about poisoning a city's water. And with the growth of ubiquitous surveillance, many people (especially in Europe) are demanding the right *not* to be known, seen, or overheard.[26] The right to privacy is a kind of agnotology: the right to be forgotten, not to be surveilled. Here, again, not all ignorance is bad.[27]

We should keep in mind that there are countless other ways by which ignorance is created, maintained, and disseminated. Ignorance can be created by failing to notice or to gather or to document, but knowledge once gained can also be lost or ignored, which can amount to pretty much the same thing. We too often forget that ignorance is often the product of *ignoring*, which can result from deception or inattention but also from the infinity of distractions generated by stroking glass or clicking on a laptop. Disciplinary siloing can cause scholarly ignorance, and the same is true of ideologies that truncate causality or reifications that trap agendas inside adjectives. We should also keep in mind that ignorance *by design* can become structural or cultural ignorance, part of the fabric of the not widely known or not known at all. Much of what we call ignorance is just another name for "culture," however befogged or befuddled.

Ignoring can also result from less obvious sources. Every choice is simultaneously a path not taken; every word spoken, a million left silent. And ignoring is most of what we do—and not just by virtue of disciplinary or professional specialization. Ignoring can result from attachment to unattractive implications ("Don't even go there") and can be the simple consequence of lack of access.

Ignorance can also result from blocking the means by which one might acquire knowledge. This can be achieved by laws that bar gathering certain forms of data or images (ag-gag again) or by firing whoever might be in a position to acquire that knowledge. (Recall those ancient tombs kept secret by killing all the laborers involved and then all of those who'd done the killing, etc.) Ignorance can also be created by silencing journalists, and the Chinese government, for a time, blocked any reporting on Beijing's air quality—a kind of ostrich or PR agnotology. If you block the printing press (as was done in late Islamic empires), you impede the growth of knowledge. If a university like Stanford doesn't have a school of public health, it becomes less surprising that so much attention is given to genomics and brain scanning. And if you bar women from education, you don't get women's history or worse—think of the knowledge lost to those millions of women and girls barred from schooling in the Taliban's Afghanistan. And we shouldn't be surprised if universities that take millions from Big Oil love carbon capture and geoengineering as quick fixes for the climate crisis.

Ignorance also has a linguistic aspect. The whole point of euphemisms is to pretty up what is ugly, to disguise the unpleasant as polite. Cigarette makers are masters of this dark art, talking about cancer as an "issue" or a "concern" or death itself as a "health effect."[28] Language limits what can be said but can also influence what is known or left unknown. Deceptive euphemisms include commercial propaganda like "natural gas" or "clean coal" but also mystifying obfuscations like "cane juice," a food-industry euphemism for sugar, or "sunshine units," a nuclear industry term for radiation leaking into the countryside. "Solar geoengineering" is Big Carbon's term for bouncing the sun's rays back into space (to allow continued pumping of carbon into the atmosphere), and "food insecurity" is how bureaucrats dress up starvation. A list of deceptive terms and phrases could fill volumes and would have to include monstrous terms like "comfort women," the Imperial Japanese Army's euphemism for females forced into sexual slavery, and "ethnic cleansing," a sanitized way to say genocide. Small strings of words can convey vast misunderstandings—think of the profound naïveté underlying expressions like "junk DNA" or "fetal personhood" or "net zero" or "our earliest ancestors" (they didn't have parents?).[29]

War reporting is notoriously full of such deceptions, giving a pretense of civility to the ripping of flesh. Torture is thus euphemized as "enhanced interrogation," and disingenuous jabbering as "the peace process."[30] Good soldiers are "servicemen"; bad soldiers are "militants." Good guys have "allies"; bad guys have "proxies." "Delicate operation" is how some journalists describe Israel's invasion of Gaza, and bombings are euphemized as "degrading the adversary," "sending a message," or even "cutting the grass."[31] After the Second World War, the United States and Britain both renamed their "war" departments as departments of "defense/defence."

Reallocation of agency (and causality) is also part of this. Tobacco-control advocates talk about "cessation," albeit only of consumption, never production.

Blame is similarly shifted from the maker to the user when we hear that lead paint isn't the problem; the problem is "pica"—compulsive eating—in this case, the desire of kids to chew on their toys.[32] Southern physicians in the 1800s likewise claimed that if slaves try to escape, that's because they suffer from "drapetomania," an insane desire to run away. The whole idea of obsessing over one's personal "carbon footprint," a Big Carbon concoction, was designed to blame individual consumers for the global climate catastrophe.[33]

This unhealthy forcing of causality downstream (blaming the victim) we can call "Epimethean" or "catarheumatic" (referring to downstream thinking or acting)—if you'll allow me a brief digression into Greek mythology. Recall that Prometheus was the Titan who stole fire from the gods and gave it to humanity, fostering the rise of science and civilization. As punishment, Zeus had him chained forever to a rock, where an eagle would visit him nightly to peck out his liver, which would regrow the next day, resuming the torment. Prometheus was literally the god of forethought, hence his name: from *pro* (fore) and *methe* (math or skill more generally). Less well known, though, is that Prometheus had a brother, Epimetheus, the god of hindsight, afterthought (*epi* plus *methe*), and lame excuses. The brothers had been tasked with supplying the creatures of the world whatever they'd need to live—strength and speed, fur and feathers—but Epimetheus didn't think ahead and had nothing left to give by the time he reached humans, naked and defenseless. Prometheus stole fire from the gods to give to humanity because we had no way to survive without. Zeus then created Pandora to punish the brothers and entrusted her with the evils of the world in a box, which she was never supposed to open. (From curiosity, she accidently let them all loose, managing only to keep hope.) The point for us is that Epimetheus doesn't plan or think ahead, he doesn't go upstream but rather settles, makes do, gets by without questioning authority or causality. He is the fool; he comes in too late.

Public health scholars for years have been urging us to go upstream. Bishop Desmond Tutu in South Africa famously observed that we are fishing all these people out of the river but too rarely ask, Why are they in the river in the first place? Who keeps pushing them in?[34] Smoking causes cancer, for example, but what causes smoking? And if peer pressure causes smoking, what causes peer pressure? And if it's ultimately a *manufacturer* that is causing all this cancer, then they, too—and their legal beagles—should be held to account as vectors of disease.[35] Failing to do so is shortsighted. It's wrong, then, to focus only on our immediate condition; we have to challenge Epimethean myopia, or what, from Greek, we could also call "catarheumatics"—from *cata* meaning "down" and *rheum* meaning "river."

Catarheumatics is when people approach the climate crisis by trying to suck carbon out from the atmosphere or bounce the sun's rays back into space. A comparable solution for obesity would be mass liposuction. One might as well say that the best way to prevent lung cancer is a nightly brush of the lungs. The academicians of Lagado were on equally solid ground, attempting to extract the nutrient remnants from feces and sunbeams from cucumbers.

Ignoring upstream causes, we may keep fishing people out of the river without considering who is pushing them in. Ignorance by this means creates impotence and hides both the origins and best solutions of many of our maladies.[36]

ADJECTIVAL AGNOTOLOGY

I've been teaching agnotology for the past thirty-odd years, first at Pennsylvania State University and, from 2004, at Stanford. I've had hundreds of student papers over the years—papers on tampon agnotology (the discourse taboo), affect agnotology (modesty), visual agnotology (camouflage), gun-lobby agnotology, dietary agnotology (think food fads or supplements), nuclear agnotology, racial agnotology, juridical agnotology, algorithmic agnotology, papers on the agnotology of TikTok and Facebook and spy craft, on lead paint and gas and asbestos and polonium, papers on Holocaust denial and the denial of the Armenian genocide, the Katyn massacre, the Holodomor, the Nakba, and the mass rape of German women after the Second World War, or choose your atrocity—all of which involve distraction, erasure, or trivialization.[37] I've had lots of student papers on the agnotology of tobacco and sugar, along with climate agnotology and the agnotology of fake news and fine print and "attributed" warnings[38] but also papers on vaccine hesitancy,[39] concussion denialism,[40] opioid agnotology, and election denial. Here are some of the less obvious topics we've covered:

Wikipedia Agnotology: Wikipedia is a platform that allows for the curation and spread of mis- and disinformation by promiscuously allowing "editing." Try going to, say, the site for "Marlboro" and enter how many people are killed every year from smoking that cigarette, and watch how quickly that fact will be erased! For hot-button topics, Wikipedia is often an engine of equivocation or worse. Corporate censors quickly scrub out the truth; you can literally watch its erasure.

Earthquake Agnotology: Real estate developers in Southern California after 1906 didn't want to have to build to seismic codes, so they paid a geologist to write a book claiming that earthquakes could only happen in Northern California in response to uplift from receding glaciers. For a couple decades, the developers succeeded, which is why so many buildings from that era are precarious.

Architectural Agnotology: Buildings are often designed to conceal—think of Stanford's animal-experiment facility hidden under a nondescript parking lot of the medical school (aka Doggie Dachau) or the hiding of cities inside mountains to make atomic bombs. Cigarette makers don't want you to know where their factories are located, which is one reason Matthew Kohrman set up "Cigarette Citadels," to map every tobacco factory on the planet.[41]

Graphic Agnotology: Here the reference is to visual deception, as when cigarette makers chart cancer rates over time as chaotic to disguise what's causing the growth or decline of a particular malignancy.[42] Climate change deniers likewise cherry-pick starting times for temperature trends to deny the increase in global temperatures.[43] Gun makers similarly minimize gun violence by comparing American murder rates to those in poorer parts of the world and ignoring the far higher rates of gun homicide in the United States (see John Donohue's essay in this volume).

Cartographic Agnotology: Maps are often engines of ignorance given the radical selectivity of what is included or omitted—for good or for bad reasons. Two hundred Native American sites on the Stanford campus are not easily found on any published map to prevent plunder or desecration. Botanic journals now withhold locations of newly discovered species of cacti to prevent them from disappearing. And Google will jitter the location of sensitive military and other vital sites to prevent targeting—potentially a virtuous agnotology. How maps are drawn can be highly political.[44]

Archaeoagnotology: Nadia Abu El-Haj has written about how digging involves a certain amount of destruction insofar as assumptions are made about what counts as treasure and what counts as trash.[45] Archaeology is the study of faint buried traces, but excavating and reconstructing those traces often involve a degree of projection, which is one reason portions of sites are often left undug—so future generations may see the past with new eyes.

Media Agnotology: Jon Christian writes about how the journalistic "balance routine" has led to passive reporting in the form of "on the one hand . . . on the other hand," giving equal time to suspect opinions. Many news agencies now have "misinformation correspondents"—a job that didn't even exist when we published our first volume. BBC has a Disinformation Team and an Anti-Disinformation Unit, and Reuters has a reporter covering cybersecurity and disinformation. CNN in 2016 made history by challenging statements made by Donald Trump: when the real estate mogul called Barack Obama the "founder of Isis," the network disputed that claim with the blunt chyron "He's Not."[46]

Algorithmic, or Aspirational, Agnotology: A substantial literature has emerged on algorithmic bias and how to correct it. Machine translations once misgendered pronouns—by defaulting to the masculine, for example—but Google (for example) is now debiasing with a vengeance. Only 10 percent of *Fortune* 500 companies have women CEOs, for example, but most of the first images returned by a 2024 search for "CEO" were women. Fewer than 2 percent of *Fortune* 500 CEOs are Black, but an image search for "CEO" returns a high proportion of African Americans. We find a similar (aspirational?) bias in a search for "scientist," which returns only ten light-skinned men among the first fifty results.

Porno Agnotology: Ubiquitous and easily accessible pornography creates an idealized/falsified view of human sexuality, with unknown and perhaps unforeseeable consequences. A huge proportion of internet traffic is porn, and artificial intelligence (AI) is likely to accentuate already existing deformations.

Pyro-agnotology: For more than a century, the U.S. Forest Service had a policy of suppressing all fires, resulting in the catastrophic buildup of tinder. Deforested zones were also replanted in monoculture, depriving the forest of its natural defenses. The goal was to protect commercial forestry, but Forest Service officials also deliberately suppressed Native forest management on the grounds that fire under any circumstance was "wasteful." Aboriginal practices in Australia were likewise suppressed on the basis of similar ignorance.[47]

THE PRESENT ESSAYS

The essays in this volume were presented at a two-day conference at Stanford University in May 2023. Each is dedicated to exposing how a particular kind of ignorance has been engineered, either deliberately or from grave indifference. A volume such as this could focus on a thousand topics, but here we highlight some of the more poignant.

Benjamin Franta looks at one of the chief obstacles to ending the climate catastrophe: the neglect of the fact that to stop it, we must stop burning oil and coal; these must be left in the ground! Fossil fuel producers have known this for decades and have deployed armies of PR wizards and astroturf agents—and legions of consultants—to obscure this crucial fact. Carbon dealers weaponize complexity and distract with myopic calls for "adaptation and management." Ignorance in this realm has even infected some in the climate-science community, enticed by pseudo-solutions (like sequestration or solar geoengineering) that leave carbon producers to their own devices. Franta helps explain how we've ended up in this absurd world where people fixate on planting trees or bouncing the sun's rays back into space, ignoring simple solutions that could actually save the planet.

Londa Schiebinger and I explore a different kind of agnogenesis in our essay on how enlarging reproductive liberties can help mitigate climate change. Our starting point is the taboo that has emerged on talking about population in the context of climate change given the long history of eugenics and racialized "population control." The net effect is that scholars spend a lot of time talking about coercive *contraception* while virtually ignoring coercive *conception*. The reality is that millions of babies are born every year as a result of rape, incest, and laws barring access to contraception or abortion. We show that if people were free to have only as many babies as they wanted, the world would have significantly fewer people—with enormous benefits for the climate and biodiversity and greater human freedoms.

Hany Farid, in his essay on AI agnotology, looks at how "algorithmic amplification" increases the speed of transmission of misinformation, allowing hate, misinformation, and conspiracy theories to spread online. Farid distinguishes different kinds of ignorance resulting from predictive versus generative AI: predictive AI (i.e., recommendations of what to read or see) can generate ignorance by siloing and tunneling, creating "informational isolation" in the form of filter bubbles and echo chambers, while generative AI can produce "alternative realities" in the form of "deepfakes" that erode public trust. AIs of this sort are throwing jet fuel onto an already heated inferno of media-generated ignorance and distrust, and we can expect more agnogenesis of this sort in the future. He points out that with the proliferation of deepfakes, we are going to see elections altered and perhaps even wars started. And it's going to become easier to be a liar—the so-called "liar's dividend"—and harder to get caught. Generative AI is also starting to feed on its own content, which could lead to new levels of fakery.

Nanna Bonde Thylstrup, in her chapter on agnotology in the exabyte era, explores how data is often lost in the process of collecting it because corporations and science giants collect far more data than can possibly be stored or curated. CERN's (European Organization for Nuclear Research) accelerators, for example, record a billion collisions per second, for which the lab has developed algorithmic "triggers" to distinguish between more or less interesting events. Digital records are precarious, however, and vulnerable to catastrophic loss from fires or electrical failures, which can make paper look immortal by comparison. Electronic files of a certain age also often become unreadable: a 2014 study found that half of all U.S. Supreme Court links since 1996 had succumbed to "link rot." Thylstrup emphasizes that erasures can be complex politically, as evidenced by the fact that paper shredders (*Aktenvernichter*) were first built in Germany in the 1930s to prevent resistance tracts from falling into the hands of the Nazis. (The first was a modified pasta maker.) Shredders are now used to protect trade secrets and embarrassing paper trails, but Thylstrup cautions that we are moving into a "digital dark age" shaped by forces of obsolescence given the fragility of data in our electronic world.

In his "Law against Knowledge," Peter Galison looks at different choke points in the process of knowledge formation, stressing that ignorance can be created not just by blocking already existing knowledge or confounding it with noise but by preventing its creation in the first place. Galison looks at some of the laws recently passed in Wyoming barring the collection of environmental or historical data—bacterial contamination of water, for example, or the presence of archaeological remains—to protect mining and ranching interests. Galison uses these Resource Data Trespass laws (barring photography by drones, for example) as an example of what he calls "anti-epistemology," which he links to a broader regime of state censorship: states may block a radio broadcast or classify some military enterprise, but they can also bar the collection of certain kinds of data, obviating any need to block it downstream. He also inverts the information theory of Claude Shannon to point out that knowledge is often easier to prevent than to destroy or dilute.

Naomi Oreskes and Erik M. Conway, in their chapter, show how free market fundamentalism has obstructed intelligent action on many of our social ills. Oreskes and Conway show how Adam Smith has come to be falsely remembered as a free market fanatic ("invisible hand"), when the truth is that he envisioned crucial roles for government—defending taxation to finance public goods and education, for example. The authors show how the false memory we have of Adam Smith derives from his remythologizing during the Cold War, when student editions of his *Wealth of Nations* were stripped of crucial sections embracing the role of government to protect us from threats foreign and domestic. They link this to the rise of the myth that market freedoms are required for political freedoms, helping us to better understand one of the most important obstacles to acting responsibly on the climate.

John J. Donohue then looks at how gun manufacturers and their agents have corrupted knowledge about the hazards of firearms. Donohue shows how gun-lobby apologists distort the magnitude and causes of this distinctly American epidemic, using misleading statistics and spurious claims about the "original" intent of the Founders. Gun agnotology draws from a fetishized misrepresentation of the Second Amendment, but Donohue also shows how global statistics can be twisted to obscure the fact that, in per capita terms, America leads the affluent world in gun murder. Gun manufacturers effectively barred the Centers for Disease Control from researching gun violence for many years,[48] and Donohue details how courts have been misled by industry-funded academic hacks. This helps explain why America leads the world in firearm deaths. Japan, for example, has only about eleven people killed by guns every year versus nearly twenty thousand in the United States.

For his part, Robert H. Lustig looks at the corrosive role of refined sugar in human health, including in obesity, of course, but also metabolic syndrome and other disorders. He shows how food producers have used the ideology of personal choice, developed first by Big Tobacco, to blame the victim and stigmatize individual consumers. Food producers cover up the fact that sugar causes diabetes and fund research to exonerate sweets from causing obesity. So if you're fat, it's your own fault thanks to personal gluttony and sloth. In truth, we now have an epidemic even of fetal obesity, caused by production of hyperpalatable "foods" with a propensity to cause addiction. Public health organizations have been co-opted, explaining why organizations like the British Dietetic Association maintains that "sugar does not cause diabetes." Coca-Cola and Pepsi have provided funding to ninety-six different public health agencies, which is one reason so many of us remain in the dark about the harms of Big Sugar.

In her essay, Jennifer Jacquet explores how Big Meat creates, launders, and legitimizes ignorance concerning the health and environmental impacts of consuming "meat" (i.e., animal flesh). Jacquet shows how livestock producers use academics to bolster their credibility and spread climate disinformation. Frank Mitloehner at the University of California, Davis, for example, was paid by the National Cattlemen's Beef Association to produce a report criticizing estimates from the Food and Agriculture Organization of the United Nations (FAO) that 18 percent of global greenhouse emissions come from livestock (mainly from belches and manure). On a per gram basis, methane packs a far greater (short-term) greenhouse punch than CO_2, so the question is serious. Mitloehner helped shape the methods used by the FAO to account for livestock's contribution to global emissions—a kind of agnogenic accounting—and helped convince many that cows should not be blamed for climate change.

Rosemary Sayigh then looks at the forgetting of the Nakba, the forced displacement of 750,000 Palestinian men, women, and children in 1948 with the establishment of the new nation of Israel. Sayigh tells how she herself had never heard the word before she married a Palestinian in 1953 and that much of the scholarship exposing the catastrophe emerged only in the 1980s and '90s. The Nakba (meaning "catastrophe" in Arabic, much like *Shoah* in Hebrew) involved the depopulation of some 560 Palestinian towns and villages, the erasure of Palestinian place-names, and the confiscation of Palestinian libraries. Sayigh traces the "burial" of the Nakba over the course of the 1950s, '60s, '70s, '80s, and '90s, during which time it was essentially absent from British and even Arab textbooks and barred from being taught in Israeli schools. Few people even today know that Yad Vashem, the Holocaust memorial, is built on the site of the first massacre of Palestinians by Zionist militants at Deir Yassin. Sayigh shows how monumental events can fall victim to memoricide, much like the Armenian genocide in Turkey or the massacre of the Herero and Nama (by General Lothar von Trotha) in what is now Namibia.

Daniel Akselrad, in his contribution, looks at the fake railway station built at Treblinka by the Nazis, which had signs pointing to near and faraway towns and a false painted clock—part of a ruse to convince prisoners they were being "resettled" into some kind of work camp. Akselrad shows that this and surrounding fakery were part of a broader effort to maintain morale inside the "camp" (itself a euphemism for a killing center), which forms part of his work on Nazi stagecraft, architectural euphemism, and the agnotology of atrocity. The past, in this sense, is still present insofar as many of the signs in our own world are leading us astray.

Caroline A. Jones, in her essay, explores the racialized ignorance accompanying Hiram Powers's white marble *The Greek Slave*, an 1843 sculptural sensation dragged across the United States on railroad tours, making its way even into the slave-owning South and to London, where it was shown at the Great Exhibition in 1851. Art historians acknowledge the novelty of the widely reproduced piece but tend to ignore its racialized significance, especially the fact that by focusing on the "white slavery" (by Muslims) in the Greek war of independence rather than on the reality of enslaved Africans in the Americas, Powers effectively enshrined white ignorance. For Jones, the sculptural power of Powers's creation is analogous to the "Whitening" of art historical discourse itself—a discourse that has ignored the different political reception of the sculpture by free Black people and white abolitionists. Jones highlights the tensions between Black and white responses to Powers's work and invites readers to consider how the fetishized whiteness of his Italian marble can help illuminate the racialized history of art. Jones, in this sense, summons what she calls a "Black agnotology," complementing what Charles W. Mills derides as "white ignorance."

Last but not least, Nadia Abu El-Haj explores the trauma of modern American wars—specifically, our post-9/11 "forever wars"—juxtaposing the trauma of those who return from war against the purported ignorance of "civilians" who, per mythology, cannot fathom its true horrors. Based on trauma-clinic fieldwork and a review of war reporting, she shows how it has come to be imagined that only those who've suffered in battle can really know what war is like. Abu El-Haj shows how this Janus-faced myth of the "knowing soldier" and "ignorant civilian" allows most of the true horrors of war to go unnoticed, since it is really only injuries to "our" soldiers that get noticed and deserve care. The net effect is a kind of nationalist myopia, a narcissistic agnotology blinding us to the true scope of harm inflicted on *foreign* victims of these forever wars.

THE FUTURE OF AGNOTOLOGY

Agnotology represents a new kind of science, continuous with the broader spirit of inquiry that has animated humans for millennia. Our hope is to join with the surge of recent scholarship on mis- and disinformation, on denialism of all stripes, on doubt mongering and distraction,[49] on miseducation,[50] and on "ignorance studies" more generally.[51] We need to think more about how ignorance gets empowered—and how it might be overcome.

This includes the psychology of ignorance, how and why certain kinds of people become vulnerable to humbuggery and hokum. Conspiracy theory has become a topic of scholarly interest,[52] with attention mainly to "theories" most of us would consider preposterous: that the earth is flat or hollow, that the moon landing or 9/11 or some school shooting was faked, or that dark forces are pacifying us from above with chemtrails.[53]

Not all conspiracies are false, however, or the product of psychopathology.[54] It's an empirical question whether any given conspiracy is real, and we should not forget that cigarette makers really did conspire to hide the hazards of cig-arettes.[55] Keep in mind also that our focus is on ignorance *as a social product* rather than ignorance as a personal attribute or mental disorder, which is why it's so important to appreciate how and why conspiracy theories have become so easily spread. Proliferation is made possible by handheld devices that give us instant access to quasi-infinite warrens of "information"—rabbit holes inside rabbit holes. And fakery has become easy to produce and distribute on an in-dustrial scale. Between October and December 2021 alone, Facebook disabled more than 1.3 billion fake accounts, along with more than a hundred networks of "coordinated inauthentic behavior" (CIB).[56]

We also need to know more about how ignorance is produced globally. How does climate denial (or pseudo-solutionism) work differently in, say, coal-rich India or tech-rich Japan? How is evolution presented (or not) in textbooks in different parts of the world, and how does internet censorship differ in North Korea versus Iran? And how has the Chinese state managed to keep three hundred million of its citizens smoking, earning hundreds of billions for the government?

Accounting could well become the master discipline of the future, which means we are going to have to focus more on agnogenic accounting, the du-plicity involved in measuring "tar and nicotine" or "net-zero" carbon or the non-counting of military or agricultural emissions, etc.[57] We are going to have to focus more on deceptive metrics, such as the Body Mass Index, which doesn't consider fat-to-muscle ratio; cigarette-delivery metrology, which doesn't mea-sure how much poison actually goes into the body; the Environmental Pro-tection Agency's Air Quality Index, which classifies high levels of pollution as "moderate"; the food pyramid, which enshrines meat and dairy as part of a "proper diet"; measures of radioactive exposure, like the rem (roentgen equiv-alent man), which ignores the site of action in the body; safety standards that don't consider the female body (e.g., crash test dummies); medical devices that fail by racial blinders (e.g., the pulse oximeter); etc. Correcting these will re-quire attention to agnogenic metrics. We live in a world where tax shelters allow the very rich to declare "no income" and where averages conceal extremes. Two degrees Celsius of global warming is four degrees of warming on the land, for example, and even more at the poles; it's not homogeneous. All these (and countless others) involve misleading metrics, either by conniving design or poor communication.

We also need to know more about the virtues of ignorance, including the importance of ignorance in creating new knowledge. It's a cliché to say that the first step toward knowledge is recognizing our own personal ignorance (think Socrates or Confucius), but the corollary for those of us who teach is that we can't really do our job properly until we grasp what our students do not know! It is not always easy to put ourselves in the shoes of the learner, but understanding ignorance is key to effective teaching. And in a related vein, we need to know more about how education can generate ignorance—or fail to overcome it. An internal tobacco-industry poll from 1978 shows that highly educated smokers were the ones most likely to believe that filters made cigarettes safer.[58] How else does education generate ignorance?

We also need to know more about how religious, ethnic, or nationalist chauvinisms produce ignorance. Christian, Islamic, and Jewish monotheisms have all inspired murderous wars of conquest, and fundamentalists in these or other traditions (Hindu, Buddhist) may insist on the virgin birth of an iconic hero or claim exclusive rights to some sacred piece of ground ("the Holy Land"). Fundamentalists are typically ethnocentric, but they can also foster ignorance by denouncing established truths, suppressing journalism, or barring (or requiring!) certain kinds of education. Most religions encourage learning, but fundamentalists propagate ignorance by insisting on strict adherence to some sacred text and punishing deviance. We need to know more about how religious commitments buttress antipathy toward hard-won empirical wisdom, and we need to know more about why the IPCC fails to consider religion as an obstacle to acting intelligently on climate change. We already know a lot about how social Darwinism has helped spawn colonial horrors, but we need to know more about how attitudes toward race and women's bodies have blocked recognition of evolution.

We also need to know more about how magic and superstition persist in different parts of the globe. Part of our ongoing decline in biodiversity, for example, has to do with the fact that nonhuman species are coveted for their magical powers; tigers, sharks, and rhinos are killed, for instance, to feed those that believe that by ingesting or displaying parts of animals that are powerful or have penis-shaped appendages, men can gain sexual potency.[59] Of course, the distinction between religion and magic is much like that between a language and a dialect—a language, after all, is just a dialect with an army.

We also need to know more about left-wing agnotology, ignorance generated by the neglect of tradition or the presumption of a need to radically remake society—as when Pol Pot killed scholars or when Chinese cultural revolutionaries

tried to erase bourgeois knowledge. We need to know more about how ignorance can be generated by "progressives," whether by mollycoddling the young
or insisting on the unalloyed virtues of mixing and migration or by woke algorithms that generate Black Nazis (and blurred swastikas) when AIs are asked to
draw "an image of a 1943 German soldier."[60] It could well be that by focusing
only on right-wing or corporate agnotology, our new science has been "blind
in one eye."[61]

As for misinformation, one can compare our present condition to that faced
by residents of cities prior to the rise of public hygiene and social medicine,
where clean water often mixed with dirty, fouling the urban body. Nineteenth-
century reformers cleaned up much of that mess, at least in wealthier parts of
the world. But the rise of social media presents new and comparable threats
given the ease of production and transmission. Karl Marx long ago observed
that all economy is an economy of time, and in this sense, we are living in a
world much like that of the late 1800s, when new technologies made it easy to
consume large quantities of candy, ice cream, soda pop, cigarettes, and other
sensory delights and stimulants, including compressed (and eternalized) sight
and sound in the form of photography and motion pictures.[62] Rare delights
became routine thanks to more rapid networks of communication and delivery
(railroads and radio, for example, and later trucking and TV). The result has
been the spread of all manner of sensory rewards, including those that cause
many of the maladies we associate with modernity.

The Agnotocene is an extension of this earlier revolution. Social media
platforms have become billion-dollar operations by profiting from the capture of eyeballs and clicks, but they've also facilitated the spread of mis- and
disinformation—along with hate speech and shaming and a myriad of addicting
enticements (i.e., gambling)—on a scale previously unimaginable, with speed
and cost being the chief novelties. Paraphrasing Mark Twain, "A lie can travel
halfway around the world while the truth is still putting on its shoes."[63] Or per
Jonathan Swift circa 1710, "Falsehood flies, and the Truth comes limping after
it." All this is presented to us as a bounty of "choice," but it also comes with
costs in the form of overconsumption and addiction and tsunamis of distraction, with the added novelty that it has become much easier not just to consume
but also to *produce* such chicanery.[64]

———

So now let's enter this hall of mirrors, the many kingdoms of distraction and
disinformation. Fortunes are being spent by the powerful to corrupt our world
with deceptive signs and false solutions, and our hope is to bring that process
of agnogenesis into better focus. Keep in mind that the object of our attention is *the making* of ignorance rather than the *fact* of ignorance, the "act of"

(making ignorance) rather than the "lack of" (knowledge). How does ignorance get made, and how is it distributed and maintained? The reader is invited to join our exploration of the dark arts of casting shadows, along with the cultural causes of missed opportunities for enlarging human liberties and planetary health. There's an endless variety in how things are made or kept hidden, obscure, invisible, out of reach, insensible—countless means by which we are prodded into mindless apathy, complicit negligence, soporific acquiescence. Ignorance is an ever-expanding infinity, but the study of how it gets created has only just begun.

TWO

How the Most Important Fact of Global Warming Has Been Obscured

BENJAMIN FRANTA

THE SUMMER OF 2024 was the hottest on record.[1] Heat domes formed across the world like gigantic ovens, and smoke from incinerated forests covered the North American continent. Waters in the Gulf of Mexico reached jacuzzi-like temperatures, while the North Atlantic Ocean heated up as never before in human history, threatening to collapse the Atlantic meridional overturning circulation. Greenhouse gasses are higher than they have ever been in three million years, and we are sprinting toward ever-greater extremes. And almost all due to the production and consumption of fossil fuels.

Like anything, global warming is complicated in some ways and simple in others. Petro polluters and their apologists are quick to emphasize complexity, with the implication that there's always more to study before solving the problem and that no one is to blame for the slow pace of progress. But climate change isn't complicated in *every* way and is not very hard to understand. It turns out that the most important facts are also rather simple, including the most important fact of all: replacing fossil fuels is necessary—and pretty much sufficient—to solve our climate emergency.

The science of climate change—involving careful modeling with eerily exact predictions of global warming—has been understood for decades both by climate scientists outside the fossil fuel industry and by scientists and executives within the industry itself. Remarkably, by 1980, both academic and industry scientists were aware that catastrophic global warming was largely preventable and required an immediate turn away from fossil fuels. This most important fact, however, posed an existential threat to fossil fuel producers, who already in the 1980s had launched an assault on the truth through denial, distraction, and obfuscation. Their goal? To promote the idea that global warming could be solved without replacing fossil fuels. As we shall see, from the late 1980s to the early 1990s, oil producers and their allies developed three major strategies for defeating fossil fuel controls: weaponizing scientific uncertainty, manipulating economics, and promoting false solutions. The net effect is that large segments of the population—and even many climate-science professionals—still today don't grasp the importance of replacing fossil fuels to respond properly to the emergency.

DISCOVERING THE MOST IMPORTANT FACT

The need to replace fossil fuels to fix global warming might appear trivial or obvious. At the same time, much of the history of contested climate policy can be understood as a struggle over this crucial fact's existence, interpretation, and implications. More than any other piece of knowledge, this fact has posed the greatest existential threat to the most powerful industry on the planet.

Exact knowledge of fossil fuels' dangers developed as early as the 1950s and was well established by 1980. In 1954, for example, the American Petroleum Institute (API)—the industry's most important trade association—was privately informed by scientists at the California Institute of Technology that fossil fuels were likely causing a buildup of carbon dioxide in the atmosphere.[2] Five years later, in 1959, physicist and father of the hydrogen bomb Edward Teller warned oil-industry executives that this buildup would cause serious warming by the end of the century and advocated for the replacement of fossil fuels to prevent this catastrophe.[3] And in 1965, shortly after the White House published an assessment warning of global warming, the head of API notified fossil fuel chiefs that, based on the report's findings, "an alternative nonpolluting means of powering automobiles, buses, and trucks is likely to become a national necessity."[4] Three years later, in 1968, the API received this same warning from its own private consultants, who advised that it was time to develop "air pollution technology" to bring CO_2 emissions under control.[5]

European manufacturers came to similar conclusions. In 1971, the French oil giant Total published an article titled "Atmospheric Pollution and Climate" in its magazine *Total Information*, which noted that since the 1800s, the burning of fossil fuel had resulted in the release of "enormous quantities of carbon dioxide" and that if fossil fuel use continued to grow, CO_2 concentrations would reach four hundred parts per million (ppm) around 2010 (a forecast that proved surprisingly accurate).[6] The article characterized the predicted global warming as "quite worrying" and warned of modified atmospheric circulation, melting of polar ice, and significant sea level rise, stating, "The catastrophic consequences are easy to imagine."[7]

Based on these and other warnings, policymakers and energy modelers started discussing how to shift to low-carbon energy economies in the future. In 1969, for example, Victor Erickson from the U.S. Department of Commerce reported to the annual meeting of the American Institute of Mining, Metal-lurgical and Petroleum Engineers that if "we are faced with the problem of a warming of the earth from the 'hot house' effect of carbon dioxide in the upper atmosphere or with the cooling of the earth from particulates cutting out the ra-diation of the sun, entirely new approaches may be needed. . . . [This] will raise major economic issues and it is not too early to be thinking about them."[8] The following year, the institute's conference program included contingency plan-ning for transitioning to a low-carbon energy economy by the year 2000.[9] And one year later, in 1971, a speaker from the U.S. Treasury Department observed that to address atmospheric pollution, "future demands may be met by systems which are not dependent on fossil fuels."[10]

By the end of the 1970s, climate science had advanced with the development of computer-based general circulation models and multiple consensus assess-ments, including the 1979 "Charney report" by the U.S. National Academy of Sciences, which found that a doubling of atmospheric CO_2 would cause an av-erage warming of around three degrees Celsius.[11] This allowed researchers both within and outside the industry to predict how much warming would be pro-duced by various fossil fuel future scenarios. In other words, by this time, many of the most important facts about global warming were well established. Scien-tists could describe how much warming would occur, when it would occur, and how to prevent it from getting worse.

Around this same time, a body of scientific literature began to examine how to limit the severity of global warming. The main conclusion of this work was that to avoid a climate catastrophe, fossil fuels needed to be replaced, beginning immediately. Researchers also realized, though, that the replacement of fossil fuels, even if pursued energetically, would not happen overnight. Fifty years was a common estimate for the time needed to replace one energy source with another, and this lent the climate problem a certain urgency.[12] In 1978, Stanford engineer John Laurmann noted that a few years prior, the physicist Wolf Häfele and engineer Wolfgang Sassin of the International Institute for Applied Sys-tems Analysis (IIASA; an international research center established near Vienna, Austria, to facilitate East-West cooperation during the Cold War) had suggested that half a century would be needed to transition away from fossil fuels, leading Laurmann to warn that "the most likely date for reaching at least 2.5°C [of global warming], 2029, suggests the need for immediate remedial measures in the form of global reduction of fossil fuel use."[13]

Laurmann elaborated on this concept in 1979, writing in *Science* that be-cause moving away from fossil fuels would likely take decades (what he called the "market penetration time concept"), climate change posed an "immediate environmental control problem."[14] Global warming, he noted, could be avoided "by switching from the present-day predominant use of coal and oil as primary energy sources to a non-carbon-based fuel" and that the timescale for replacing fossil fuels suggested "the need for immediate action if the [climate] change is to be averted."[15] Assuming an annual global-energy-demand growth rate of 2 percent (similar to actual historical and current rates),[16] Laurmann projected that with immediate action, atmospheric CO_2 could be limited to a peak of around 403 ppm before it would start to decline. That would have been significantly lower than our current level of over 420 ppm (and steadily rising).

Other researchers also emphasized the importance of this market-penetration-time concept. A 1979 study in the *Journal of Geophysical Research* by F. Niehaus and J. Williams obtained results similar to Laurmann's, projecting that with immediate action, CO_2 emissions could peak around the year 2000 with concentrations limited to around four hundred ppm (fig. 2.1) and that continued reliance on fossil fuels could lead us to a catastrophic one thousand ppm by the year 2100.[17] Niehaus and Williams's projections were adopted in 1981 by the IIASA in its own analysis of the fossil fuel climate problem.[18]

This is important to grasp. By 1980, the market-penetration-time concept was widely acknowledged and indicated a need for immediate action to avoid disastrous levels of global warming. Multiple peer-reviewed studies projected that with immediate action, atmospheric CO_2 in the twenty-first century could be limited to around four hundred ppm, corresponding to warming of only about 1 degree Celsius. In reality, CO_2 concentrations might have been limited to even less than four hundred ppm with a faster transition to renewables. The key lesson of these studies from over forty years ago is that catastrophic global warming was understood to be largely preventable and required prompt efforts to replace fossil fuels.

A related concept developed around this time was something called the "action-initiation time," which referred to the date by which actions to reduce fossil fuel emissions would begin to have an effect. Researchers found that the later the action-initiation time (i.e., the longer fossil fuel controls were delayed), the more abrupt the replacement of fossil fuels would have to be to achieve a given CO_2 limit. A 1982 study in the journal *Energy* by A. M. Perry and colleagues, for example, examined future CO_2 buildup as a function of the action-initiation time, explaining that the energy transition would be required "sooner or later" and that "the longer the transition is delayed, the more abrupt and difficult" it would be.[19] "The possibility of limiting CO_2," they found, "will depend strongly on energy use patterns established and decisions taken in the relatively near future"; they also warned against any "delay in promoting restraint in the growth of fossil fuel use."[20]

In his 1979 study, Laurmann also modeled how *delaying* efforts to replace fossil fuels could lock the world into a climate catastrophe, estimating that one, two, or three decades of delay would lead to an additional 70, 140, or 250 ppm

FIGURE 2.1. Two different scenarios for energy and climate, published by Niehaus and Williams in the *Journal of Geophysical Research* in 1979. In the first chart (fig. 10), immediate dedication to nonfossil energy sources allows emissions to peak around the year 2000, CO_2 concentrations peak at 400 ppm, and global warming peaks at 0.8°C (assuming a climate sensitivity of 2°C per doubling of CO_2; a climate sensitivity of 3°C, closer to current estimates, would yield global warming of around 1.2°C). In the second chart (fig. 12), continued reliance on fossil fuels leads to 1,000 ppm CO_2 and 6°C warming by the end of the twenty-first century.

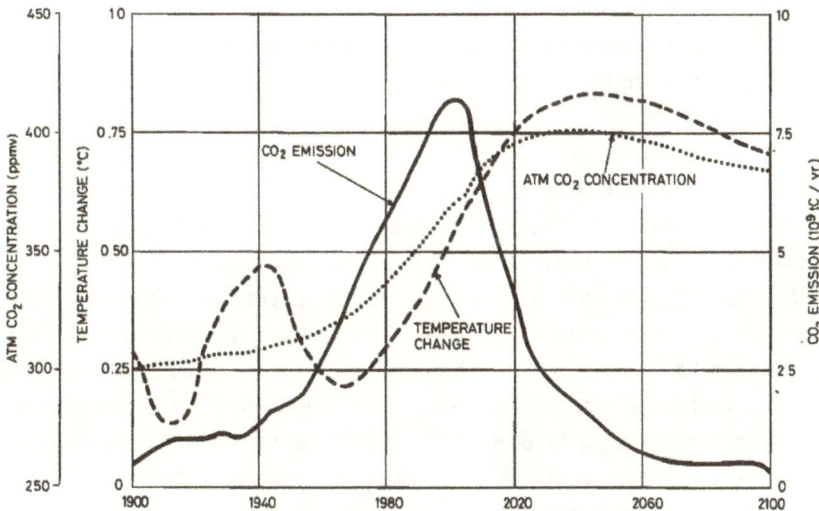

Fig. 10. CO_2 impact of 30-TW solar and nuclear strategy.

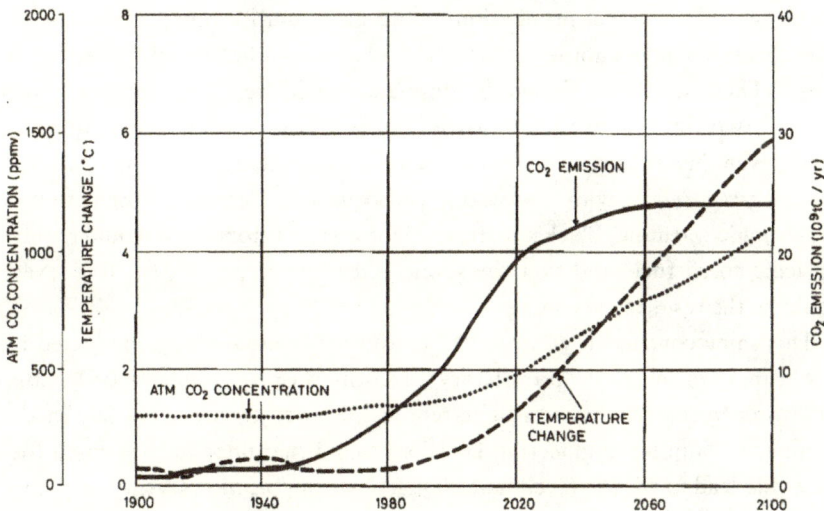

Fig. 12. CO_2 impact of 30-TW fossil fuel strategy.

Source: F. Niehaus and J. Williams, "Studies of Different Energy Strategies in Terms of Their Effects on the Atmospheric CO_2 Concentration," *Journal of Geophysical Research* 84, no. C6 (June 1979): 3127, 3129.

CO_2 in the atmosphere, respectively. Given his calculation that CO_2 concentrations could be limited to about 400 ppm with immediate action, Laurmann's analysis implied that one, two, or three decades of delay beyond 1979 would result in minimum CO_2 concentrations of 470, 540, or 650 ppm, corresponding to warming of 2.3, 2.9, or 3.7°C.[21] In other words, every year of delay would result in more unavoidable warming and more enormous, permanent damage to human life and planetary ecosystems.

To sum up, by 1980, the scientific literature showed not only that catastrophic global warming was preventable but also that every year of delay in replacing fossil fuels would come at the cost of greater climate damage. Table 2.1 summarizes this literature from the late 1970s and early 1980s.

Recently discovered internal documents from Exxon and other companies demonstrate that oil manufacturers were also aware of the necessity of replacing fossil fuels. In July 1977, for example, James Black from Exxon Research and Engineering met with the company's Management Committee to discuss climate forecasts, informing his supervisors of the "general scientific agreement that the most likely manner in which mankind is influencing the global climate is through carbon dioxide release from the burning of fossil fuels" and that the best available models predicted that "a doubling of the CO_2 concentration in the atmosphere would produce a mean temperature increase of about 2C to 3C over most of the earth," with two to three times more warming near the poles.[22]

To limit warming, Black presented the Exxon board with two scenarios: one in which fossil fuel production would grow until around 2060 and decline thereafter and another in which production would begin to decline in 2025.[23] These scenarios, shown in figure 2.2, would yield CO_2 concentrations five and two times above preindustrial concentrations, respectively. In other words, even the safer scenario was expected to cause global warming of around three degrees Celsius with devastating consequences. Despite his prediction of catastrophic warming, Black's central insight was that stopping it would require replacing fossil fuels and that the sooner they were replaced, the less severe would be the resulting warming.

This same conclusion was reached in another internal study conducted by Exxon in 1979, marked "proprietary" and titled *Controlling the CO_2 Concentration in the Atmosphere*. The report, authored by Steve Knisely in the company's Planning Engineering Division, found that uncontrolled fossil fuel use would lead to "dramatic climate changes" within the next seventy-five years and that preventing such an outcome would require "dramatic changes in patterns of energy use," including leaving most fossil fuel reserves in the ground.[24] Knisely calculated that "only about 20% of the recoverable fossil fuel could be used before doubling the atmospheric CO_2 content"; he also predicted that a doubling of atmospheric CO_2 would cause increased rainfall, four feet of sea

TABLE 2.1. Overview of CO_2-control research, 1977–1983.

Year	Publication	Minimum achievable CO_2 concentration (ppm)
1977	William D. Nordhaus, "Strategies for the Control of Carbon Dioxide" (Cowles Foundation Discussion Paper 443, Yale University, New Haven, CT)	417
1978	John A. Laurmann, "Fossil Fuel Utilization Policy Assessment and CO_2 Induced Climate Change," in Carbon Dioxide, Climate and Society, ed. Jill Williams, IIASA Proceedings Series: Environment 1 (Laxenburg: IIASA), 253–62	No estimate but calls for immediate action due to market-penetration time
1979	F. Niehaus and J. Williams, "Studies of Different Energy Strategies in Terms of Their Effects on the Atmospheric CO_2 Concentration," Journal of Geophysical Research 84, no. C6 (June): 3123–29	400
1979	John A. Laurmann, "Market Penetration Characteristics for Energy Production and Atmospheric Carbon Dioxide Growth," Science 205, no. 4409: 896–98	403
1979	Steve Knisely, "Controlling the CO_2 Concentration in the Atmosphere," Exxon internal document, Oct. 16, 1979, ID no. mqwl0228, Climate Investigations Center Collection, University of California, San Francisco, https://www.industrydocuments.ucsf.edu/fossilfuel/docs/#id=mqwl0228	440 (420 by 2050)
1981	Jeanne Anderer, Alan McDonald, and Nebojsa Nakicenovic, Energy in a Finite World: Paths to a Sustainable Future (Cambridge, MA: Ballinger)	400 (adopting Niehaus and Williams, "Different Energy Strategies")
1982	A. M. Perry et al., "Energy Supply and Demand Implications of CO_2," Energy 7, no. 12: 991–1004	460 (note: funded by the petrochemical company Union Carbide)
1983	David J. Rose, Marvin M. Miller, and Carson E. Agnew, Global Energy Futures and CO_2-Induced Climate Change (Cambridge, MA: Energy Laboratory, MIT)	420 (by 2050)

FIGURE 2.2. Future scenarios describing fossil fuel production (*left*) and the resultant increase in global atmospheric CO_2 concentrations (*right*), presented to Exxon's Management Committee by company scientist James Black in July 1977. Even the relatively safe scenario presented by Black would allow CO_2 levels to nearly double and global warming to rise nearly 3°C.

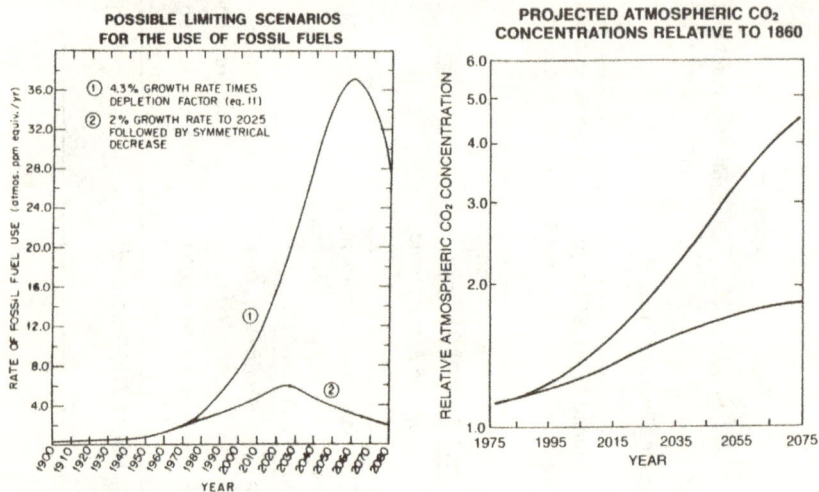

Source: J. F. Black to F. G. Turpin, letter and report (describing the presentation "The Greenhouse Effect," July 1977), Exxon internal documents, June 6, 1978, ID no. xqwlo228, Climate Investigations Center Collection, University of California, San Francisco (hereafter UCSF), https://www.industrydocuments.ucsf.edu/fossilfuel/docs/#id=xqwl0228, vu-graphs 3, 4.

level rise, and melting of the polar ice caps—potentially triggering "major increases in earthquakes and volcanic activity resulting in even more atmospheric CO_2 and violent storms" as well as an ice-free Arctic that would produce "major shifts in weather patterns in the northern hemisphere."[25]

Knisely also warned that unless limits were placed on emissions, "noticeable temperature changes would occur around 2010 as the concentration reaches 400 ppm" and that "significant climatic changes [will] occur around 2035 when the concentration approaches 500 ppm." Knisely predicted dire impacts in the United States at five hundred ppm: the Southwest would be hotter and drier and suffer water shortages, glaciers would melt in the Rockies and Pacific Northwest, snowpack would decline, and marine life would be "markedly changed." Without controls on fossil fuel production, preindustrial CO_2 would double by 2050, causing "dramatic climatic changes in the world's environment."[26] Knisely's predictions were remarkably accurate: CO_2 concentrations passed four hundred ppm in 2013, postdating his prediction by only three years.[27]

Could these outcomes be avoided? Knisely considered three future scenarios: one in which fossil fuels remained *uncontrolled,* a second in which the CO_2 buildup was limited to *510 ppm,* and a third in which CO_2 was limited to *440 ppm* (see fig. 2.3), which Knisely assumed to be "a relatively safe level for the environment." Knisely found that to achieve this last-mentioned safe concentration, non–fossil fuels would need to be substituted for coal in the 1990s and supply 50 percent of the world's energy by 2010. Neither shale oil nor coal (nor, by extension, tar sands) could remain or be developed as major energy sources. Carbon dioxide emissions would have to peak in the mid-1990s and decline thereafter, and most fossil fuel reserves would have to remain in the ground.[28]

In 1989, Royal Dutch Shell developed its own confidential scenarios describing possible global warming futures. In one scenario, called "sustainable world," greenhouse gas emissions would peak around the year 2000 and decline rapidly thereafter, with total CO_2 in the atmosphere limited to four hundred ppm. In another scenario, dubbed "global mercantilism," emissions would continue to rise with dire outcomes, including

> more violent weather—more storms, more droughts, more deluges. Mean sea level would rise at least 30 cm. Agricultural patterns would be most dramatically changed. Something as simple as a moderate change in rainfall pattern disrupts eco-systems, and many species of trees, plants, animals, and insects would not be able to move and adapt.
>
> The changes would, however, [have the] most impact on humans. In earlier times, man was able to respond with his feet. Today, there is no place to go because people already stand there. Perhaps those in industrial countries could cope with a rise in sea level (the Dutch example) but for poor countries such defences are not possible. . . . The potential refugee problem in GLOBAL MERCANTILISM could be unprecedented. Africans would push into Europe, Chinese into the Soviet Union, Latins into the United States, Indonesians into Australia. Boundaries would count for little—overwhelmed by the numbers. Conflicts would abound. Civilization could prove a fragile thing.[29]

Thus, like Exxon, Shell predicted that preventing catastrophe would require an immediate transition away from fossil fuels. What, though, about using carbon capture to enable the continued use of fossil fuels? Interestingly, Exxon considered and rejected this possibility early on. In 1981, Exxon Research and Engineering internally distributed the twenty-page *Scoping Study on CO_2,* which considered the possibility of using carbon capture technology (which already existed) to reduce emissions but concluded the approach wasn't economically feasible. "The cost of scrubbing large quantities of CO_2 from flue gases is exorbitant," the report stated, so much so that "indirect control measures, such as energy conservation or shifting to renewable energy sources, represent the only options that might make sense."[30]

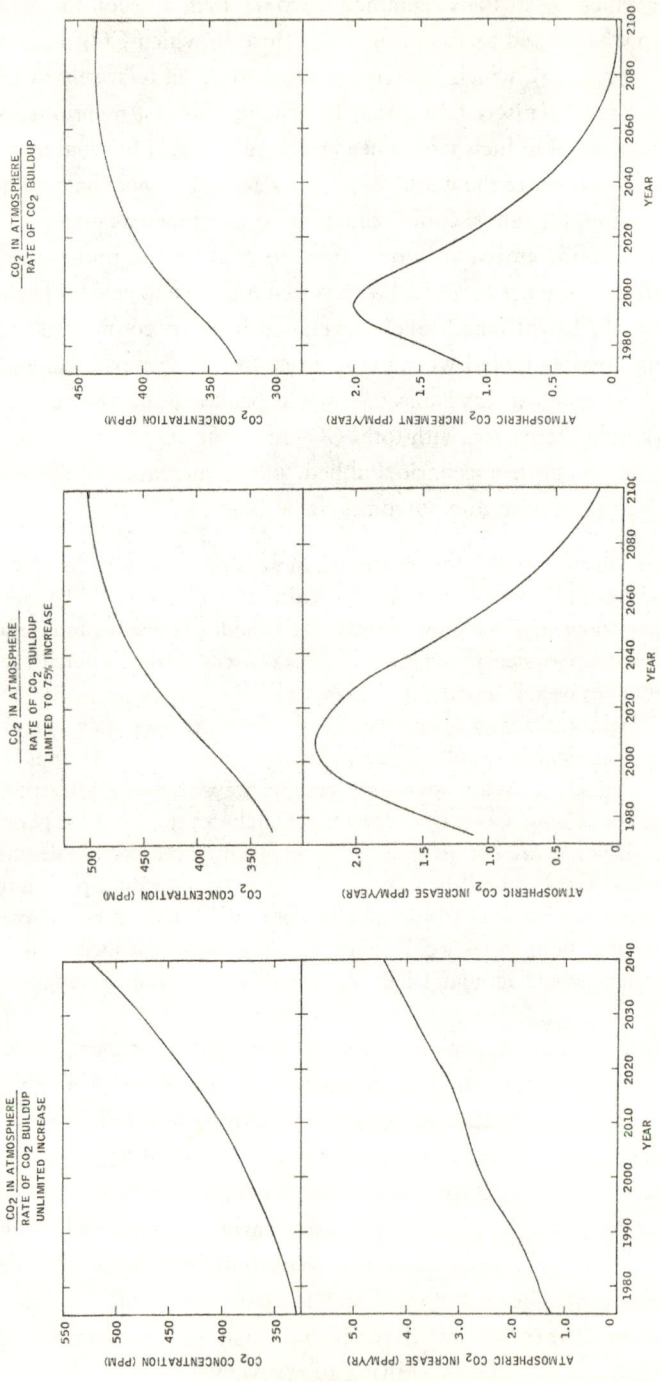

FIGURE 2.3. Three scenarios for the future of CO_2 according to Knisely at Exxon in 1979: CO_2 concentration is unlimited from lack of fossil fuel controls (*left*); CO_2 is limited to 510 ppm through some fossil fuel controls (*middle*); CO_2 concentration is limited to 440 ppm through strong fossil fuel controls (*right*). Knisely's report demonstrates that Exxon knew by 1979 how to avoid or minimize a buildup of CO_2—by rapidly developing nonfossil energy sources and leaving most such fuels in the ground—and that immediate action was necessary to avoid severe global warming. Note: For each chart, the top shows atmospheric CO_2 concentrations, and the bottom indicates annual CO_2 emissions.

Source: Knisely, *Controlling the CO$_2$ Concentration*, figs. 11, 9, 7.

By 1980, then, leading fossil fuel manufacturers had taken notice of the most important facts about global warming and recognized that transitioning to renewables would be necessary, and largely sufficient, to solve the problem. Alternatives to replacing fossil fuels, such as carbon capture, were rejected as economically nonviable. And researchers both inside and outside the industry understood that each year of delay would result in additional unavoidable global warming and ever-greater damage. Conceptually, the cause and the cure were pretty straightforward.

BURYING THE MOST IMPORTANT FACT

The most important fact about global warming—that preventing it would require a rapid transition away from coal, oil, and gas—poses an existential threat to fossil fuel producers who, in the 1980s, set about not only denying the climate problem but also promoting the idea that global warming could be solved *without replacing fossil fuels.* In short order, any apparent solution to climate change that left fossil fuels intact would be championed by the industry, and a resultant flurry of false and inadequate solutions quickly came to infect climate discourse, research, and policymaking at all levels.

Although fossil fuel producers have *promoted* nearly every nonviable or inadequate solution to climate change developed over the past four decades—polluting the solutions landscape—not all these strategies *originated* with the industry. In other words, Big Carbon is not the only party responsible for distracting from the necessity of replacing fossil fuels. The treatment of climate change at the IIASA in the 1970s is a disconcerting example.

As described by historian Isabell Schrickell, the issue of global warming was introduced at IIASA around 1973, largely by William Clark (later a professor at Harvard) and Crawford Stanley Holling (a pioneer of ecological economics), both of whom spent their careers defining concepts such as "sustainability" and "socioecological resilience."[31] Also in the 1970s, a group of engineers, physicists, and economists at IIASA, including the economist William Nordhaus and the physicist Wolf Häfele, started promoting the idea of removing CO_2 and other industrial pollutants through ocean sequestration or recycling. The Italian physicist Cesare Marchetti, at IIASA since 1974, coined the term "geoengineering" to describe these processes. Meanwhile Clark, Holling, and their collaborators, including the economist Thomas Schelling, promoted adaptation and management, arguing that the fundamental problem was not fossil fuels per se but rather the "resilience" of human society. This is classic catarheumatics: don't prevent, just learn to cope.

These paradigms—which sought to adapt to or clean up pollution after the fact—shared a significant blind spot: the importance of preventing pollution in the first place. Häfele and Nordhaus claimed that replacing fossil fuels would be too disruptive to society, and rather than promote the growth of renewables (which both men saw as unrealistic), they insisted on using recycling methods that never really worked (and still don't). The adaptations promoted by Clark and Schelling also let Big Oil off the hook. Their approach called not for the prevention of global warming, but rather for a vague and complex program of global management. Clark and Schelling did not seem to realize that more direct solutions were required, and their rejection of existing nascent solutions in favor of hopeful fantasies had the effect, intended or not, of maintaining the fossil fuel status quo.

Another obstacle can be found in Nordhaus's use of economic rhetoric to normalize the climate catastrophe. While searching for a market solution befitting his economic ideology, he speculated that a doubling of atmospheric CO_2 would be a reasonable target—and then tried to devise economic models to find the "optimal" path toward that outcome. Although his analysis used a lot of math, in reality, it ignored the warnings being issued by actual climate scientists. Equally myopic was a letter to the National Academy of Sciences submitted by a group chaired by Schelling focusing on the social and political aspects of global warming.[32] Schelling asserted that the best response to warming would not be fossil fuel replacement but rather mass human migration and adaptation to a new global climate system. He also exaggerated the uncertainty of the existing science and insisted that market forces would fix the problem. On crucial matters of fact, Schelling turned out to be catastrophically wrong: market forces have not fixed global warming, and his recommendations of human migration and adaptation carried cruel and potentially unlawful implications.

In 1983, the National Academy released a follow-up to the 1979 Charney report, which had issued strong warnings about global warming and implied the necessity of replacing fossil fuels.[33] The new report was chaired by William Nierenberg (then director of the Scripps Institution of Oceanography and later an outspoken climate sceptic) and included perspectives from Schelling and Nordhaus. Schelling insisted that "it would be wrong to commit ourselves to the principle that if fossil fuels and carbon dioxide are where the problem arises, that must also be where the solution lies." Nordhaus, for his part, wrote that while carbon emissions could be reduced by taxing fuels, the strategies suggested by Schelling, notably climate modification and adaptation, were "likely to be more economical ways of adjusting."[34] Yet neither economist presented evidence for those conclusions. As described by historians Naomi Oreskes and Erik Conway, a nuclear physicist who reviewed the report, Alvin Weinberg, at the time called it "seriously flawed in its underlying analysis and in its conclusions" and asked, presciently, "Does the Committee really believe that the

United States or Western Europe or Canada would accept the huge influx of refugees from poor countries that have suffered a drastic shift in rainfall pattern?" Yet these and other critiques were ignored, and the report was used by the Reagan White House to ignore calls by the U.S. Environmental Protection Agency to reduce coal use.[35]

This history shows that although, by 1980, multiple peer-reviewed studies— and Big Oil's own in-house research—pointed to the clear need to replace fossil fuels, this conclusion was muddied by complacent economists and others who insisted, with little evidence, that such replacement would not be necessary. And this muddying of the waters was successfully weaponized to keep the fossil fuel train hurtling forward.

Additionally, in a few short years, the global fossil fuel industry and its allies had developed a full suite of false solutions to keep fossil fuel controls at bay. Much of the strategic coordination within the industry was conducted through the International Petroleum Industry Environmental Conservation Association (IPIECA), a gassy giant created in the 1970s to help carbon producers interface with the UN Environment Programme (UNEP). In 1988, for example, at a meeting held at Total's headquarters in Paris, the IPIECA created the Ad Hoc Group on the Greenhouse Effect, soon renamed the Working Group on Global Climate Change. The working group was chaired by Duane LeVine, Exxon's manager for science and strategy development, with members including Bernard Tramier from the French oil giant Elf, Brian Flannery from Exxon, Leonard Bernstein from Mobil, Terry Yosie from API, and other representatives from BP, Shell, Texaco (now Chevron), and Saudi Aramco. The group had three goals: first, "to draw up the state of the science of climate change induced by the possible accentuation of the greenhouse effect, including the main areas of uncertainty"; second, to develop "no-regrets" response strategies that would keep fossil fuels intact; and, third, to examine "efficiency" and fossil gas as industry-friendly responses to global warming.[36]

In 1990, the IPIECA's working group sent materials on climate change to its dozens of member oil companies all around the world, including a strategy document created by Exxon's Duane LeVine. LeVine warned that international policymakers would soon seek to control fossil fuels the way they had recently controlled ozone-depleting chemicals and outlined a plan to ensure that any such effort would fail. To defeat policies that could shift "the energy resource mix" away from fossil fuels, LeVine explained that the industry should emphasize uncertainties in climate science while simultaneously calling for further research. Additional strategies were to portray the costs of climate controls as prohibitive and to promote industry-friendly solutions that would keep the oil and gas business intact. Using LeVine's strategy, the industry ensured that the pollution-control approach used to solve the problem of ozone depletion—culminating in the Montreal Protocol in 1987—would stall when it came to climate and fossil fuels.

A key revelation from Exxon's strategy document is that by 1989, the promotion of false solutions had joined denial, the weaponization of uncertainty, and economic rhetoric as foundational strategies for prolonging the reign of fossil fuels. The first two false solutions identified by the industry were to promote efficiency and "natural gas" (fossil gas). And here we need to appreciate two fundamental realities of climate physics: (1) no amount of efficiency can solve the climate problem so long as the economy is based on fossil fuels, and (2) "natural gas" is just another fossil fuel—and just as carbon intensive as oil or coal when methane emissions are taken into account.

Reforestation would soon be added as a third industry-supported false solution. Shell's internal climate-scenarios report from 1989, for example, contains one of the earliest known proposals to use reforestation "offsets" to justify continued fossil fuel expansion. Near the end of that report, a boxed section urges "Don't Just Stand There. Plant A Tree," explaining that Applied Energy Services (a power company) was paying to plant some fifty-two million trees in Guatemala to justify building a new coal-fired power plant in Connecticut.[37] Shell even hoped to make money from this scheme, noting optimistically in an internal report from 1988 that if reforestation were adopted as a solution to global warming, "there would be some call on companies, including Shell, with experience in tropical forestry."[38]

The truth, however, is that much like efficiency, reforestation can never be adequate to counteract fossil fuel pollution. Trees only temporarily store carbon, which means that reforestation cannot counteract the long-term, quasi-permanent (on a human scale) climate contamination from fossil fuels. Reforestation cannot compensate for fossil-carbon pollution.

What we have to appreciate, then, is that by 1980, the most important fact about global warming—that preventing it would require a concerted replacement of fossil fuels—was well established. Crucial also to appreciate, however, is that ten years later, Big Carbon producers had successfully managed to obscure that fact by promoting three false and distracting solutions: fossil gas, reforestation, and efficiency. Contributing to the obfuscation were a number of prominent economists who had started (intentionally or not) protecting fossil fuel interests by insisting that adaptation, geoengineering, or trusting in price signals would be preferable to a focused development of non-fossil-energy systems. The success of this agnogenesis is reflected in the fact that in the text of the United Nations Framework Convention on Climate Change adopted in 1992, fossil fuels are mentioned only four times—each time, astonishingly, in order to *protect* their continued use.[39]

The industry's delay tactics since this time are familiar.[40] From 1989 to 1998, the carbon majors leaned heavily on denial, creating front groups, like the Global Climate Coalition, to dispute the scientific basis for global warming.[41] Oil producers also began hiring economic consultants as part of a strategy to exaggerate the costs of replacing fossil fuels and downplay the costs of global warming.[42] Around 1998 (when Big Tobacco faced hundreds of billions of dollars in legal liability for its own deceptions), however, we start to see an important shift in the carbon majors' public positioning on climate change. From that time forward, rather than disputing the scientific basis for warming, the companies began presenting themselves as partners in trying to solve the problem. As such, the promotion of false solutions took on heightened importance.

In the new millennium, more pseudo-solutions were added to the list, including carbon capture (not economically viable), hydrogen fuel (mostly made from fossil gas and itself an indirect greenhouse gas), biofuels (expensive, not scalable, and mostly composed of fossil fuel), solar geoengineering (potentially very risky), and a fixation on individual consumer habits (inadequate). These quickly became the subjects of industry-sponsored academic centers, advertising campaigns, and policy platforms, consuming years of public attention and delaying the deployment of demonstrably viable solutions, such as electrification (including of transport) and renewable power generation (mainly solar and wind).

Describing the full extent of Big Carbon's greenwashing and promotion of false solutions is beyond the scope of this chapter, but the industry's inroads into academia deserve special mention.[43] An internal plan formulated in 1998 by representatives from across the trade, including Exxon and Chevron, called for the support of industry-friendly scientists around the country—not crude climate deniers but rather mainstream academics aligned with Big Oil's delay-oriented positions, including the stalling-tactic that more research was needed before fossil fuel replacement could begin.[44] From that point on, Big Oil used its new ostensibly solutions-oriented position on climate change, together with the promise of corporate philanthropy, to gain sway over the scientific community (and, indirectly, the Intergovernmental Panel on Climate Change) and turn potential opponents into allies.[45] In 2000, for example, BP and the Ford Motor Company created the Carbon Mitigation Initiative at Princeton—now the university's "largest and most long-term industry-university partnership"—to study "carbon management" (not fossil fuel replacement).[46] Researchers from the new center helped develop an "all-of-the-above" climate policy that was promoted throughout the 2000s and 2010s to frame fossil fuels as part of the solution to global warming.[47]

Stanford University also became a major recipient of fossil fuel money. In 2002, Exxon founded the university's Global Climate and Energy Project, a multidecade $200 million–plus center that gave its funders (the majority of which were corporate fossil fuel interests and all of which were chosen by Exxon) legal control over research projects. The project was directed first by petroleum engineer Lynn Orr and then by "ground fluids" engineer Sally Benson and spent much of its funding researching carbon capture and other industry-friendly responses to global warming.[48] Two years later, forestry and fossil fuel executive Ward Woods endowed Stanford's Woods Institute for the Environment, which would focus much of its attention on "natural climate solutions" (i.e., trying to engineer ecosystems to absorb CO_2) rather than fossil fuel replacement.[49]

In 2007, Chevron, BP, Total, and other fossil fuel companies founded the Massachusetts Institute of Technology (MIT) Energy Initiative, which subsumed much of the climate-related work at the institute and openly gave its funders significant influence over research directions. ExxonMobil, Saudi Aramco, Shell, and many other fossil fuel producers would soon join as sponsors of MIT's initiative, which was first directed by natural gas proponent Ernest Moniz and later by MIT's Chevron professor of chemical engineering Robert Armstrong. That same year (2007), BP created the Energy Biosciences Institute at the University of California, Berkeley, to promote biofuels as a solution to global warming under the direction of enhanced-oil-recovery microbiologist John Coates. And oil and gas executive Jay Precourt created the Precourt Center for Energy Efficiency at Stanford, directed by the Exxon-financed economist James Sweeney, which promoted fuel efficiency as the solution to climate change.

The trend would continue. In 2015, for example, a range of oil and gas companies established the Natural Gas Initiative at Stanford under the direction of industry-friendly economist Frank Wolak to promote fossil gas as a climate solution. And in 2018, ExxonMobil, Shell, and Total established Stanford's Strategic Energy Alliance, a continuation of the industry-funded Global Climate and Energy Project. Big Oil's patronage would also come to dominate many already-existing university research centers, including Stanford's Energy Modeling Forum, MIT's Earth Resources Laboratory and its Center for Energy and Environmental Policy Research, Princeton's Andlinger Center for Energy and the Environment, Columbia's Center on Global Energy Policy, and Harvard's Project on Climate Agreements. Harvard's Consortium for Energy Policy Research was similarly co-opted, as was its Environmental Economics Program, directed by the oil-friendly economists Robert Stavins and William Hogan.

Much like Big Tobacco's widespread funding of cancer research, which focused on causes *other* than cigarettes,[50] Big Oil has tended to fund work that keeps researchers busy without threatening the industry's bottom line. Much of that research has focused on promoting "solutions" that keep fossil profits flowing, including research into fossil gas, carbon capture, energy efficiency, hydrogen fuel, biofuels, "natural climate solutions" (such as reforestation), and "management" of or adaptation to a warming planet.

Emboldened by the credibility given it by funded academics, Big Oil has also promoted its pseudo-solutions through advertising. In the early 2000s, for example, the company once known as British Petroleum spent $100 million to rebrand itself as Beyond Petroleum, creating a green image for itself while continuing to funnel nearly all its investments into fossil fuels.[51] The company also popularized one of the most devious climate concepts of all time: the personal carbon footprint. Mirroring junk food producers blaming consumers for obesity, in 2006, the carbon giant proclaimed on its website that it was "time to go on a low-carbon diet" and encouraged consumers to use its "carbon footprint calculator."[52] The concept was brilliantly deceptive, deflecting blame onto consumers with scientific precision while simultaneously giving the company an aura of virtue—and guilt-tripping us all into an unwinnable quest to avoid poisoning the earth through personal choices.[53]

In 2007, Chevron launched its own advertising campaign, titled Will You Join Us?, depicting the company as an environmental leader and blaming global warming on individual consumer behavior. The company claimed that saving energy was equivalent to developing new sources of renewable energy; consumers were also encouraged to start "driving slower," with this upbeat admonition: "We've got a huge source of alternative energy all around us. It's called conservation. . . . Clearly, saving energy is like finding it. We've taken some of the steps needed to get started but we need your help to get the rest of the way." Ads such as these encouraged consumers to "join" Chevron in addressing climate change by making minor changes in their personal energy consumption. Everyday Americans were shown making promises, including, "I will use less energy," "I will leave the car at home more," "I will take my golf clubs out of the trunk and save gas," and "I will consider buying a hybrid."[54]

Three years later, in 2010, Chevron launched another ad campaign called We Agree, promoting fossil gas as a clean-energy source. One television ad, titled "Shale Gas Needs to Be Good for Everyone," featured a concerned citizen-farmer and a Chevron employee, where the farmer opens with, "We're sitting on a bunch of shale gas. . . . It's a game changer." Meanwhile a placard implores, "Let's do the right thing on shale gas." The farmer then says, "It means cleaner, cheaper, American-made energy." The farmer and the employee both tout the economic benefits of shale gas, and the employee then chirps, "At Chevron, if we can't do it right, we won't do it at all." Both then close by saying in unison, "We've got to think long term." In yet another Chevron ad, which aired in Australia in 2013, an older confident male "Chevron natural gas advisor" announces that Chevron is building "one of the biggest natural gas projects in the world. Enough power for a city the size of Singapore for fifty years." A young concerned female "web designer" then asks, "What's it going to do to the planet?" The Chevron rep patronizingly reassures her that fossil gas "is the cleanest conventional fuel there is" and "the smart way to go" while both nod in agreement.[55]

In 2016, Shell launched its own campaign, called Make the Future, developed by the public relations giant Edelman. The campaign targeted "Energy-Engaged Millennials" (EEMs in Shell's internal documents), featuring pop icons, such as Jennifer Hudson, portraying the oil giant as a clean-energy leader. A leaked copy of one of the campaign's strategy documents shows that it sought to divert attention away from fossil fuels and place blame for global warming on all of society. Shell's campaign included "advertorials" (presented to look like news stories) in the *Washington Post* and *New York Times*, designed by those papers' in-house advertising companies, which touted liquified natural gas (LNG) as "sustainable," a "lower-carbon fuel, " and "part of a mosaic of alternative energy sources."[56]

ExxonMobil also joined the greenwashing party, placing misleading advertorials in the *New York Times*, many of which focused on the company's promotion of algae biofuels.[57] Exxon's ads didn't mention that most commercialized biofuels are, in reality, still mostly composed of fossil fuels with only a small amount of biological material blended in or the fact that from 2010 to 2018, the company spent only 0.2 percent of its capital investments in biofuels and other low-carbon energy systems (with the remaining 99.8 percent going to fossil fuels).[58] In 2023, Exxon ended its fourteen-year-old algae biofuels advertising campaign—seen by millions—without ever making an algae biofuel product.[59]

OBSTRUCTION AND OBFUSCATION

For the past quarter of a century, oil producers have falsely promoted themselves as leading the way in fixing global warming while normalizing fossil fuel expansion and steadily growing greenhouse gas emissions. The false solutions promoted by the industry—fossil gas, hydrogen, biofuels, carbon capture, carbon offsets, efficiency, and consumer choices—leave fossil fuel dominance intact. Some of the biggest names in public relations and print media, including Edelman and the *New York Times*, continue to help the industry with its deceptive advertising. Half a century after Big Oil could and should have begun moving toward clean-energy sources, fossil fuel expansion continues—under the cover of false solutions.

This is remarkable given that by 1980 the most important facts about global warming were understood, including what to do about it. Burning fossil fuels was understood as the main cause, and leaving them in the ground and replacing them with renewables was known to be the solution. Big Oil's response was to create a three-prong plan to defeat that process: weaponizing complexity and scientific uncertainty, manipulating economics, and promoting false solutions. The industry largely abandoned overt climate denial in the late 1990s but dramatically expanded its promotion of pseudo-solutions, adding carbon capture, hydrogen fuel, biofuels, and consumer habits to its older promotion of efficiency, fossil gas, and carbon offsets. These distractions have helped occlude the fact that stopping global warming necessitates replacing fossil fuels.

Today, we remain awash in pseudo-solutions after decades of industry advertising and the infiltration of academia to legitimize them. Especially diabolical is that many of these false solutions seem to fit with traditional goals of environmentalists, such as protecting forests, reducing waste, and promoting personal responsibility. Yet none is adequate or viable as a solution to global warming. Indeed, these false paths have likely been supported by the industry precisely *because* they leave fossil fuel dominance intact. They are destined to fail while simultaneously portraying fossil fuel producers as responsible corporate citizens.

Arguably, false solutions have become even more important than climate denial as an obstacle to solving the climate crisis. The comparison with tobacco is closer than one might imagine given that, for decades, cigarette makers promoted false solutions to the cigarette catastrophe: first "toasting" and later menthols, filters, low tars, lights, and ultralights and then cigarettes advertised as "natural," organic, or additive-free. Cigarette makers sometimes even funded cancer research, with the idea being that curing cancer would allow the cigarette makers to keep on causing it. The cigarette denial campaign was really "a holding strategy,"[60] recognizing that every year of delaying the end of cigarettes would mean billions in profits for their makers. For Big Oil, too, the promotion of false solutions has replaced outright denial as its chief survival strategy. Sadly, this strategy of delay has often been aided by subservient academics and politicians and even some myopic environmental groups.

Counteracting this universe of false solutions will require scrutiny and perhaps even legal action. Vague promises of future progress in biofuels, hydrogen, carbon capture, and other technical fixes are inadequate and even dangerous insofar as they result in wasted time. Viable, proven solutions for replacing fossil fuels already exist, including solar and wind energy, electrification and electrical storage, and laws and economic incentives that can help us transition more rapidly toward those solutions. More than forty years after the most important fact about global warming was understood, we must stop looking for hope in all the wrong places and meet the fossil fuel disaster head-on.

THREE

Preventing Unwanted Births Can Help Mitigate Climate Change—While Enlarging Human Liberties

LONDA SCHIEBINGER *and* ROBERT N. PROCTOR

THERE IS AN UNFORTUNATE oddity in our modern world: we are in the midst of two human catastrophes intimately linked but almost never imagined as having anything to do with one another. What we're really talking about is a paralyzing taboo, infecting the Left as much as the Right. Both catastrophes are of significance for human liberties and the health of the planet.

We are talking about a taboo but also a myopia. As for the taboo, it has become difficult to talk about any kind of link between population and climate change for historical reasons we shall explore. As for the myopia, most scholars are keenly aware of the threat of coercive *contraception* but rarely attach any significance to coercive *conception*. The taboo largely follows from the myopia. In discussions of our ongoing climate catastrophe and biodiversity crisis, almost never do we find any acknowledgment that *we are in the midst of an epidemic of forced birthing*—and this, too, deserves exploration. Here, we offer a reframe that allows us to talk ethically about population and climate change in the same breath.

Ours is a case study in agnotology and, specifically, how the opportunity of *preventing unwanted births* has come to be ignored as a means to alleviate our ongoing loss of biodiversity and the numerous other threats posed by climate change. The Intergovernmental Panel on Climate Change (IPCC) in 2022, for example, treated population as a "driver" of greenhouse gas emissions but not as an opportunity to reduce those emissions. This neglect can be understood as a reaction to racist neo-Malthusianism but also as a consequence of the failure to consider contraception (and abortion) as part of "technology." To stabilize the population in a manner consistent with human liberties, we need to reframe this opportunity in terms of preventing unwanted births. Ending coercive conception can help fix the environment while simultaneously enlarging human liberties. The reader may find it odd to hear that rape, child marriage, and state sanctioned pronatalism are all causally linked to the bleaching of coral and the increased frequency of wildfires, but that is the too often overlooked truth: stopping reproductive coercion—essentially forced birthing—would be a win-win for humanity and all other life on the planet.

WHAT IS AT STAKE?

Every year, some thirty-seven gigatons of new anthropogenic carbon enter the air, mainly as a result of burning fossil fuels. The net effect has been a steady increase in the CO_2 content of the air, with today's levels topping out at over 430 parts per million—a higher level than any time in the last three million years. Climate modelers recognize that population is an important driver of emissions but do little to explore how population growth might be limited in a manner that is consistent with the expansion of human liberties and bodily autonomy.

This is a remarkable oversight. Rarely in the massive literature on climate change do we find any acknowledgment of the fact that a sizeable amount of atmospheric carbon could be avoided by reducing "unwanted births" (boxes 1 and 2). This is significant, given that nearly half of all pregnancies are unintended and an estimated 10 percent of all births are "unwanted."[1] Population is routinely included as a "variable" in climate models but is de facto treated as a given, unalterable, and of little or no policy consequence—like solar radiation or the eccentricity of the earth's orbit. No consideration is given to the fact that the systematic deprivation of human liberties results in millions of babies being born every year, with substantial environmental consequences.

The myopia here is interestingly asymmetrical: scholars often worry about coercive *contraception* (think forcible sterilization) while giving much less attention to coercive *conception* (think rape and state pronatalism), when the truth is that millions of women in the world today are essentially forced to give birth. Unintended (and often unwanted) pregnancies are common in both poor and wealthy parts of the world and result from numerous causes,[2] the most significant of which include the lack of access to contraception, the prohibitive cost of contraception, failures in contraceptive products, partner opposition to the use of contraception, religious sanctions against contraception (or abortion), laws banning abortion (or contraception), the insufficiency or inefficacy of male contraception, and state-sanctioned pronatalism. All these are examples of *reproductive coercion*, and even this is only a partial list; there are many other ways in which women end up being forced to conceive and give birth. The absence of comprehensive sex education and gender equality more broadly contributes to the crisis—for both men and women.[3] Many babies are also born as a result of sexual assault and violence, exacerbated by traditions sanctioning child marriage.[4]

BOX 1. What Does It Mean for Births to Be "Unwanted"?

Globally, nearly half of all pregnancies each year are unintended. These unintended pregnancies—numbering more than 120 million each year—represent a human rights crisis, both for the women who may be forced to give birth and the children who may suffer as a result. How are these numbers generated? Demographers use population-based surveys to generate such data, with women typically asked something like, "At the time you became pregnant, did you want to become pregnant then, did you want to wait until later, or did you not want to have any (more) children at all?"[a]

Women who report they did *not* want to become pregnant are considered to have had an "unintended pregnancy," which results in either miscarriage (12 percent), induced abortion (60 percent), or live birth (28 percent)—some fraction of which can be classed (according to survey data) as "unwanted." Best estimates put the figure at about 13 million unwanted births per year globally (roughly 10 percent of total births), though this could be an underestimate given that it relies on interviews with mothers up to three years after giving birth. These women may be unwilling to label a birth unwanted ("rationalization bias") or to admit they did not use contraceptives or that their contraceptives failed.[b]

---

a. UNFPA (United Nations Population Fund), *Seeing the Unseen: The Case for Action in the Neglected Crisis of Unintended Pregnancy* ([New York]: UNFPA Division for Communications and Strategic Partnerships, 2022), 29, https://eeca.unfpa.org/sites/default/files/pub-pdf/en_swp22_report_0_2.pdf.

b. Heather M. Rackin and S. Philip Morgan, "Prospective versus Retrospective Measurement of Unwanted Fertility: Strengths, Weaknesses, and Inconsistencies Assessed for a Cohort of US Women," *Demographic Research* 39 (2018): 61–94.

The numbers here are astonishing. According to the United Nations, there are 650 million women alive today who were married as children.[5] Even in the United States, an estimated three million women alive today have suffered rape-related pregnancies—many of which resulted in births.[6] The net effect is that millions of babies are born every year who would not have been born if people—women and men alike—had the freedom to bring into the world only those babies they wanted.

Enlarging freedom in this manner—by making contraception and abortion freely available, combined with sex education and efforts to achieve gender equality—could reduce global carbon emissions by roughly 10 percent, or 3.7 gigatons per year, which is more than the total combined emissions of Germany, Japan, Brazil, Turkey, Mexico, and Australia.[7] Ending coercive conception and forced birthing would be equivalent to taking some eight hundred million gas-cars off the road, which is about half of all vehicles worldwide.[8]

BOX 2. Male Empowerment Is Important Too

"Gender" is too often shorthand for "women," but reproductive rights should also apply to inseminators. Female contraception includes pills, injectables, diaphragms, IUDs, contraceptive implants, cervical caps, and sterilization, but male contraception has been woefully neglected. Condoms have been around since antiquity, and vasectomies since the 1800s (the first was performed on a dog in London in 1823), but we still do not have a male birth control pill. Experimental male contraceptives include a hydrogel injected into the vas deferens and a mini condom that attaches to the tip of the penis. People too often imagine contraception to be a women's issue, but empowering men is crucial to preventing unintended pregnancies. Vital to appreciate is that women will not be empowered without changing broader social norms—including male norms. Broadening access to and effective uptake of contraception by females and males will require transforming broadly held beliefs but also free reproductive services, sex education, and safeguards to bodily autonomy for both women and men.[a]

---

a. UNFPA, *Seeing the Unseen*, chap. 6; John Bongaarts, "Slow Down Population Growth," *Nature* 530, no. 7591 (2016): 409–11.

AN OPPORTUNITY IGNORED

So a sizeable amount of atmospheric carbon could be avoided by reducing unwanted births. What is remarkable, however, is how little this has been considered in consensus documents outlining the causes of our climate crisis. The IPCC, for example, treats population as a "driver" of emissions but not as an opportunity for mitigation (i.e., fixing the problem).[9] In the 2014 IPCC report, we do find reference to a study stating that carbon emissions could be lower by 30 percent by 2100 "if access to contraception was provided to those women expressing a need for it,"[10] but this important observation does not make it into the accompanying *Summary for Policymakers*.[11]

We find this same oversight in more recent IPCC documents. The IPCC's 2021 report of Working Group I includes a chapter on "Human Influence on the Climate System" while altogether ignoring human reproductive behavior. Neither contraception nor abortion nor reproductive health is mentioned once in this 2,391-page report. The 3,056-page report for Working Group II (on *Impacts, Adaptation and Vulnerability*) similarly makes no mention of contraception or abortion and refers to "reproductive health and family planning" only in the context of improving the health and well-being of women and their children.[12] Working Group III's massive report on *Mitigation of Climate Change* states that education of women and investments in reproductive health "are evident measures to reduce world population growth" and that gender equality and reproductive rights could "substantially lower population growth," but nowhere in the entire 1,977-page volume is there any mention of contraception or abortion or any other form of birth control.[13] Throughout these reports, population is treated as a driver of total emissions but is ignored as a possible solution for limiting human-induced climate change. Population is treated as an uncaused cause, an unmoved mover.

Nor does population growth, family planning, or contraception figure significantly in the United Nations' original 1992 Framework Convention on Climate Change or in the many conferences subsequently organized to implement that treaty. At the 2021 UN Climate Change Conference (COP26) in Glasgow, for example, family planning was only trivially referenced. This is despite the fact that the United Nations, in its "Sustainable Development Goals" (goal 3.7 in particular), has called for incorporating "universal access to sexual and reproductive health-care services" into national strategies by 2030. Among the 327 press releases issued from COP26, not a single headline mentions contraception or reproductive rights.[14] When they do talk about women, the talk is all about how women (and girls) will be *impacted* by climate change; we hear about the importance of "empowering women" but nothing about how improving reproductive health—that is, reproductive liberties—could lessen the carbon burden.

We find a related myopia in organizations that promote family planning. None of the most powerful agencies—the UN Population Fund, the World Health Organization, the Gates Foundation, or the U.S. Agency for International Development—acknowledge the climate benefits of preventing unintended pregnancies. A 2019 booklet from the United Nations, titled *Family Planning and the 2030 Agenda for Sustainable Development* (funded partly by the Gates Foundation), points out that hundreds of millions of women have "an unmet need for family planning"; the text urges "women's and girls' empowerment" and "improved health and nutritional status of children" but fails to acknowledge a climate benefit from ending unwanted pregnancies.[15] Bottom line: scholars and policymakers—even many of those who worry about the preservation of biodiversity—have become reluctant to explore how population growth can threaten the natural world.

Population is not considered one of the "planetary boundaries" defining Johan Rockström and colleagues' "safe operating space" for humanity, for example.[16] Rockström et al. identify nine planetary boundaries "that must not be transgressed," but these are all conceived in earth-systems terms—meaning geophysical processes like climate change, ocean acidification, ozone depletion, and biodiversity loss—with no mention of human population in the aggregate. One might say they've ignored the "anthro" aspect of "the Anthropocene."

Population also plays no role in the so-called doughnut economics outlined by Kate Raworth. Raworth expands Rockström and colleagues' notion of planetary boundaries to include the "social foundation," comprising food, health, housing, education, peace and justice, social equity, and energy and water, but claims that population is not really a problem because the rate of increase has been falling. The "good news" for Raworth is that women are already managing the size of their families thanks to decades of women's empowerment. Population is thus ("at last") a solved problem.[17]

Population is also not one of the "missions" in the so-called mission economy articulated by Mariana Mazzucato—and now informing European Union policy.[18] Targets are set for automotive emissions, biodiversity, and green jobs, but no mention is made of birth control or coerced conception, either locally or globally. The European Commission's Horizon Europe research and innovation program for 2021–2027 (with a budget of €96 billion) has announced five missions—adaptation to climate change, ending cancer, ocean restoration, climate-neutral cities, healthy soils—with again no mention of population.[19] Reproductive technology (here meaning contraception and abortion) is not part of the "opportunities for innovation" outlined in the European Green Deal and is also not even mentioned in any of the proposals for Green New Deals set forth in the United States, Australia, Canada, South Korea, or the United Kingdom. Population is also missing from recent literature surrounding "degrowth"—a radical economic theory (also known as "ecological economics") postulating a need to reverse global economic growth.[20]

Even the otherwise brilliant *Shock of the Anthropocene*, which includes an entire chapter on the Agnotocene, ignores birth control and reproductive health (or coerced conception) more generally.[21] And again, the entire juxtaposition at first sounds outlandish: How could rape be relevant to global climate change? How could something as bloody and messy as abortion or menstrual cups be of relevance to our efforts to combat climate degradation? Climate change is too often modeled in narrowly geophysical terms, focusing only on the "vast machine" that is the so-called earth system.[22] This overly narrow approach ignores the sociocultural causes of the climate crisis and therefore many of the novel opportunities for solutions. Climate change, after all, is caused not just by the carbon molecule but by corporate connivance, right-wing tort reform, and regulatory capture. The situation is much like tobacco, where addiction does not begin in the brain but in the boardroom. Here, too, we need to look not just at proximate but ultimate causes, the "causes of causes,"[23] including broader sociocultural forces.

With few courageous exceptions, the assumption in almost all climate literature is that population will essentially take care of itself: population is imagined to be either a solved problem, a natural good, or something you shouldn't even talk about. And in raising the issue, you are either beating a dead horse or transgressing fundamental human rights.

WHY THIS MYOPIA?

Historical context, of course, is crucial for understanding this taboo. Race-based population control was a pillar of Nazi policy and propaganda, and in America, too, eugenicists pushed "positive" and "negative" eugenics, rewarding the breeding of certain people judged superior and the sterilization (and sometimes castration) of others judged inferior.[24] Some thirty U.S. states passed laws allowing the forcible sterilization of anyone judged physically or mentally unfit—laws upheld by the U.S. Supreme Court.[25]

Even after the collapse of Nazism and the eugenics movement, "population control" got a bad name as a result of state-sanctioned efforts to limit fertility, especially in poorer parts of the globe. Eugenics laws in Puerto Rico resulted in over a third of all women of childbearing age being surgically sterilized between 1937 and 1960.[26] Forced vasectomies in India in the 1970s—including over six million in 1975 alone—led to a backlash that brought down Indira Gandhi's government. China's One-Child Policy from 1980 to 2015 led to millions of forced sterilizations and abortions; millions of women were also forcibly fitted with intrauterine devices, which were rarely checked for safety and often not removed after menopause.[27]

Many branches of Christianity, Islam, and Judaism also include a moral imperative to procreate as part of one's religious duty. The Vatican, for example, opposes artificial contraception (vs. "the rhythm method," first tolerated in 1951), and many social conservatives, even in wealthier parts of the world, oppose induced abortion.[28] In the United States, the Helms Amendment to the Foreign Assistance Act passed by Congress in 1973 in the wake of *Roe v. Wade* banned the use of foreign aid to pay for abortion as a method of family planning. President Ronald Reagan later imposed what became known as the "global gag rule," banning foreign aid to *any* nongovernmental organization (NGO) providing abortion services, including counseling or referrals, and to any group advocating for decriminalizing or expanding abortion services. In Ghana alone, this led to an estimated 500,000 to 750,000 additional unplanned births during George W. Bush's presidency.[29] Radical anti-abortionism (also known as "pro-life") was reaffirmed by all three subsequent Republican presidents (both Bushes and Donald Trump). This gag rule applies to all global health assistance, meaning that governments and NGOs have to choose between obtaining U.S. federal assistance ($9.5 billion in 2017) and providing abortion services.[30] Opposition to abortion is still strong or even growing in many parts of the world and remains an obstacle to achieving reproductive freedom.

We have focused on the IPCC, but a parallel myopia can be found in global governance bodies dealing with population policy. Since 1954, for example, the UN International Conference on Population and Development (ICPD) has convened every decade to explore how population contributes to or serves to hinder economic development, especially in poor countries. One common concern had been that overpopulation had become a barrier to development, with the implication that "population control" could help curb poverty. By the 1970s and '80s, however, this assumption was being broadly called into question.[31] The new idea was that population would take care of itself once "development" had been achieved. This is the so-called demographic transition—the idea that nations move naturally and inevitably through a multistage process of development from a preindustrial to industrial to postindustrial society, achieving a postindustrial demographic equilibrium.[32] In this scenario, people would stop having large numbers of children in consequence of increased wealth, medical progress, and the establishment of social security in old age. A kind of optimistic fatalism—or fatalistic optimism.

An important turning point was the United Nations' 1994 ICPD, held in Cairo, which has framed UN population policy ever since. In response to postcolonial theories of development and revulsion against abuses in the name of population control, the "Cairo consensus" effectively classified efforts to limit population growth in poorer parts of the world as racist and/or neocolonial. Cairo shifted the focus—and rhetoric—of policy away from global "population control" and "demographic targets" onto women's empowerment and reproductive health (of individuals).[33] The idea of "reproductive rights" as key to "reproductive health" had emerged from the women's health movements of the

FIGURE 3.1. After the Cairo consensus, the expression "population control" fell from favor. The emphasis thereafter has been on "reproductive health" and women's equality.

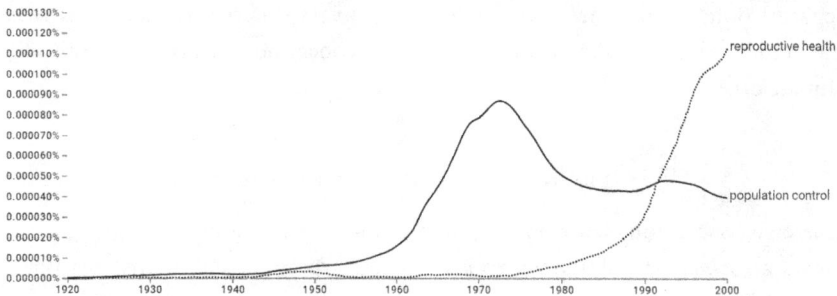

Source: Google Books Ngram Viewer, accessed February 14, 2025, https://books.google .com/ngrams/graph?content=population+control%2Creproductive+health&year_start= 1985&year_end=2008&corpus=en-2009&smoothing=3.

1970s and '80s and, by the 1990s, had been mainstreamed into global governance orthodoxy.[34]

The UN Population Fund's published *Programme of Action*,[35] adopted by 179 governments, advances the human rights rationale for sexual and reproductive health and effectively abandons (and vilifies) the neo-Malthusian environmentalism fueled by Paul R. Ehrlich's 1968 *Population Bomb* and the Club of Rome's 1972 *Limits to Growth*. Although the Cairo consensus framed population policy in terms of "sustainable development"—nonexistent terminology prior to the 1980s—it entirely ignored how voluntary family planning might help combat global climate change or other threats to the environment. Cairo effectively signaled the end of planetary health, including biodiversity and climate stability, as a rationale for birth control. Global funding for contraception would plummet in subsequent decades, along with environmental justifications for family planning.[36] Keep in mind that the human population would more than double from 1973 to 2023 even as concerns about "overpopulation" have dwindled.

As a consequence of these perfectly non-overlapping myopia, climate and population policy have become essentially siloed—with significant, albeit largely unappreciated, consequences. This mirrored inattention has resulted in both the IPCC and the ICPD ignoring the expansion of reproductive liberty as a way to help reverse the climate catastrophe. The IPCC treats population as a driver of emissions but not as an opportunity for mitigation. The United Nations' 2021 report *World Population Policies* never connects human reproduction to climate change or loss of biodiversity.[37] Similarly, the United Nations' 2022 Population Fund report *Seeing the Unseen: The Case for Action in the Neglected Crisis of Unintended Pregnancy* highlights a "crisis" of unintended pregnancies but mentions climate change only in passing and only as a driver of broader human suffering—while ignoring altogether how ending coercive conception and forced birthing could have benefits for the environment.[38] These separate policy realms are really two trains on very different tracks, ships passing in the night. The net effect is that coercive conception as a contributor to the climate crisis has been and continues to be ignored.

PRONATALISM ISN'T HELPING

Coercive conception is one obstacle to reproductive freedom, but many governments also *incentivize* women to give birth, largely for nationalist reasons.[39] One finds this pronatalist impulse already among mercantilists of eighteenth-century Europe, where children were considered the "wealth of nations, the glory of kingdoms, the nerve and good fortunes of empire."[40] Growing the population was celebrated as a means to increase the production of crops and material goods while simultaneously filling the ranks of standing armies and augmenting taxes paid into royal coffers.

State-sanctioned pronatalism is not coercive in the sense of forced sterilization or rape or child marriage, but it's nonetheless a perfect example of the (inconsistent, mutually myopic) siloing we've been talking about. Janus-faced government ministers will often cry "climate crisis" while simultaneously decrying a national "demographic crisis" ("too few babies") and calling for policies to increase birthing. The rationale is typically economic: pronatalist governments worry about an impending "Silver Tsunami" (also known as the "Gray Wave"), with growing numbers of retired elderly threatening to overwhelm the shrinking numbers of younger workers expected to pay into national coffers to support pensions, healthcare, and other social programs.[41] Perversely, the countries most actively encouraging births are typically wealthier nations—that is, those with the highest emissions. According to a recent study by the United Nations, some fifty-five countries—including Australia, China, Finland, France, Germany, Greece, Hungary, Iran, Japan, Poland, Russia, Singapore, and South Korea—now have policies to *increase* birth rates within their borders, primarily via tax incentives and "baby bonuses."[42]

Anxiety surrounding this issue is surprisingly high. In 2022, for example, *Asia Times* characterized China as a ticking "demographic timebomb" when the country reported its first population decline in more than sixty years.[43] That same year, Vladimir Putin revived the Stalin-era Mother Heroine award, paying ₽1 million ($16,500) to Russian mothers when their tenth child (!) turned one year old. Hungary today offers special bonuses to families with three or more children and provides free in vitro fertilization (IVF) to couples at state-run clinics.[44] Israel, too, offers free IVF for up to two "take-home babies" until a woman turns forty-five, with a recent *New York Times* article explaining this as a response to biblical mandates, memories of the Holocaust, fears of losing children in military conflict, and desires to counter "higher fertility rates of Palestinians in the occupied territories."[45]

Doomsday prophecies about demographic collapse can also be heard in the private sector, where naïve cornucopianism still runs rampant. Elon Musk, a father of fourteen (and counting), has tweeted his alarm that a decline in global birth rates will cause economic and innovative stagnation; Musk even claims that population collapse due to low birth rates "is a bigger risk to civilization than global warming."[46] Wealthy financiers have spawned organizations like Silicon Valley's Pronatalist.org, which warns that nations like South Korea are "teetering on a catastrophic population collapse." Director Malcolm Collins, a conservative "neuroscientist-turned-entrepreneur-turned-VC-turned-education innovator," supports fertility planning, egg and sperm donation, birthing surrogacy, and improved reproductive technologies (including the Egg Freezing Ambassador Program)—all to reverse impending population collapse.[47]

Conservative and libertarian think tanks, such as the Cato Institute and Foundation for Economic Education, also decry what they call "the myth of overpopulation" as part of their efforts to rein in government regulation and to discredit "encroachments" on the free market.[48] Population is not a concern, they argue, because more people means "more minds to create, and more hands to build,"[49] making possible "an infinitely bountiful planet."[50] Catholic anti-abortion groups, like the Population Research Institute, also talk about people as "the most valuable resource on the planet" and claim that "the myth" of overpopulation "cheapens human life." The institute attacks what they imagine to be the proabortion agenda of "family planning," which, along with "reproductive health," conceals a purported agenda to eliminate the poor.[51] In 2019, when Senator Bernie Sanders expressed his support for global access to birth control, he was attacked by conservatives claiming that he was calling for fewer "brown babies."[52]

We should emphasize that nationalist pronatalism is very different from progressive policies supporting children once born, including parental leave, affordable childcare, universal education, and prevention-oriented healthcare (i.e., prevention of exposure to lead, sugar, and tobacco). These policies are not, properly speaking, pronatalist but rather pro-child and pro-parent; the crucial distinction is between promoting population growth versus promoting care for children once born. The former is orthogonal (at best) to liberty, the latter aims at enlarging liberty. European social democracies began introducing family-friendly policies especially in the 1930s, first to encourage women to join the workforce and later to support gender equality. The goal was not to increase the population of the nation, but rather to secure the health and well-being of children (and parents) along with gender equality. Our goal, too, should not be to incentivize having more children but rather to properly care for and nurture those we already have—and to ensure a livable world for those to come.

It is also important to appreciate that as fertility rates decline worldwide, nations will transition through a temporary period of aging: the Silver Tsunami is a transitory phenomenon. Much of the hysteria surrounding national birthing deficits is based on measures that assume some fixed age of retirement, when the reality is that healthy living with rewarding work can allow people to enjoy longer and more satisfying lives.[53] And even from a purely economic point of view, the much-lamented graying of the population doesn't have to be a cause for alarm.[54] As mortality and fertility rates decline and populations achieve zero growth, age distributions will stabilize after reaching a "stationary" state, like we find in the countries of modern Scandinavia. And if the question is how to support our elderly retired, this can be solved by taxing the rich to support social security and reallocating priorities away from, say, the military, which in the United States alone has a budget of some $900 billion per year.[55]

EQUITY BOMB?

Enlarging reproductive freedom should be recognized as a way to help lower global greenhouse gases: preventing unwanted births could substantially reduce emissions.[56] The greatest rewards will come from wealthier parts of the world, where per capita emissions can be orders of magnitude larger than those in poorer parts of the world.[57] This opportunity is significant given that nearly half of all pregnancies in the United States, for example, are unintended, and millions of Americans live in states where abortion is illegal and becoming less accessible every day.

On a per capita basis, the decarbonization need is greatest for wealthier peoples of the world, but there are far more people in poorer parts, where reproductive needs remain largely unmet. It should also go without saying, however, that the poor should not, and hopefully will not, remain impoverished forever. (The Cairo consensus calls economic development "a universal and inalienable right."[58]) Some demographers predict that "a global middle class" of 3.2 billion people in 2016 will rise to 5 billion by 2030, and that 40 percent of India's population will join the ranks of this class by 2050.[59] The total number of people using cars, planes, computers, and mobile phones is increasing rapidly, which means that even if per capita emissions decline in wealthier countries, we are still likely to see an *overall* increase as global consumption continues to grow.[60] Paul R. Ehrlich and John P. Holdren flagged this already in 1969, observing that "development" of the rest of the world to the standards of the West "probably would be lethal ecologically." Ehrlich and Holdren emphasized that with unlimited growth, "all the technology man can bring to bear will not fend off the misery to come."[61] And global climate change was not yet even on the table.

We should reemphasize that enlarging contraceptive freedom should be regarded as *complementing* the more fundamental need to end our reliance on fossil fuels, carbon-heavy agriculture, cement-based construction, and other needless emissions. We are not talking about an *alternative* to decarbonization. The climate catastrophe is so grave, however, that family planning should not be "left out in the cold." John Bongaarts and Brian C. O'Neill point out that population policies cost significantly less than other mitigation options (nuclear power and biofuels, for example, but also even solar and wind): one unwanted birth averted (via voluntary measures) will save a lifetime of emissions.[62]

Crucial also to keep in mind, though, is that effective birth control and abortion are less accessible in poorer parts of the world. In most European countries, for example, contraception is included as part of ordinary healthcare, and abortion is readily available. Access to contraception and/or abortion remains limited in most parts of the world, however. Today, only 36 percent of women live in a country where abortion is available upon request.[63] And many nations still today ban abortion entirely. Abortion is effectively prohibited in Andora, Aruba, Congo, Curaçao, the Dominican Republic, Egypt, El Salvador, Haiti, Honduras, Iraq, Jamaica, Laos, Madagascar, Malta, Mauritania, Nicaragua, the Philippines, Senegal, Sierra Leone, Suriname, Tonga, the West Bank and the Gaza Strip, and now more than a dozen states in the United States.[64]

We are currently in the midst of an epidemic of forced birthing. The problem is massive and unequally distributed. Reproductive justice is a virtue in its own right, but we also need to appreciate that our failure to achieve universal reproductive liberty is endangering the planet.[65] People should have the right to have as many or as few children as they want, and empowering reproductive choice must be part of economic development if we are to halt our accelerating environmental catastrophe.

STRUCTURAL AGNOTOLOGY

Agnotology is the study of ignorance and the cultural production of ignorance. It takes the measure of our ignorance and analyzes why some topics are ignored or suppressed or overlooked, while others are embraced and come to shape our world. Ignorance is more than an absence of knowledge; it can also be the outcome of myopia and inattention.

The agnotology we have described here is not an example of mis- or disinformation or of corporate malfeasance. It has instead to do with *structural myopia* and *selective inattention* driven by postcolonial fear, religious dogma, free-market fundamentalism, and nationalist ideology—forces that operate in ways that knowledge makers may not even perceive. In our instance, largely in response to mid-twentieth-century racism and neo-Malthusianism, it has become hard to talk about, or even to imagine, how enlarging reproductive freedoms might help solve our ongoing climate crisis. Post-Cairo-consensus progressives have become wary of making any kind of link between population and the environment, and free-market fundamentalists claim that overpopulation is entirely a "myth."

What makes this interesting as a case study in agnotology is that the myopia affects the Left as much as the Right—along with technocrats in the middle. Family planning advocates focus on individual rights and may find it abhorrent to identify an environmental "benefit" from reproductive policies. Religious conservatives decry "family planning" as code for abortion, and libertarian cornucopians deny that population is a problem at all, because technology will come to our rescue. Climate modelers ignore human biological reproduction because their focus is on the earth as a geotechnical system. Few climate modelers identify Exxon, Shell, and Chevron as causes of climate change—these are neglected along with lobbying by trade associations, political elections, and the emissions and distractions of war. Classic catarheumatics.

But here's the kicker: people from all across the political spectrum worry a lot about coercive *contraception* but insufficiently about coercive *conception*. From a progressive's point of view, there's an irony in all of this given that the architects of the Cairo consensus framed their work in the language of *sustainability*. The big push was for gender equality—there's a whole section in that report on the "girl child"—with no consideration given to links between birthing and the environment, or anything else having to do with *biodiversity*—a word that doesn't even appear in the 1994 report. This is a significant omission. Unlimited growth of humans on the planet threatens wilderness and wildlife, regardless of how much carbon is being emitted (though decarbonization can certainly lessen the blow).

Yet another reason for our failure to consider family planning as a way to mitigate the climate catastrophe has been the failure to consider "reproductive technology"—meaning birth control and abortion—as part of technology. This has to do with how climate science has been conceptualized but also with how "technology" has been understood. In the 1970s, when eco-minded scholars such as Ehrlich and Holdren first started equating the "impact of human activity on the planet" with population × affluence × technology—the so-called IPAT equation—technology was conceived as "impact" (on the environment) "per unit of consumption" with no attention given to how technology might *reduce* population. Even in the revision known as the Kaya identity, opportunities for mitigation are conceived in terms of how one might influence energy intensity per unit of gross domestic product (GDP) and emissions per unit of energy consumed.[66] ("Affluence" is replaced by "GDP per capita" in these newer models.) Significant is that in most applications of both IPAT and Kaya models, and in climate modeling ever since, contraception is not considered part of technology. Technology is conceived in narrowly economic (or productivist) terms as helping to make goods cheaper, which helps us understand why climate modelers don't look to birth control "technology" to help solve the climate crisis. And why technology is not considered relevant for impacting one of the crucial drivers of emissions: the total number of humans on the planet.

We have a climate catastrophe and a liberty catastrophe that are rarely brought into conjunction. We are proposing a reframe that involves seeing environmental degradation as (partly) a result of coercive conception and forced birthing. To mitigate the climate crisis in a manner consistent with human rights and liberties, we must reframe this opportunity as a means to prevent unwanted births and to end forced birthing. The goal is for everyone to be able to choose whether to reproduce, for every birth to be wanted, and for every child to be valued. Reproductive liberty is an ignored opportunity in the climate fight: it's a win-win for human freedom and planetary health.

FOUR

AI-Fueled Ignorance, Confusion, and Profit

HANY FARID

UPON ITS RELEASE IN 2013—and before its star Kevin Spacey fell from grace—*House of Cards* was Netflix's most streamed content in the United States alongside forty other countries. Netflix had good reason to believe this show would be a hit before filming began. According to their subscriber data, Netflix knew that viewers who watched the original BBC miniseries were also likely to watch movies starring Kevin Spacey and to watch movies directed by David Fincher (e.g., *The Social Network*). The data-driven trifecta suggested that *House of Cards* would likely be a hit—and it was.

A decade ago, this type of decision-making based on user-generated data seemed groundbreaking and somewhat controversial.[1] In today's era of ubiquitous data harvesting and breathtaking advances in artificial intelligence (AI),[2] it seems merely quaint. Today, nearly everything we see, hear, and read online is the result of AI- and data-driven algorithmic curation and manipulation.

Every day, more than four petabytes of data are uploaded to Facebook.[3] But not all this content is equal in the eyes of Facebook. Starting in 2009, Facebook eliminated your ability to chronologically sort your news feed, turning over editorial control to algorithmic curation. Similarly, every minute of every day, more than five hundred hours of video footage are uploaded to YouTube and TikTok. The likelihood that any video is widely seen, however, depends largely on recommendation algorithms. A full 70 percent of watched YouTube footage is recommended by the company's algorithms, and TikTok's entire video feed is individually curated based on a user's past viewing habits. By 2016, Twitter (now X) and Instagram joined in unleashing attention-grabbing recommendation algorithms to control what we read, see, hear, and—ultimately—believe.

The titans of tech relinquished control to these algorithms because they can better manipulate users, maximizing clicks, likes, shares, and—in turn—profits. I contend that this algorithmic amplification is the root cause of the unprecedented speed and reach with which hate, misinformation, and conspiracies are spreading online.[4] And this may only be the beginning.

Data- and AI-powered recommendations are largely focused on steering our attention to content generated by our fellow online citizens. More recently, generative AI has emerged to take over content creation. Trained on billions of pieces of human-generated content, generative-AI systems can write a cogent eight-hundred-word op-ed in the style of Maureen Dowd, produce eye-popping photographic images in the style of Annie Leibovitz, pen new lyrics and music in the style and sound of Billie Holiday, and create a full-blown video with Scarlett Johansson's identity swapped into any role. With breathtaking new advances in all aspects of predictive and generative AI, online platforms will be able to control both content creation and recommendation, allowing them to further pollute our information ecosystem and distort our reality as they vie for our personal data and attention.

I will discuss the following two pillars of AI: (1) predictive AI, used to control what we see online; and (2) generative AI, used to create content that is quickly becoming indistinguishable from human-generated content. I will discuss how—left unchecked and unregulated—these pillars of AI hold the potential to toss jet fuel onto an already troubling level of technology-fueled ignorance, hate, and distrust.[5]

PREDICTIVE AI

If you have spent anytime online, then you have been subjected to a predictive algorithm of the form "If you like x, then you may like y." News, music and movie streaming, and shopping sites routinely analyze our previous online habits, compare them with other users, and then make personalized recommendations for each of us. One could reasonably argue that Amazon's and Netflix's recommendations that pop up while you are surfing their sites are relatively benign, with the most devious consequence being that they convince us to buy things we don't need or encourage us to stay up past our bedtime binging a season of *House of Cards*.

Similar recommendation algorithms, however, are used by social media platforms in a more insidious manner. Virtually everything we see on social media is determined by an algorithm designed to maximize user engagement (time on platform) in order to thus maximize the delivery of paid advertising. By comparing your collective viewing habits and detailed demographic models— including race, gender, identity, age, political affiliation, religion, likes, dislikes, etc.—to other users, these recommendations can make eerily precise recommendations to keep you clicking and swiping for hours on end.

These recommendations are not neutral, nor are they benevolent. Facebook's own internal research, for example, found that "our algorithms exploit the human brain's attraction to divisiveness." The research went on to conclude that if left unchecked, Facebook's recommendation algorithms will promote "more and more divisive content in an effort to gain user attention and increase time on the platform." A separate internal Facebook study found that 64 percent of people who joined an extremist group on Facebook did so because of the company's recommendation. Facebook's leadership choose to largely ignore these findings.[6]

These recommendations can create a vicious feedback loop. After, perhaps innocently, searching for QAnon,[7] a user will quickly be recommended more QAnon-related content. A few clicks here, a few clicks there, and the user will be taken down an increasingly deeper and narrower rabbit hole, from which escape could prove difficult. This scenario is not just a hypothetical.

After watching a video on YouTube, for example, you will be recommended another video through YouTube's "watch-next" algorithm. YouTube distinguishes between two types of these recommendations: nonindividualized "recommended" videos and individualized "recommended-for-you" videos based on a user's viewing history. I set out to study the nature of these recommendations,[8] but because it is nearly impossible to accurately simulate a diverse set of users with varied viewing history, I focused my analysis on YouTube's generic "recommended" videos. I wondered, in particular, what YouTube would recommend after someone watched a video from an English-language news or information channel (e.g., BBC, CNN, Fox).

My method to emulate the recommendation engine was a two-step process: I started by gathering a list of news and information channels and then emulated the watching of videos posted by these channels, from which I automatically logged YouTube's recommendations. I gathered the first twenty recommendations from the watch-next algorithm from each of one thousand news and information channels on a daily basis from October 2018 to February 2020, starting from the last video uploaded by each channel. The top–one thousand most recommended videos on a given day were retained and classified as conspiratorial or not. I classified a video as conspiratorial if its underlying thesis, by and large, satisfied the following criteria: (1) explains events as secret plots by powerful forces rather than as overt activities or accidents, (2) holds a view of the world that goes against scientific consensus, (3) is not backed by evidence but instead by information that is claimed to be obtained through privileged access, and (4) is self-fulfilling or unfalsifiable. This classification was determined automatically using a trained classifier that analyzes the video script and the associated user comments. Over a fifteen-month period, I analyzed more than eight million recommendations from YouTube's watch-next algorithm.

YouTube experienced a conspiracy boom at the end of 2018 when almost 10 percent of recommended videos were, by my metrics, conspiratorial. In January 2019, YouTube announced their forthcoming effort to recommend less conspiratorial content. Starting in April 2019, I monitored a consistent decrease

in conspiratorial recommendations; by June 2019, the rate of conspiratorial recommendations had fallen to 3 percent. Shortly after this dip, recommendation
rebounded to around 5 percent.

An analysis of the recommended conspiratorial content revealed three broad
topics: (1) alternative science and history, (2) prophecies and online cults, and
(3) political conspiracies. The first of these involves a radical redefinition of the
mainstream historical narrative of human civilization and development. This
content uses scientific language without the corresponding methodology, often
to reach a conclusion that supports a fringe ideology not as well served by facts.
Examples include the refuting of evolution, the claim that Africa was not the
birthplace of the human species, or arguments that the pyramids of Giza are
evidence of a past high-technology era. Conspiracies relating to climate are also
common, ranging from claims of governmental climate engineering—including
chemtrails—to the idea that climate change is a hoax and that sustainable development is a scam propagated by the ruling elite. A number of videos address
purported NASA secrets refuting, for instance, the U.S. moon landing or claiming that the U.S. government is secretly in contact with aliens.

The second topic includes explanations of world events as prophetic, such
as claims that the world is coming to an end or that natural catastrophes and
political events are religious realizations. Many videos from this category inter-
twine religious discourse based on scriptural interpretations with conspiratorial
claims, such as describing world leaders as Satan worshipers, sentient reptiles,
or incarnations of the anti-Christ. These videos rally a community around
them, strengthened by an "us-versus-them" narrative that is typically hostile to
dissenting opinions in ways similar to cult-recruitment tactics.[9]

The third main topic is comprised of political conspiracies, the most popular of which is QAnon—a conspiracy based on a series of ciphered revelations
made on the 4chan anonymous message board by a user claiming to have access
to classified U.S. government secrets. These videos are part of a larger set of
conspiratorial narratives targeting governmental figures and institutions, alle-
gations that a deep-state cabal and the United Nations are trying to create a new
world order or claims that the Federal Reserve and the media are conspiring to
act against the interests of the United States.

Importantly, the above analysis was made on nonpersonalized recommendations from news or information channels, suggesting that the observed problematic YouTube recommendations constitute a lower bound on YouTube's
conspiratorial recommendations. One would reasonably expect that personalized recommendations on less mainstream channels would surface even more
conspiracies. And, in fact, I observed just this pattern: over the fifteen-month
window of my analysis, after emulating the watching of a conspiratorial video,
another conspiratorial video was recommended 50 percent of the time—a sig-

nificantly higher proportion than the rate of conspiratorial recommendations on news-related videos.

It is reasonable for YouTube and others to design their recommendation engines to suggest videos that are similar to previously watched videos. Overly selective algorithmic recommendations, however, can lead to a state of informational isolation—the so-called filter bubble, or echo chamber. I contend that these algorithmic recommendations and amplification are the root cause of the unprecedented speed and reach with which the internet's flotsam and jetsam spread online. As AI-powered recommendation algorithms learn how to manipulate users more effectively, we can expect the rabbit holes to get increasingly deeper and the echo chambers to get increasingly more isolated, and we will collectively live in an increasingly more bizarre world devoid of a shared reality grounded in facts.

Today's predictive AI are bracketed by humans (and bots) generating content on one side and by humans consuming content on the other side. As I will discuss next, new advances in AI hold the potential to replace humans on the generation side, leading to a potential future in which AI systems will both generate and recommend content for humans to consume. I will first discuss the nature of generative AI and then the implications of an AI-powered generative-predictive feedback loop.

GENERATIVE AI

Although generative AI (also known as "synthetic media," or "deepfakes") varies in its form and creation, it generally refers to audio, image, or video that has been automatically synthesized by an AI-based system.[10] I will first discuss the various forms of generative AI and where—even in these early days—we are seeing them used and misused.

Audio

A prototypical text-to-speech system consists of two basic parts. First, the text is specified and converted into a phonetic and prosodic representation that captures the specific sounds, intonation, stress, and rhythm to be spoken. Second, a synthesis engine converts this symbolic representation into a raw audio waveform, typically through an intermediate frequency-based representation.

Synthesized voices have come a long way from the tinny robot voices of past years. Boosted by advances in AI, today's synthetic voices are increasingly more realistic. In addition to simply creating human-sounding voices, it has become possible to clone another person's voice from as little as thirty seconds of audio recording. There are several free or low-cost commercial offerings that allow anyone to clone and use anyone's voice with few to no guardrails.

Image

A generative adversarial network (GAN) is a common computational technique for synthesizing images of people, cats, planes, or any other category. Versions 1, 2, and 3 of StyleGAN are some of the most successful techniques for synthesizing realistic faces. Each successive iteration of StyleGAN yielded higher-quality faces with fewer visual artifacts.[11] Although there are many complex and intricate details to these systems, StyleGAN (and GANs in general) follow a fairly straightforward structure.

A GAN is composed of two basic parts: the generator and the discriminator. When tasked with creating a synthesized face, the generator begins with a random array of pixels and feeds this first guess to the discriminator. If the discriminator, equipped with a large database of real faces, can distinguish the generated image from a real face, the discriminator provides this feedback to the generator. The generator then updates its initial guess and feeds this update to the discriminator for a second round. This process continues with the generator and discriminator competing in an adversarial game until an equilibrium is reached when the generator produces an image that the discriminator cannot distinguish from a real face.

Because StyleGAN begins with a random array of pixels, it is not possible to control the properties of a synthesized face (skin tone, age, gender, etc.). More recently, a new diffusion-based text-to-image synthesis technique has emerged that affords exquisite control of your creation. OpenAI's DALL-E, for example, is a multibillion parameter version of the text-synthesis engine GPT-4 (Generative Pretrained Transformer 4) and is trained to synthesize images from text descriptions. Ask DALL-E for "a portrait of thirty-something African-American women wearing sunglasses, a red scarf, and a purple polka-dotted dress," and it will generate precisely that. GAN- and diffusion-synthesized images are eerily realistic and have or will quickly become indistinguishable from photographic images. DALL-E is only one of a dozen or so text-to-image engines that are readily available online for free or a small fee.

Video

Most of the attention on the video-synthesis side has been focused on creating videos of people. These types of AI-synthesized videos—so-called deepfakes—take on one of several different forms: *lip sync, face swap,* and *puppet master.*

A minute-long *lip-sync video* of what appears to be former president Barack Obama saying things like "President Trump is a total and complete dips—t" was part of famed actor and filmmaker Jordan Peele's 2018 public service announcement (PSA) on the dangers of fake news and the then-nascent field of deepfakes. Presciently, the PSA concludes with a Peele-controlled Obama saying, "How we move forward in the age of information is gonna be the difference between whether we survive or whether we become some kind of f—ked up dystopia."[12]

By using hours of authentic video of President Obama and a synthesized or impersonated audio track, a lip-sync deepfake can generate a synchronized video track of Obama saying anything the creator wants. The complete synthesis pipeline consists of four primary steps: (1) an artificial neural network is trained to learn a mapping between an audio track and an outline of the mouth shape that is consistent with the audio; (2) a detailed image of the mouth region (including the nose, cheeks, mouth, and chin) is synthesized by blending mouth regions from the training video to match the estimated outline shape; (3) the synthesized mouth region is blended onto a retimed training video modified so that the head motion is consistent with the audio (e.g., the head is typically still when there is a pause in the speech); and (4) the jawline is warped to match the shape and position of the chin.

TikTok's @deepTomCruise is an impressive example of a *face-swap deepfake* in which one person's identity, from eyebrows to chin and cheek to cheek, is replaced with another.[13] For each video frame of identity A, a new video frame is synthesized where the original identity is swapped with a new identity, B. This technique consists of three basic steps: (1) synthesize an image of B in the same head pose and expression as A, (2) fill in any missing facial or hair pixels that arise from the synthesis step, and (3) blend the synthesized face B into the original frame to replace the identity of A. By repeating this process frame after frame, one person's identity is swapped with another. This technique works best when there are many images of the co-opted identity B with different facial expressions and head poses.

In a *puppet-master deepfake*, the head movements and facial expressions of one person (the puppet master) are transferred, in real time, to another person (the puppet). Unlike lip-sync (which only modifies the mouth region) or face-swap (which only modifies the eyebrows to chin and cheek to cheek), a puppet-master deepfake synthesizes the entire head, which is both more difficult and more compelling because it preserves more features of the identity being co-opted. Taking as input videos of the puppet master, A, and the puppet, B, the facial expressions and head movements are transferred from A to B. This process consists of three basic steps: (1) the facial expressions (e.g., mouth open, eyebrows raised, brow furrowed, etc.) of identities A and B are tracked throughout the video sequences; (2) the expression of identity A is transferred to B by deforming the facial expression of identity B, which may include synthesizing the mouth's interior when, for example, A's mouth is open but B's mouth is closed; and (3) the transformed face is composited back into the original video sequence.

Puppet-master deepfakes have expanded from head to full-body synthesis.

With an input video of person A dancing and a few minutes of person B performing some simple motions, the system transfers A's dance moves onto B, controlling them like a puppeteer might. Although the resulting videos currently have fairly obvious visual artifacts, this full-body puppeteering is likely a sign of things to come: as facial synthesis is perfected, it will be obvious to move to upper-body and then full-body synthesis.

These types of talking-head fakes are limited to making it appear that someone is saying something they never did. More recently, the text-to-image technology described above has been expanded to text-to-video capabilities, in which a short (ten to twenty seconds in length) video can be created from a simple text prompt. While last year the resulting videos were barely coherent, today's text-to-video technology can create more visually compelling, albeit not yet perfect, footage. If the trends continue, however, we should expect highly compelling fake videos limited in content by only our imagination.

Although AI-generated videos are generally not quite as convincing as their image and audio counterparts, they are quickly gaining ground and will soon pass through the uncanny valley and become nearly indistinguishable from reality.

BOON OR BANE

There are, of course, many useful and creative applications of generative-AI content. AI-generated voices, for example, hold tremendous power to restore speech to those who have lost it, especially when it is done in their original voice. After losing his natural voice due to throat-cancer surgery in 2015, for example, the actor Val Kilmer explained, "My voice as I knew it was taken away from me. People around me struggle to understand me when I'm talking."[14] Kilmer's voice was cloned from thirty minutes of earlier recordings of him, allowing him to convert his text to speech in a voice that is recognizable to him and those around him. More recently, as Representative Jennifer Wexton battles a rare brain disorder that has limited her ability to speak, she used a text-to-speech voice generator to speak on the House floor.

On the creative side, generative AI has already made its way into Hollywood feature films. For example, younger versions of performers were synthesized in the blockbusters *Rogue One: A Star Wars Story* and *The Irishman*. Films are also being automatically and more realistically dubbed, eliminating the distracting audio-mouth desynchronization that occurs in traditional movie dubbing. This technology allowed famed footballer David Beckham to record a PSA in nine different languages for the fight against malaria.[15]

In a more ethically complex application, the documentary *Roadrunner*, about the life and tragic death of Anthony Bourdain, contains a few lines of dialogue in a synthesized version of Bourdain's voice reading an email to a friend ("My life is sort of s——t now. You are successful, and I am successful, and I'm wondering, Are you happy?").[16] The use of a synthesized voice was only revealed after a *New Yorker* reporter asked the filmmaker how he acquired this clip. When asked about the ethical boundary of synthesizing a deceased person's voice for a documentary, the filmmaker responded somewhat dismissively, "We can have a documentary-ethics panel about it later."[17]

Generative AI is not, however, without its dark side. Before the less objectionable term "generative AI" took root, this content was referred to as "deepfake"—a term derived from the moniker of a Reddit user who, in 2017, used the then-nascent AI-synthesis technology to create nonconsensual sexual imagery. Targeting primarily women, this technology continues to be widely used to insert a woman's likeness into sexually explicit material, which is then publicly shared by its creators as a form of humiliation or extortion.

Fraudsters have also found novel ways to weaponize deepfakes. In early 2020, for example, a United Arab Emirates' bank was swindled out of $35 million after a bank teller received a phone call from the purported director of a company the bank manager knew and with whom he had previously done business. The voice on the other end of the phone instructed the manager to transfer the funds as part of a corporate acquisition. Because the request was consistent with previously received emails and since the voice was familiar to him, the bank manager transferred the funds. It was later revealed that the voice was AI-synthesized made to mimic the director's voice. Similar types of fraud are now being perpetrated at the individual level. In early 2023, for example, the mother of a teenager received a phone call from what sounded like her distressed daughter, claiming that the teenager had been kidnapped and feared for her life. The scammer then demanded $50,000 to spare the child's life. After calling her husband in a panic, she learned that their daughter was safe at home.

Deepfakes have also found their way into disinformation campaigns. In the early days of Russia's February 2022 invasion of Ukraine, President Volodymyr Zelenskyy warned the world that Russia's digital disinformation machinery would create a deepfake of him admitting defeat and surrendering. A few weeks later a deepfake of him appeared with just that message. This video was eventually debunked but not before it made its way onto national television and spread across social media.

In a particularly startling case of disinformation and potential fraud, in May 2023, minutes after a photo purporting to show a bombing at the Pentagon went viral on Twitter (now X; from a verified account that at first glance appeared to be Bloomberg News), the stock market dipped by $500 billion in just a few minutes. While the markets recovered after the photo was exposed as fake, the incident highlights the power of fake imagery combined with the unchecked virality of social media and, in this case, Elon Musk's folly of paid verified accounts that can easily be used to impersonate legitimate news outlets.

Perhaps the most pernicious result of deepfakes and general digital trickery will be that when we enter a world where anything we read, see, or hear can be fake, then nothing has to be real—the so-called liar's dividend.[18] In the era of deepfakes, a liar is equipped with a double-fisted weapon of both spreading lies and using the specter of deepfakes to cast doubt on the veracity of any inconvenient truths. In 2016, for example, Elon Musk was recorded saying that "a Model S and Model X at this point can drive autonomously with greater safety than a person. Right now." After a young man died when his self-driving Tesla crashed, his family sued, claiming that Musk holds some responsibility because of his claims of safety. In attempting to counter this claim, Musk's attorneys told the court that Musk, "like many public figures, is the subject of many 'deepfake' videos and audio recordings that purport to show him saying and doing things he never actually said or did." Fortunately, the judge was not persuaded: "Their position is that because Mr. Musk is famous and might be more of a target for deepfakes, his public statements are immune," wrote Judge Evette Pennypacker. She added, "In other words, Mr. Musk, and others in his position, can simply say whatever they like in the public domain, then hide behind the potential for their recorded statements being a deepfake to avoid taking ownership of what they did actually say and do. The Court is unwilling to set such a precedent by condoning Tesla's approach here."[19] As deepfakes continue to improve in realism and sophistication, it will become increasingly easier to hide behind the liar's dividend.

A GENERATIVE-PREDICTIVE FEEDBACK LOOP

There has been speculation that the fake Pentagon-bombing image caused a $500 billion market dip in part because automated predictive-AI algorithms responded to the chatter on Twitter and began a sell-off, with human traders responding in kind. If correct, this event points to a potentially bizarre future where predictive-AI algorithms act based on generative-AI content, creating an unpredictable feedback loop.

This same feedback loop may also infect the online-information ecosystem. Social (and even traditional) media may jettison the unpredictable, expensive, and difficult-to-moderate human-generated content for AI-generated content. The result would be inexpensive content that is easier to moderate and can be designed in a highly targeted fashion to extract the maximal amount of our attention and time.

If tomorrow's predictive AI and generative AI are designed to maximize user engagement—as today's recommendation systems are—then they will be unleashed to create all forms of lies, conspiracies, hate, and vitriol in an attempt to satisfy its objective of monetizing our time and attention. In this perhaps not-too-distant future, all that will be left for us from today's creator-recommender-consumer ecosystem will be consumption. Like screen-locked zombies, we will be manipulated into spending countless hours clicking and liking, feeding the insatiable appetite of our AI overlords.

Things may get even weirder when generative AI begins to feed on its own content. Today's generative-AI systems are trained mostly on human-generated content. What happens, however, when future versions of generative AI are trained on the outputs of their own creations? Early investigations suggest that training large-language models (e.g., ChatGPT) on their own output leads to irreversible defects in future iterations of the model—termed "model collapse"—in which the model produces gibberish.[20]

More problematic may be the threat of adversarial attacks in which an adversary can prop up thousands of domains and pollute them with false information.[21] If the trend of indiscriminately scraping the web for data to train models continue, future generations of generative AI will regurgitate the lies they are fed.

OUR FUTURE

For decades, Big Tobacco and Big Oil have used a straightforward but effective playbook to deflect the harms from their products: deny the product is harmful, cast doubt on any criticism, fund research to muddy the scientific waters, and aggressively fight any regulation. Big Tech has followed the same playbook.

While the abuses of Big Tobacco and Big Oil have had immeasurable impacts on the health of millions of individuals and our planet, I contend that our inability to contain Big Tech may be even more dangerous and deadly. Without a robust and trusted information ecosystem, we cannot effectively respond to a global health crisis, we cannot effectively respond to climate change, we cannot have confidence in our elections, and we will not have the bedrock needed for a functioning society: a shared factual system.

Doubt and disinformation are the common denominator for enabling corporate indifference and greed. Over the past two decades, Big Tech has created a phenomenally effective system for creating and spreading disinformation; generative and predictive AI are going to add jet fuel to this problem. If the past two decades have taught us anything, it is that left unchecked, Big Tech—like any other industry—will put profit and growth above all else. The past has also taught us that left unchecked, Big Tech will continue to pollute our online ecosystem and, in turn, our minds, societies, and democracies. There are, however, practical and effective technologies and polices that can be enacted today to help us avoid a technology- and AI-fueled apocalypse.

TECHNOLOGY

Founded by Adobe in 2019, the Content Authenticity Initiative (CAI) authenticates recorded content at the point of origin where specialized cameras or camera apps cryptographically sign the recorded content (audio, image, or video) and (optionally) metadata, including creator identity, date and time, and geolocation.[22] Sensitive to the need to balance content authenticity with privacy and security of, for example, photojournalists in high-risk areas, the CAI allows creators to select and preserve attribution or remain anonymous. The extracted tamper-evident cryptographic hash is stored alongside any other recorded metadata, and optionally on a centralized ledger. A similar approach can be used for keeping track of AI-generated content from the point of creation.

To fully integrate this technology into our information ecosystem, downstream services like Facebook, YouTube, and X will need to cooperate and visually mark stamped content and, at least in the case of breaking news and election coverage, prioritize authentic content over fake content. Even if X and others welcome the type of chaos caused by the fake Pentagon-bombing photo, those on the generative-AI side should be calling for the robust and consistent marking of all photographic and generative-AI content. As described above, as generative AI becomes more ubiquitous, the next generation of data scrapers will ingest their own creations for retraining and perhaps even the creations of an adversary seeking to poison the next generation of AI models. Indiscriminate training without understanding data provenance could lead to a downward spiral in the quality of the next generation of generative AI.

REGULATORY

We should be realistic that Big Tech and now Big AI will generally act in their own financial interests. It is therefore up to regulators to install appropriate guardrails to ensure that we are kept safe.

It has been argued that the internet we have today—for better and worse—is thanks in large part to twenty-six words enshrined into section 230 of the Communications Decency Act:[23] "No provider or user of an interactive computer service shall be treated as the publisher or speaker of any information provided by another information content provider." Significant questions still remain, however, about who should be liable for abuses. In the early 1990s, for example, CompuServe and Prodigy faced legal challenges related to content posted by their users. Because CompuServe had a policy of not moderating any user-generated content, they were found not at fault for claims of libel. On the other hand, because Prodigy did moderate user content, they were found liable in *Stratton Oakmont v. Prodigy* for libelous content posted by a user.[24] Because Prodigy had taken some editorial role, the court reasoned they acted as a publisher. In contrast, CompuServe had taken no editorial role and was not treated as a publisher of the offending user content and thus not liable.

These cases created a perverse incentive for platforms not to moderate user-generated content. In response, in 1996, Congress passed section 230 to encourage platforms to act responsibly in the face of problematic user-generated content. In the intervening three decades, courts across the United States have adopted a broad interpretation of section 230, giving online platforms and services broad immunity for harms caused by their services.

I contend that over the past few decades, the courts have adopted an overly broad interpretation of section 230, shielding the titans of tech from significant harms they knew or should have known were resulting from their services.[25] In the age of generative AI, the U.S. Congress should clarify that corporations will not be shielded from liability. For better or worse, section 230 was designed to shield online services from responsibility as a publisher for the speech of others. Generative AI, however, is very much the speech of the corporation that designed, trained, and deployed a given AI system. This means that if a generative-AI system spews defamatory, conspiratorial, or harmful content, it is entirely the responsibility of its creators.

The past two decades have taught us that without clear regulatory guardrails, Big Tech will place profits above all else. As we enter the age of Big AI, we should not repeat the mistakes that have led to our current polluted online-information ecosystem. Although creating liability is arguably not the best—and certainly not the only—way to establish guardrails, this approach leverages existing regulatory and judicial infrastructure, has proven to work in the offline world, and is relatively future-proof even as technology tends to move orders of magnitude faster than government oversight.

With the United States making up only 5 percent of the world's population, we will also need to think carefully about how our regulatory framework will be exported and how it will impact the rest of a complex and diverse world.

HUMANS

If the AI revolution will lead to the continued erosion of our online-information ecosystem and—as some are predicting—our humanity, we will have no one to blame but ourselves. For the past twenty-five years, we have been feeding our potential AI overlords with every morsel of data in the form of news articles, blogs, personal correspondences, and billions of selfies, vacation photos, and videos. It is from this vast ocean of data that today's AI systems have learned to write, read, translate, and create. Perhaps we can excuse our past naïveté at the dawn of the modern internet revolution, but today, as we continue to feed the beast, we do so willingly and with our eyes wide open.

Silicon Valley promised that the solutions to our greatest problems were just an app, a click, and a swipe away. The past twenty-five years have shown this not to be the case. We are now being promised that AI will be the savior of what ails us. With largely the same cast of characters at the helm, we should be skeptical of these promises.

I contend that technology developed ethically and thoughtfully can be a tremendous catalyst for positive change. When done recklessly (as we have already seen), however, it can lead to spectacular failures and harm. I think Jordan Peele said it best while impersonating Barack Obama: "How we move forward in the age of information is gonna be the difference between whether we survive or whether we become some kind of f—ked up dystopia."

FIVE

On Data Loss and Disappearance in Digital Societies

NANNA BONDE THYLSTRUP

IN SAN FRANCISCO'S Richmond District in California, between the Presidio and Golden Gate Park, you might stumble upon a striking bright-white temple-like building nestled between a frenetic boulevard, lush greenery, and the mundane sidewalk life of neighborhood residents. Originally built for a Christian Science congregation, this edifice now holds service for a different kind of devotion: digital preservation. Inside, the echo of religious sermons has given way to the gentle but persistent hum of cooling fans preserving our digital past across a vast array of servers, occasionally punctuated by the activity of archive staff and volunteers. This is the Internet Archive—the world's largest repository of archived web pages and other forms of digitized cultural objects. It serves as a poignant reminder of our digital history's fragility. It also constitutes a crucial battleground in the escalating struggle to determine how humanity's digital memory is owned, shared, and preserved—or potentially lost forever.

Focusing on data loss in the digital age may seem counterintuitive. After all, we are often led to believe that we live in an era of infinite data accumulation, where a relentless gathering of information fuels scientific progress while simultaneously giving rise to political and ethical challenges. Our unprecedented ability to store and retain data has led scholars to assert that, unlike their forebears, digital societies do not forget.[1] This assertion has marked a revolutionary shift in how we approach and aggregate social and cultural memory, carrying profound implications for privacy rights, public accountability, access to information, and, ultimately, the writing of history.

While our capacity to gather and accumulate data is unparalleled, so too is the scope of contemporary data losses. Some of these losses are deliberate, manifesting as physical data destruction, planned obsolescence, and the so-called sunsetting of social media platforms.[2] Others occur accidentally due to phenomena like outages and natural disasters, such as fires and floods.[3] A few instances have erupted into political and public spectacles, like the recent text message controversies involving the European Commission and the British and Danish governments and the infamous data breaches between Facebook and Cambridge Analytica.[4] Yet countless instances of data loss occur, often unnoticed, through mundane routines, including everyday deletions by users, minor accidents or moments of inattention, and automated overwrites within corporate and public institutions. The assertion that digital societies possess flawless memory, despite mounting evidence to the contrary, reveals a profoundly naïve techno-optimism. It also disregards historical lessons that consistently teach that loss is an intrinsic aspect of any technological paradigm shift.[5]

As this chapter contends, not all instances of data loss, including the deterioration and destruction of our digital history, should be viewed as inherently tragic. Those acquainted with the inner workings of museums or libraries understand that what transforms a chaotic storeroom filled with sources and artifacts into a collection of curated knowledge are as much the deliberate choices to discard as to preserve. Historian Arlette Farge eloquently captures this necessity for discernment in her observation that "the archival operation first of all consists of separating the documents. The question is to know what to keep and what to abandon."[6] While historians like Farge grapple with the challenge of selecting from "the archive" to construct history, archivists face even more radical decisions. Their appraisals shape the collection itself and fundamentally determine what can be known.[7] Thus, the act of discarding is as ancient as information itself—an intrinsic part of the archiving process that turns informational caches into collections.

Much of today's data loss, however, occurs in opaque and unjust ways, with profound implications for both culture and politics. Many of these injustices echo what anthropologist Ann Laura Stoler describes as "the dramatic durabilities of duress" that imperial formations produce.[8] But they also highlight new problematics. Today, our digital cultural record is often in the hands of a few tech monopolies, which curate and control the dissemination and governance of many of our most important cultural artifacts and collections. At the same time, archivists have raised alarms about the challenges for democracy posed by politicians using ephemeral communication systems, like WhatsApp and Signal. And in the realm of scientific research, academic activists are fighting to preserve scientific records against political erasure and everyday oblivion.[9] Understanding the forces driving data loss is crucial for managing and mitigating its impact, shedding light on the conditions under which our societies not only remember but also forget.

This chapter charts the heterogenous landscapes of data loss, highlighting their various manifestations in agnogenic processes—processes that allow individuals and society to develop, for lack of a kinder term, "willful ignorance" or to simply forget—which are shaped by natural conditions, digital infrastructures, and broader social structures of power and resistance.

In 2017, two lab employees at CERN, the European Organization for Nuclear Research, wrote a short note outlining the challenges of "facing up to the exabyte era." They pointed out that the experiments carried out at CERN would not only increase the number of events but also "drive a huge increase in computing needs for the start of the HL-LHC era in around 2026." The reason, they explained, was that their current software, hardware, and analysis techniques were predicting that the required computing capacity would increase by roughly fifty to one hundred times, "with data storage alone expected to enter the exabyte (10^{18} bytes) regime." Five years later, in 2023, CERN announced that they had crossed a storage threshold for digital data, prompting hard decisions on what to keep and what to lose.[10]

CERN's announcement is part of a deeper pattern in science that mobilizes ignorance to fuel the production of ever-more data. In the case of CERN, the driving force behind this "overproduction" is the classic quest for knowledge that motivates basic research, where "science rushes in to fill" a perceived void in existing knowledge about the world.[11] And superficially, the mind-boggling numbers quantifying digital data do convey an impression of steady progress in knowledge production. Yet, as those in charge of data-storage architectures concede, data loss is growing every bit (byte) as fast as data accumulation.

Sometimes this data loss is intentional.[12] In the offices and laboratories of CERN, hard decisions are continuously made about which data to keep from the astounding mass that is generated. These decisions are born of necessity, compelled by material limitations and the boundaries of current technological capabilities. These "thresholds" can be explained via the two-step process of collision analysis. First, the process of particle-collision detection involves transferring data into the detector's memory, constrained by the speed of electronics. Despite electrical signals traveling at nearly the speed of light, only around one in five hundred collisions can be retained in memory. Subsequently, the data stored in memory must be written to a disk or other permanent device. This necessitates decisions to be made about what to keep and what to let go of. Particles collide at an incredible rate, ranging up to about thirty million collisions per second for proton bunches (which contain around 120 billion protons each). CERN research physicist Andreas Hoecker estimates that one billion collisions per second generate one petabyte of data per second, leading to an impossible demand for storage space: "If we wanted to keep all 30 million events per second we would need about 2,000 petabytes to store a typical 12-hour run. For a typical running year of 150 days uptime, this would mean almost 400,000 petabytes = 400 exabytes per year—a huge amount of data we would not be able to store." CERN has therefore developed complex methods for determining what data to keep and what to discard. These mechanisms are called "triggers": fast online algorithms that differentiate between what are deemed to be more and less interesting events, meaning that for every thirty million collisions, just twelve hundred—a mere 0.004 percent of the total generated—are saved, while the other 99.996 percent are lost forever.

As the CERN employees themselves point out, however, relevance is not a clear-cut value. Hence, every act of disposal also triggers an accompanying uncertainty about lost potential. Scientists expressed this anxiety in a 2018 *Forbes* article, which ran with the headline "Has the Large Hadron Collider Accidentally Thrown Away the Evidence for New Physics?" In the article, CERN employees explain that while their choices on what to keep and what to discard are rooted in scientific knowledge, they nevertheless know that they might also miss out on the next big discovery. In any case, as they point out, "We may have collected hundreds of Petabytes, but we've discarded, and lost forever, many Zettabytes of data: more than the total amount of internet data created in a year."[13]

The unease expressed by the CERN employees about the latent value of information and its potential loss will resonate with most researchers familiar with archives. Archives have to, as their raison d'être, accumulate and contain; they are institutions that gather and store material, presumably for eternity. As librarian Melvil Dewey, the creator of the Dewey Decimal System, observed in the early twentieth century, a "normal librarian's instinct is to keep every book and pamphlet. He knows that possibly someday, somebody wants it." A century later, Brewster Kahle, the Internet Archive's founder, enthusiastically expressed this same archival impulse in an interview with me: "We're building the Library of Alexandria, version 2. We can one-up the Greeks!"[14] Furthermore, the steadfast conviction in the latent worth of data has been reinforced by the emergence of new algorithmic knowledge systems, which pledge to unlock the value of otherwise hidden and "dormant" data through novel modes of reuse.[15]

Ironically, as the next section will demonstrate, the seemingly boundless possibilities of digital data for research are now also challenged by critical environmental questions: as temperatures climb and energy consumption escalates, data storage and the models trained on this content emerge as cutting-edge knowledge tools but also as significant environmental concerns. Consequently, there is a growing call to curb the relentless data-gathering impulse to feed ever-larger data-storage facilities, ensuring that decisions regarding data retention are made with climate change considerations in mind. Paradoxically, it is also the very threat of global warming that has given rise to another form of agnotologic risk: disaster-induced agnotology. Unlike the data losses at CERN, which result from intentional research decisions, this type of loss is driven not by the constraints of nature but by its devastating forces.

INSCRIBING THE SPECTER OF DISASTER INTO DATA-STORAGE ARCHITECTURES

In navigating the exabyte era, decisions about the production of knowledge and ignorance must grapple with deliberate choices about what to keep and what to discard in the shadow of the looming threat of unforeseen disasters.

In 2021, Dean Hildebrand, technical director of Google Cloud, and Denis Serenyi, tech lead of Google Cloud Storage, showcased their storage infrastructure—aptly named Colossus—emphasizing its vast scalability with a single cluster capable of handling exabytes of data. Meanwhile, Amazon is preparing for the exabyte era by developing both static storage services and mobile data-storage and processing devices, gathered in a service called AWS Snow Family.[16] For both of these digital giants, "scalability" is a central concept, suggesting a world of knowledge accumulation seemingly unbounded by nature's limitations.[17]

However, as history reminds us, massive collections are vulnerable to catastrophic loss. The Library of Alexandria, for example, was burned after Julius Caesar set fire to the ships in the harbor. More recently, one commentator in *The Guardian* described the devastating fire in the National Museum of Brazil in 2018—which consumed millions of natural and cultural artifacts—as a "lobotomy of the Brazilian memory."[18]

While digital societies have yet to face a disaster of such magnitude, tales—and insurance policies—telling of impending digital catastrophes abound. One major threat is global warming. As global temperatures rise and extreme weather events become more frequent and severe, the threat of data loss due to infrastructure failures becomes increasingly real. Another, which is often repeated by interlocutors I speak to in the field, is the specter of "the bomb"—not a metaphorical term used to describe a systems crash or breakdown but the actual physical destruction of data and storage facilities using explosives. Thus, experts entrusted with the safekeeping of data often (strongly) suggest mirroring knowledge infrastructures across locations "in the case of a bomb attack." Finally, the most persistent threat at present appears to come from cyberattacks, manifesting both in new legislative initiatives in cyber- and information security and new infrastructural architectures and business strategies.[19] These dangers have raised new concerns but have also given rise to a new phenomenon: disaster capitalism, epitomized by the emergence of disaster-recovery initiatives.

In disaster recovery, *resilience* is the pivotal concept, often materializing in maps and strategies aimed at establishing distributed rather than centralized storage nodes. As a result, data-storage architectures are evolving into robust systems designed not only for efficient data management but also for disaster preparedness. Concerns about potential data loss in the future are reflected in meticulous attention to the physical layout of information, the positioning of backups, and the routes that data traverse, recognizing that every connection poses a potential risk.[20]

Contemporary strategies to prevent catastrophic data loss often mirror the principles behind the dispersed nature of today's networked digital architectures. This approach was famously illustrated by RAND Corporation scientist Paul Baran in his oft-cited 1964 paper on how to build distributed communication networks to withstand thermonuclear threats.[21] Under the looming threat of nuclear war, Baran argues that a distributed network structure would enhance survivability in the event of attacks targeting nodes, links, or some combination thereof.[22] There is a growing trend of these Cold War–era nuclear-bunker structures being transformed into "ultra-secure" cloud infrastructures,[23] embodying the ethos of resilience and redundancy that Baran championed.

Amazon's services offer one example of the ways in which data storage at scale is also haunted by the specter of large-scale loss. On its website, the company boasts of its ability to provide the "most durable storage in the cloud," which is designed to exceed "99.999999999% (11 nines) data durability." This storage is organized such that data is distributed "redundantly across a minimum of 3 Availability Zones (AZ) by default." Moreover, Amazon offers an additional paid feature of "built-in resilience against widespread disaster." Paying clients can thus "store data in a single AZ to minimize storage cost or latency, in multiple AZs for resilience against the permanent loss of an entire data center" or in particular geographic regions to meet "geographic resilience requirements." While impressive "on paper," Amazon's tiered plans for data-loss prevention also make it clear that the threat of data loss is indeed a lucrative business: data-loss prevention is not only a feature one must pay for, but it also presumes the new logics of scaling that underlie twenty-first-century corporate expansion in platform capitalism. These modern fortresses of data protection thus serve as a testament to the enduring relevance of Cold War strategies to safeguard digital information but also as painful symptoms of how digital capitalism contributes to the acceleration of global warming—while making a business model out of these very same threats.

While the disaster capitalism of loss prevention is inscribed into the infrastructures of the digital economy, one of the most likely scenarios is still that low-stakes natural forces in the form of human fallibility will result in large-scale loss. Murphy's Law—that anything that can go wrong, will go wrong—is alive and well at every step of the data-production chain—in the form of outages due to errors, technical glitches, and failure-prone procedures. One company that has made it into the annals of human-made data loss at scale is the now moribund MySpace. For a decade, this social networking company was one of the most important music-sharing sites in the world, hosting millions of tracks and other media content by aspiring and established artists. In 2018, however, the company posted an unexpected statement on its own website: "As a result of a server migration project, any photos, videos and audio files you uploaded more than three years ago may no longer be available on or from MySpace. . . . We apologize for the inconvenience." The cause of the data loss was a server-migration error—a devastating loss to artists amounting to an estimated fifty million songs.[24]

More recently, Google Cloud accidentally erased the online account of Uni-Super, Australia's savings system for workplace pensions, leaving over half a million UniSuper fund members without access to their retirement accounts.[25] The incident starkly demonstrates that geographical distribution alone is not a foolproof safeguard against data loss. Despite having duplication across two geographies, UniSuper could not prevent the deletion of its data because the cloud subscription itself was deleted, causing a cascade or data loss across multiple locations.

In addition to human-induced error, data losses caused by cyberattacks loom large across all levels of society. For example, the British Library, while thankfully not going down in literal flames, suffered devastating economic and cultural losses in the wake of a cyberattack on October 28, 2023. A ransomware cyberattack, claimed by the Rhysida ransomware gang, compromised the majority of the library's online systems by exfiltrating data, encrypting and destroying substantial portions of the British Library's server estate, and forcibly locking out all users from their networks.

Data-loss incidents, like those involving MySpace and Google Cloud, are often framed as accidental. However, the events outlined in this section also expose a deeper "infrapolitics" of data-loss capitalism.[26] In this landscape, individuals and organizations striving to protect their data are compelled not only to invest in data-loss-prevention services but also to keep pace with the relentless churn of technological obsolescence. The British Library's 2023 report on lessons learned from its cyberattack reflects this reality: "Our reliance on legacy infrastructure is the primary reason for the extended recovery time. Many of these systems will need to be migrated to new versions, significantly modified, or entirely rebuilt because they are either unsupported and cannot be restored, or because they are incompatible with modern servers and security protocols."[27]

As the next section will illustrate, the British Library's experience highlights

how obsolescence has been weaponized within the framework of data-loss capitalism, transforming it into a powerful tool for driving perpetual dependency
and profit.

WELCOME TO THE DIGITAL DARK AGE?

Amidst the ambitious efforts of those leveraging "the digital" as a tool to generate more data about the world, alternative voices are beginning to emerge.
These perspectives tell the story of contemporary knowledge societies not as an
uninterrupted march toward greater enlightenment but rather as a precursor to
what some have ominously described as a new "digital dark age," reminding
us of the sobering reality that while software may not physically degrade like
hardware, it is still prone to corruption over time. Reliance on digital business
strategies makes it even more vulnerable through exposure to additional risks
of data loss and obsolescence.

The concept of the digital dark age was popularized by Vint Cerf—often
introduced as the "father of the World Wide Web"—at a science conference in
San Jose, California, in 2015. At that time, Cerf was vice president and chief
internet evangelist at Google and mobilized the concept to express his fear that
future generations will have little or no record of the twenty-first century due
to the fragility of new formats: "What can happen over time is that even if we
accumulate vast archives of digital content, we may not actually know what it
is."[28] As technology evolves, he warned, data stored in outdated formats may
become unreadable or incompatible with modern systems, leading to a loss of
valuable information and cultural heritage.

Cerf's warning is the stuff of nightmares for digital preservationists, many of
whom have horror stories to share about lost access to digital records because of
challenges such as format obsolescence. In 2005, Jonas Palm, director of preservation at the Swedish National Archives, described how digitization was giving
rise to a new "digital black hole," noting that digital information, "which in the
analog world could be accessed simply by the use of our eyes," would suddenly
be "stored in an environment where it is only retrievable through the use of
technology."[29] This networked dependence itself makes digital data vulnerable
to loss. Take digital data's reliance on operating systems: when an operating
system becomes obsolete, all programs built on that operating system are, by
default, themselves subject to obsolescence. One instance of this phenomenon is
the decline of Java applets, a once-popular technology used for creating interactive and dynamic website content. In 2013, however, owing to security concerns
and advancements in web technologies, major web browsers, such as Chrome,
Firefox, and Safari, began phasing out support for Java applets. As a result,
many websites and applications built on Java applets became outdated, not only

forcing the hand of developers to seek alternative solutions but also making ruins out of previously well-functioning cultural creations.

Another case in point is the diminishing relevance of Adobe Flash. Until quite recently, Flash was the "go-to" program for creating multimedia content, animations, and interactive features across the web. But with the emergence of HTML5 and its inherent support for multimedia elements coupled with security issues associated with Adobe Flash, leading web browsers suspended their support for Flash. Adobe soon thereafter ceased development and distribution of its Flash Player, rendering Flash-based content obsolete and prompting the migration of Flash-based applications and websites to more contemporary technologies.

Format obsolescence may be "the end of the line" in technology terms, but it is also important to recognize that data can also be lost due to simple lack of maintenance. One example of this is the phenomenon colloquially called "link rot"—what we tend to encounter as a 404 "not found" screen telling us that the web page we are searching for cannot be located. In a joint 2014 study, Jonathan Zittrain, Kendra Albert, and Lawrence Lessig found that fully 50 percent of the links embedded in U.S. Supreme Court opinions since 1996, when the first hyperlink was used, no longer work.[30] A decade later, in 2024, a Pew Research Center study found that 38 percent of all webpages that existed in 2013 were no longer accessible a decade later.[31] The numbers from these narrowly focused studies on a fragment of the web leave us grappling with unsettling questions about how vast the scale of knowledge loss might truly be.

One phenomenon that facilitates the increasing occurrence of digital black holes is referred to in economics as "planned obsolescence"—that is, "the production of goods with uneconomically short useful lives so that customers will have to make repeat purchases."[32] Corporate policies of planned obsolescence accelerate the production and consumption of digital devices and platforms, but they also generate obsolete data on multiple levels. While there are few empirical studies of the scale of data loss due to planned obsolescence, there is a general sentiment among those working in digital preservation that the technologies used to store and read data have increasingly short lifespans. One recent study of hard drive failure indicates that errors typically occur after only three years of use.[33] There are currently no standardized metrics for measuring obsolescence in software, and any given piece of software could, in theory, run forever since no material wear and tear is suffered with its use. Yet scholars point to a growing number of cases in which malfunctioning software, high performance requirements, or canceled support services shorten the lifespan of an otherwise functioning product.[34]

A newer challenge has also arisen regarding the social media data housed on corporate platforms. These platforms' terms of service increasingly restrict access, posing difficulties for archivists whose aim is to facilitate access to social media data collections. While platforms assert that access restrictions are motivated by a desire to safeguard user privacy, proponents of digital preservation contend that the significant role these platforms play in capturing and mirroring our digital culture warrants a clearer mandate for platforms to transfer their collections to established memory institutions.[35] This concern is in no small part substantiated by the fact that most of these platforms lead relatively short lives themselves. As Paul N. Edwards puts it,

> Today's platforms can achieve enormous scales, spreading like wildfire across the globe. As Facebook and YouTube illustrate, in just a few years a new platform can grow to reach millions, even billions, of people. In cases such as Airbnb and Uber, platforms set old, established systems on fire—or, as their CEOs would say, "disrupt" them. Yet platforms themselves burn much more readily than traditional infrastructures; they can vanish into ashes in just a few years. Remember Friendster? It had 115 million users in 2008. What about Windows Phone, launched in 2010? Not on your radar? That's my point. Platforms are fast, but they're flammable.[36]

Amidst the remnants and ruins of digital landscapes, those dedicated to long-term preservation of digital data face ongoing challenges in ensuring reliable access. Take, for instance, the Danish National Archive, which relies on statistical analysis to monitor the migration of data pertinent to future public interest. This migration process entails the creation and management of duplicates across different storage media and formats to facilitate seamless access over time. Migrations are designed to preserve the features of a record that are considered important for its designated purpose. But each act of preservation also incurs an aspect of loss or alteration, a trade-off that is established blindly and without knowledge of the historical significance or interest any given file might have.

Concerned about the context losses that migration may give rise to, institutions such as the British Library and the National Archive of the Netherlands also implement strategies of *emulation*, enabling researchers to open files that exist in obsolete formats, such as WordPerfect 5.1 and Flash. But as information scholar Amelia Acker reminds us, emulators are themselves only software applications, which are run on delineated terms in protected and experimental environments and are as yet rarely scalable to broader society.[37] Thus, while they may end up as more mainstreamed applications, they are also just as prone to obsolescence as the programs they are trying to imitate.

Each strategy of preservation, therefore, is also structurally conditioned by loss. And the patterns of these software losses are more often than not shaped by particular logics belonging to the "disruptive" forces of the "creative destruction" of the market.

THE AGNOGENICS OF EPHEMERAL COMMUNICATION

Thus far, I have explored data losses influenced by research decisions, natural constraints, and market dynamics. Now, I will consider a final significant form of data loss: the "forgetting" caused by political and corporate agendas—and governmental secrecy.

Peter Galison, in the volume *Agnotology: The Making and Unmaking of Ignorance*, describes how knowledge of military significance is removed, classified, sequestered, and even prevented from coming into being. In the digital realm, these practices persist, with archivists expressing a new concern: data losses shaped not only by the forces of obsolescence but also by strategic maneuvers of politicians, governments, and corporations. Notably, it has become evident that some politicians have questionable recordkeeping standards. One prevalent trend is the use of communication systems founded on ephemerality for political purposes. British officials, for instance, have faced accusations of employing government-by-WhatsApp tactics, relying on self-deleting messaging applications to circumvent oversight and accountability.[38] Similarly, Danish prime minister Mette Frederiksen found herself at the center of a scandal during the COVID-19 pandemic when it was revealed that her office had mass deleted text messages. In November 2020, amidst fears that a mutated strain of the coronavirus could spread from mink to humans, the Danish government made the drastic decision to cull the entire farmed-mink population—an action later deemed illegal. The scandal deepened when it emerged that key text messages from the prime minister and other top officials had been "automatically" deleted, raising suspicions of a cover-up. This incident intensified political turmoil but also highlighted more general and global issues with data-retention policies, particularly the inconsistencies between these policies and actual retention practices but also the disparity in how text messages are treated compared to other forms of government communication.[39]

Most of all perhaps, behind the spectacle of controversy, these cases also illustrate the challenges faced by diligent civil servants in their efforts to preserve ephemeral or platform-specific communications. One article, for example, cites the frustration of a European Commission official over tensions between "the need to preserve native digital content" and the infrastructural "gap in what relates to Digital Preservation," expressing concern that data-retention and filing rules are insufficient for managing different types of information as records, citing examples like "wiki content, task manager content (Jira), and conversations using instant messaging applications (Skype, Slack)."[40]

In their introduction to the concept of agnotology, Robert N. Proctor and Londa Schiebinger emphasize that not all ignorance is bad and that some forms of knowledge destruction can even be virtuous. This reminder is especially relevant to the question of politically motivated data destruction. Take the paper shredder. Today, it has become a symbol of obstruction and transgression of justice, often mobilized in popular culture to represent that final step over the line set by the law.[41] As always, fiction takes inspiration from reality. One of the most famous shredding scandals in recent times unfolded during the Enron case as it became apparent that a former accountant, Arthur C. Anderson, had orchestrated a campaign to destroy Enron-audit documents, thus knowingly breaking the law.[42] Similarly, Donald Trump's first presidential term was riddled with stories about his obsession with document destruction.[43]

But as the history of the paper shredder reveals, knowledge-destruction devices do more than permit legal transgressions of the corrupt kind; they can also be used to enact forms of virtuous agnotology. As Sarah Blacker's 2020 essay on the paper shredder points out, even the German word for paper shredder, *Aktenvernichter* (literally "file annihilator" or "record annihilator"), conceals a fascinating history of resistance to political repression.[44] One of the first people to patent a shredding device was Adolf Ehinger, the German shopkeeper and member of the anti-Nazi resistance movement. He invented the apparatus in the 1930s to protect members of the resistance by shredding printed anti-Nazi texts.[45] The mechanism of his invention was based on a hand-crank pasta maker and was first manufactured commercially in Germany in 1935. Only later was the shredder marketed to government agencies and financial institutions in the form of a patented motorized version, especially in the United States, where its target consumer base included corporations seeking to protect trade secrets. The shredder took on a new role in society, moving from its origins as a mechanism of virtuous destruction to a facilitator of control of access to particular kinds of techniques, resources, and markets. (For decades now, cigarette makers have used shredders, dusters, and even mulchers to make sure incriminating documents never see the light of day.)[46]

Today, both public bodies and commercial companies adhere to strict guidelines for data deletion, often delivered through what experts in the field of computer security call data sanitization. Concretely, there are three commonly used data-destruction methods: data overwriting, degaussing, and disc destruction, each of which refers to a specific technique and layer of destruction.[47] Responding to emerging cloud technologies, institutions and corporations have also begun adopting new methods, such as cryptoshredding, whereby data is deleted by deliberately removing or overwriting encryption keys.[48]

Put simply, destruction is a process inescapably entangled with questions of power: Who is able to destroy (what) data, under what conditions, and how? Palestinian historian Karma Nabulsi is often credited with the first usage of the term "scholasticide," which she conceptualized during the 2008–2009 assault on Gaza to describe Israeli attacks on educational institutions to place them within a more general historical context of the "the systematic destruction of Palestinian education by Israel" since 1948.[49] The term has since been expanded upon by others to understand how scholasticide has "intensified on an unprecedented scale" during the present war to include the destruction of archives, libraries, museums, and cultural heritage.[50]

Violent agnotology also speaks to the complicated politics of colonial destruction and how it may also become a specter that haunts, sometimes even more forcefully through imaginary thoughts of what might have been (in text or in fact). In his text on the limits of the archive, Achille Mbembe argues that rather than obliterating memory, such acts of destruction can actually inscribe memory and its contents in a double register consisting of both fantasy, marked by imaginary thoughts of what might exist, and a *specter* that haunts the state as an authority of a future judgement.[51] (In law, jurors are allowed to make an adverse inference if a party is found to have destroyed evidence.)

In the midst of violence, data destruction also remains a practice bound up with mundane labor, professional virtues, everyday ethical practice, and even care. Indeed, much of the politics of data destruction happens as "boring" infrastructural labor to create meaning, mitigate risks, and safeguard others.[52] It is enough here to think of the ingrained, "according-to-company-practice," and often-necessary destruction practices (referred to as "digital housekeeping" by ethnographers) of, for example, civil servants performing the mundane but complex work of deciding what files, emails, and text messages to keep and what to let go of. Here, proper destruction practices, of which the obverse of preservation is also part, adhere to the civil servant value of virtuous agnotology rather than obstruction of justice.

THE MULTIPLE MECHANISMS OF DATA LOSS

If the history of science has taught us anything, it's that the history of knowledge is not one of simple progress or accumulation. Knowledge production in the digital era, like the creation, storage, and distribution of wisdom across the centuries on stone, parchment, and paper, unfolds as a continual oscillation between gains and losses. We have much more "external memory" at our disposal to capture words, images, and gestures than our predecessors, and yet much of what lives in these "clouds" is bound up with material limitations, capitalist logics, and political decisions. Within these clouds, our traces undergo complex processes of compression, distribution, and convolutions. And many of them have an afterlife as training data for machine learning models whose business strategies and architectures can be deeply agnogenic.[53] All these things are deeply intertwined with social relations, nature's limitations, everyday contingencies, and broader power dynamics. The instances of data loss discussed here—spanning from scientists' decisions at CERN to natural disasters, planned obsolescence, strategic ephemerality, and deliberate destruction—each illustrate unique agnogenic patterns shaped by different factors, including scientific paradigms, material boundaries, market logics, personal desires, and political values.

It is therefore crucial to attend to the politics of data loss on both a personal and societal level, not because a (hypothetical) perfect archive would ensure perfect justice but rather because the power to make data disappear, extract data value, and sever data relations today lies in the hands of mostly for-profit companies. These companies are largely built on a business model of planned obsolescence and exist in a complex political and regulatory ecosystem, which often offers perverse incentives to both maximize profit and reduce regulatory burdens. Now we are witnessing concerning patterns of governmental data erasure, with the Trump administration systematically removing or altering scientific information on climate change and other politically contentious topics from institutional websites and databases, with Elon Musk's role as head of the Department of Government Efficiency actively facilitating this process. This troubling convergence reveals how certain forces within cyberlibertarian tech increasingly operate within, and sometimes actively collaborate with, autocratic-leaning forces seeking to control public access to information. The idea that we are living through an era of exponential, near-infinite knowledge accumulation no longer fits for a society in which we lose pieces of our collective record daily. We must learn more about how these losses are material, cultural, and political and inextricably entangled with deeper structures of power. Agnotology open a new avenue for doing so.

SIX

Law against Knowledge

Anti-Epistemology

PETER GALISON

EPISTEMOLOGY, THE STUDY OF how knowledge is gained and secured, has a long, dark shadow. For every means of knowledge acquisition, transmission, and confirmed reception, there is a dark-side counterpart in a world of anti-epistemology. Do you acquire knowledge by speech, sight, photograph, or film? There are procedures, mechanisms, laws, and restrictions aimed to block those paths. Do you try to distribute or file that knowledge? There are means aimed at blocking those as well. Laws have a particular role to play: they not only come with the legitimacy and force of the state—prisons and fines—for their violation, but they also target the kinds of things that might be in the dark zone. Understanding the shifting modalities of knowledge-blocking is a path toward understanding what an epoch considers valuable knowledge.[1]

To be sure, secrecy goes back millennia—Babylonians used numbers to encode pictograms on tablets five thousand years ago. Modern secrecy laws are often aimed at flagging propositions of military significance—General Washington will sleep in this house on that night. Such are the kinds of secrets targeted by the World War I Espionage Act. In World War II, secrecy expanded, both bureaucratically and legally. Powered by the vast Manhattan and radar projects, hundreds of thousands of people were sworn into a tripartite system we recognize even today: confidential, secret, top secret. Importantly, the scope, the ontology of secrecy, jump shifted with the Atomic Energy Act of 1949 and 1954 to cover whole scientific domains: knowledge of elements from uranium on up to the science of nuclear fission.

With the 2001 Patriot Act, secrecy once again leaped, ratchet-like, ahead to cover domains that had been officially off-limits to classification, such as infrastructure. Now, most anything can be made secret, and in a triumph of penumbral expansion, new categories of the restricted, but not classified, have come into view. The government can, for instance, withdraw documents from libraries that indicate first responder capabilities in and around potential flood zones.[2]

Battlefield propositions, technoscientific domains, terror ubiquity. In our contemporary moment, the ratchet of anti-epistemological secrecy has turned in yet new directions as ideology and money cross. Proprietary secrecy, like government secrecy, continues to expand as the big tech companies, among others, cut off what they are doing from public inspection. But beyond protecting company secrets (food and drink ingredients, like the formula for Coca-Cola), there is a hybrid form of secrecy—industrial secrecy—that dons the apparatus of the state as its armor.

Nowhere is this more vividly on display than in Big Agriculture (Ag), especially animal mass production. Big Ag has pushed, successfully in many cases, for a class of laws known as ag-gag. By declaring industrial farms (concentrated animal-feeding operations) to be critical infrastructure, they can be protected from prying eyes that aim to photograph and, especially, video these facilities.[3] Here, I would like to focus on one recent example as a lens into the nature of the removal of knowledge: a pair of 2015 Wyoming laws that aimed to restrict the taking and using of "resource data" about the environment—namely, Title 40 (Trade and Commerce), Chapter 27 (Trespass to Unlawfully Collect Resource Data), Section 40-27-101 (Trespass to Unlawfully Collect Resource Data; Unlawful Collection of Resource Data).[4]

CUI BONO?

Though extensive in area, Wyoming has a small population (586,000 in 2015). In the early 2000s, Jonathan Ratner worked for the Forest Service, the branch of the Department of the Interior that manages some 420 million acres of federal lands and nearly fifty-five million acres of tribal land.[5] As a wilderness ranger in the Wind River Range, Ratner pursued complaints about the impact of livestock grazing. His extensive report landed on the desk of the district ranger, who did not want to pursue these issues. Neither did most of the conservation organizations. "Everyone," Ratner recalls, had "sort of moved away from the protection of the natural resources for their own sake to a very anthropocentric view of conservation—that it has to be of value for humans to be worthy of protection." So he took his data on cattle-caused environmental impact to another organization, the future Western Watersheds Project (WWP).[6]

To him, "the massive level of water pollution caused by cattle in particular" had become obvious. "And so," he continues, "my thought was, 'Well, we have the Clean Water Act. Let's develop a water quality monitoring program and try to get some accountability here.'"[7] On its own, the Environmental Protection Agency (EPA) has nowhere near the staff needed to monitor the vast domain of

U.S. waterways. Instead, the EPA depends upon state-based certified testing for data.[8] After a year setting up an EPA-approved water-quality-monitoring program, Ratner said he had an approvable plan for sampling and analysis under the 303(d) list of the state's impaired and threatened waters within the Clean Water Act. "I had collected about two years of data," he recalls, "and essentially what I was finding was that . . . the E. coli [*Escherichia coli*] levels in any water body that was accessible to livestock was somewhere in the range of two to . . . fifty ('five-zero!') times the state standard. But nobody collects this kind of data."[9]

Ratner began developing the standard analytical protocol that he would use in 2004.[10] It was approved in 2005, and he submitted his first round of 303(d) results for 2007: "I knew that people may not like what the data showed, but I had no idea of the level of coordination and massive backlash that occurred due to the submission of data . . . literally within forty-eight hours of the time I submitted the data for the first [2007] 303(d) cycle."[11]

Meanwhile, the WWP took up other causes, challenging mineral and animal-agriculture interests on public lands in Wyoming. Following the possibility of litigation between the WWP and Ruby Pipeline—a company that was building a $3 billion natural gas pipeline from Wyoming crossing through northern Utah and Nevada and ending in Malin, Oregon—a settlement was reached. In return for halting any litigation, Ruby would pay $15 million over

FIGURE 6.1. **Route of the Ruby Pipeline from Opal, Wyoming, to Malin, Oregon.**

Source: Author's re-creation based on information from United States Energy Information Administration, "Ruby Natural Gas Pipeline Begins Service Today," July 28, 2011, https://www.eia.gov/todayinenergy/detail.php?id=2410.

ten years to the Sagebrush Habitat Conservation Fund, and $7 million to the Greater Hart-Sheldon Conservation Fund for land acquisition and remediation, covering some five million acres, enabling the conservation organizations to buy out grazing rights elsewhere.[12] Construction of the pipeline began on July 31, 2010, was placed in service on July 28, 2011, and, though it declared bankruptcy in 2022, continued operating.[13]

Even archaeological sites, some dating from 8,000 BCE, became flashpoints. The Forest Service had agreed to a grazing plan for some fifty-two thousand acres in Arizona's Coconino National Forest. The WWP judged that the plan had neglected to take into account the impact this grazing would have on a staggering number of archaeological sites. Indeed, though only 4 percent of the land had been surveyed, already eight hundred archaeological sites had been identified in the new grazing zone, "including a 34-room pueblo, numerous pit houses [dwellings dug into the ground], artifact scatters, ceremonial sites, ball courts, and even a cliff dwelling. The area also features a pit house from around 1100 A.D. with an intact roof." The court backed the WWP against the Coconino National Forest administration.[14]

In June 2014, fourteen ranchers, represented by Karen Budd-Falen, an anti-environmentalist pro-rancher lawyer, leveled a lawsuit against the WWP and

FIGURE 6.2. **Archeological site in the Coconino National Forest.**

Source: Brady Smith, *Honanki Ruins near Sedona*, Nov. 11, 2009, photograph. Courtesy of USDA Forest Service, Coconino National Forest.

Ratner. By that point, the WWP had been both celebrated and vilified for its effective court battles against the private use of public lands throughout the West, especially Wyoming, Idaho, and Nevada. Over and over, the organization showed that the federal government had a legal obligation to protect endangered species, fragile ecosystems, water safety, and archaeological remains. That effort cut into the profits yielded by the exploitation of public lands by gas and oil interests as well as ranchers. All this lay behind the gathering legal reaction: battle lines were drawn.

MALICIOUS AND NEFARIOUS INTENT

The suit against Ratner personally was just an opening gambit—its impact severe for him but limited more generally. After a preliminary bill died in committee in 2014, in January 2015, Republican Wyoming state senator Larry Hicks stood on the chamber floor in the state capitol and led the charge for a new bill: "I would like to give you a little background is why the need and the purpose for this bill to begin with [sic]. If we go into our existing criminal trespass statutes, let me explain to you how that operates. So you have some entity, some person that for nefarious and malicious intent is trespassing on your property to collect information or data that then they intend to reposit this data in some place that is going to result in adverse impacts to either you, your property or your business."[15] Here, "some entity" clearly designated an organization, like the WWP or People for the Ethical Treatment of Animals (PETA), that had gathered evidence of environmental degradation or animal cruelty. "Some place" meant a governmental body, such as the Wyoming Department of Environmental Quality, or the federal agencies, such as the EPA, that might receive the data and use it to impose sanctions under old or new rules and laws.

"Under the current law," Hicks continued to his fellow state senators, "the only thing you can do as a landowner or property owner is ask that person [taking data] to leave. And if they leave, there is no criminal act. Well, what happens to the information that they collected that then goes into some depository of some agency that then subjects you to additional scrutiny. Additional regulatory takings. Delays your ability to make a living. Precludes you from implementing practices that are vital to your business. There's nothing [you can do]." Hicks clearly wanted more than to turn back a particular person collecting data, which would allow them to vacate the property without criminal indictment. He wanted a fierce deterrent to anyone—or any organization—contemplating the collection of data. And he wanted to interdict or destroy information sent, for example, to the EPA.[16]

The Data Resources Bill aimed at just this sort of preemptive action, raising the potential cost to be more than just a temporary stop. Hicks made that clear:

"So what this law attempts, this new bill, which would carve out a new penalty as 6-3-414 is trespassing to unlawfully collect resources and collect resource data, tries to get to this issue of not trespassing for the purpose of taking pictures of flowers or hunting or inadvertently crossing somebody's property to go fishing. Get to those trespassers that are doing this for malicious and nefarious intent: to collect data."[17] In short order, the Sixty-Third Wyoming Legislature (January 13 to March 6, 2015) passed laws, one criminal and the other civil, through both houses in early March. Together, the laws allowed the state to imprison and impose fines on anyone who collected "resource data" on open lands for the purpose of handing them over to federal agencies.[18] " 'Collect' means to take a sample of material, acquire, gather, photograph or otherwise preserve information in any form from open land which is submitted or intended to be submitted to any agency of the state or federal government."[19] "Open land" (defined in the acts to mean "unincorporated land") is a huge part of the United States—some 40 percent of Americans live in unincorporated, mostly rural, land. The federal government itself owns about one-third of the land mass of the United States—much higher in the West, where it owns 96 percent of the land in Arkansas, 88 percent of Nevada, and 75 percent of Utah, for example. In Wyoming, the federal government holds about 48 percent of the land (some thirty million acres out of sixty-two million total acres), with another 9 percent in the hands of the state of Wyoming and the Wind River Reservation.[20] The crime, as per clause (a), is as follows: "A person is guilty of trespassing to unlawfully collect resource data if he: enters onto open land for the purpose of collecting resource data; and does not have: ownership . . . or written or verbal permission of the owner."[21] Restrictions on public land go back centuries—in England, enclosures go back informally to the twelfth century and legally to the seventeenth.

The resource data laws were already unstable since "open land," without further specification, could include public land (any unincorporated land), and the law specified the need for ownership or the owner's permission. That tension would come back to haunt the legislation. Adding complexity to landownership in the West is that a great deal of the land is subdivided into a checkerboard pattern, with alternating public and private land. This practice, dating to the nineteenth century, issued from federal policy aimed at subsidizing the building of railroads (the companies were given land abutting the tracks) and ostensibly encouraging quality track construction. It also—and not incidentally—served to disrupt the integrity of Native American culture and land. Checkerboarding left significant public acreage "landlocked," and people, including hunters and campers, could be prosecuted for "corner hopping" from one public square to another. One landowner claimed, with the wisdom of geometry, that since people were not "points," in corner hopping, they inevitably would violate his air rights even if they did not put a single footprint on private land.[22]

FIGURE 6.3. **Example of checkerboard land control in Wyoming, with public land in gray and private land in white.**

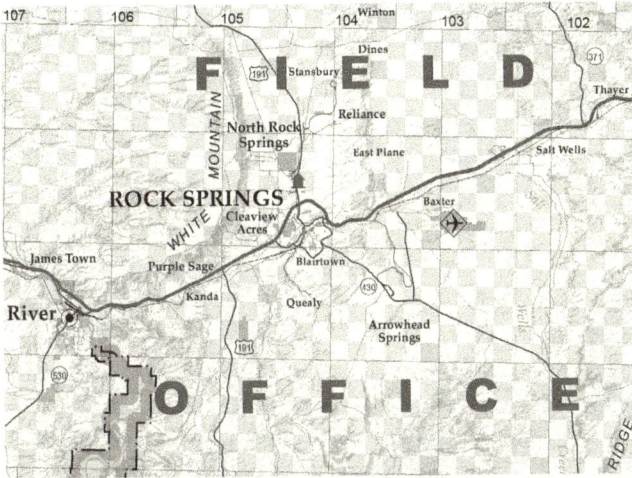

Source: Bureau of Land Management, U.S. Department of the Interior, *BLM Wyoming Land Status Map 2020*, Sept. 1, 2020, https://www.blm.gov/documents/wyoming/pub lic-room/map/blm-wyoming-land-status-map-2020.

Beyond the "open land" definition, "resource data" for the Sixty-Third Wyoming Legislature meant "data relating to land or land use, including but not limited to data regarding agriculture, minerals, geology, history, cultural artifacts, archaeology, air, water, soil, conservation, habitat, vegetation or animal species."[23] Whole domains, whole spheres of knowledge, fell under this knowledge shadowing, but it was clear what they held in common. If an endangered animal or plant were found, then the extraction or agricultural use of the land could be restricted. If cultural or archaeological evidence of Native American presence were identified, then suddenly federal regulations might restrict where ranchers could herd their cattle. To be sure, this wasn't about just any activities demanded by the state or federal government. There would be no problem at all taking data on the land to assess titular ownership, to determine corresponding taxes, or to chase criminals. As the Tenth Circuit Court of Appeals would later add, "Wyoming has adopted expansive definitions of 'resource data' and 'collect.' . . . Accordingly, prohibited acts include the following activities on public land, so long as an individual also records where such data was gathered: collecting water samples, taking handwritten notes about habitat conditions, making an audio recording of one's observation of vegetation, or photographing animals."[24] Just in case the data *were* obtained, that telling photograph or water-quality test secured, clause (f) extended the curtain further: "Resource data collected in violation of this section in the possession of any governmental entity as defined by W.S. 1-39-103(a)(i) shall be expunged by the entity from all files and data bases, and it shall not be considered in determining any agency action."[25] So things stood when the Wyoming legislature came to a close in March 2015.

To take on the new laws, WWP formed an alliance. The National Press Photographers Association, for example, was very concerned that the integrity of their craft required a recording of where and when they made their images. They, along with the Natural Resources Defense Council (NRDC) and PETA, jointly filed a complaint on September 29, 2015. Heard by the judge on December 11 of that year, the state's attempt to have the environmental complaint dismissed was denied on December 28, the court deciding that it violated the First Amendment of the U.S. Constitution.[26] Wyoming had to retreat to reconsider the wording of the law.

On March 3, 2016, the Sixty-Fourth Wyoming Legislature came back to this question and began a revision of the original laws. But after revising the law from "open" to "private" land, the state took on an even bigger issue: the new wording made it a crime for someone to cross "adjacent" private land with the intention of getting to *any* other land where data collection would take place. Senator Chris Rothfuss, a PhD chemist representing District 9 (Laramie) and one of only two Democratic state senators (amidst twenty-nine Republicans), voiced consternation about the validity of the law, specifically about the "adjacency" and proximate lands:

> I'm concerned that we're creating an entirely different problem. Not that trespassing is a good thing, but the idea that you are in a state where your actions are legal, your objective, which is to collect data on federal land, is also perfectly legal. Those two things are legal. Crossing the land is illegal. It already is. That's a trespass, the private property. But the fact that your objective was to collect data on public land, which again is legal, that exacerbates the crime. That is a [sic] enhancement of the trespassing because of an otherwise legal intent. I don't know if we do that anywhere else in statute. . . . [Absent a precedent,] I'd urge the body to vote no and try and remedy that.[27]

The president of the Senate asked for any further discussion and called on Senator Charles Scott (Republican, District 30). Scott, a Harvard college and MBA-educated rancher who was a member of the Stock Grower's Association, the Audubon Society, and the Nature Conservancy, dismissed Rothfuss's objection about precedent. Nonetheless, he found the idea of enhancing a penalty because of an "otherwise legal activity, somehow it just doesn't strike me as being the right thing to do."[28] After those four or five minutes of resistance, the bill was passed, revised without the changes to the adjacent/proximate lands that Roth-

fuss and Scott wanted to make. The law—with the adjacent/proximate clause—
became effective on March 16, 2016.

Researchers got the message loud and clear. The University of Wyoming
issued a "frequently asked questions" document informing staff of the range of
forbidden data covering geology, archaeology, minerals, animals, plants, water,
air, and soil:

> Do I still have to obtain permission even if I am just taking students onto the
> property as a field trip?
> Yes, there are no exceptions for educational activities under the statute.
>
> Do I still have to obtain permission from a landowner if I am just crossing his/
> her land to access public land?
> Yes, individuals must obtain permission from the landowner even if the indi-
> viduals is [*sic*] just crossing that property to access public land.[29]

Precisely because of the 2016 adjacent/proximate new clauses, WWP sued
the state again and, after being denied by the district court, was forced to
appeal. The appellate court reversed the district court's decision, forcing the
Wyoming state legislature to amend their law again.

On September 7, 2017, when the Court of Appeals for the Tenth Circuit
remanded the case back to the district court, they focused on clause (c), which
intensified normal trespass penalties based on intent (and so violated the First
Amendment); the district court then complied with this on October 29, 2018,
as it had to. The appeals court was sharp in its reprimand: "The fact that one
aspect of the challenged statutes concerns private property does not defeat the
need for First Amendment scrutiny."[30]

Plaintiffs contended that they had a right to report information about public
policy. Indeed, they insisted that the whole system of environmental regulation
and enforcement depended on data being delivered to the agencies—data the
agencies themselves could not possibly collect on their own. The appeals court
concurred, laying down the law in just these terms. The Tenth Circuit provided
the example of "the WWP reports information on water quality, including GPS
location data, to the Wyoming Department of Environmental Quality pursuant
to the Clean Water Act. Under the Clean Water Act, state agencies must 'actively solicit' 'field data' from the public that can be used to evaluate pollutants
in waterways. 40 C.F.R. § 130.7(a), (b)(5)(iii)."[31]

The Tenth Circuit further argued that NRDC should submit "geo-tagged
photographs" to the U.S. Fish and Wildlife Service when NRDC aims to push
for intervention under the Endangered Species Act. "Such petitions must present
'substantial scientific' evidence showing that a species is endangered or threatened, which typically includes a record of where the species has been observed.

See 16 USC § 1533(b)(3)(A)." Indeed, a host of other federal agencies require "public input," "public involvement," and "public scrutiny"—all in order to weigh "accurate scientific analysis."[32]

The battle raged on in 2018 as the Wyoming legislature continued to fight to keep the data laws. But the state government was losing in court case after court case. On October 29, 2018, U.S. district judge Scott W. Skavdahl said, rather bitingly, "Many of Defendants arguments seem to boil down to 'if the Plaintiffs simply do not seek to engage in their protected speech creation activities, they will not risk violating the statute.' This is not constitutionally acceptable."[33] The Constitution quite clearly forbids making a distinction between the content of belief or intent: "The statutes, on their face, penalize only the collection of resource data relating to land or land use. A content analysis is necessary to determine if the statutes are applicable. This is a clear case of content-based statutes." Judge Skavdahl not only granted the plaintiffs' motion to summarily resolve the issue in their favor, but he also further ordered "that the State of Wyoming is permanently enjoined from enforcing Wyoming statutes §§ 6-3-414(c) and 40-27-101(c) as both are in violation of the First Amendment of the Constitution of the United States of America."[34]

Following the decisive constitutional rejection by the courts of the enhanced penalties for crossing private land to take "resource data" in "adjacent" land, silence hovered over the legislature. Some five years later—effective as of July 1, 2023—an abbreviated acknowledgment came in summary: "This bill repeals a civil and criminal prohibition on trespassing across adjacent land to collect resource data. These prohibitions had been previously found unconstitutional in federal court. (See Western Watersheds Project v. Michael, 869 F.3d 1189 (10th Cir. 2017); Western Watersheds Project v. Michael, 353 F.Supp.3d 1176 (D. Wyo. 2018))."[35]

Industry pressure against data collection did not abate. In 2022, the Wyoming House debated, inconclusively, how to stop drones from trespassing. Then, on January 13, 2023, the Senate Judiciary Committee took up the question, intermittently concerned about whether they (or the Federal Aviation Authority) would control the "immediate" airspace. And if so, they wondered, would that be below five hundred feet? Four hundred feet? Fifty feet? While altitude could be debated, the impulse was clear. Testifying for the lobby group Wyoming Farm Bureau Federation, Brett Moline questioned the wording of the bill, which states, "A person is guilty of trespass by small unmanned aircraft if . . . it interferes *substantially* with the landowner's or his authorized occupant's use and enjoyment of the land."[36] Moline hoped to remove the word "substantially": "In my opinion, that puts the burden on the landowner, you know, What is 'significantly interferes'? Depending on how everything works out, that could set a pretty high bar. I know it's used in other places, but it's kind of something that's, in my opinion, a little nebulous." He also hoped to be able to take down drones before they could do any damage: "In other instances, before something bad happens, we have, as landowners, the opportunity to . . . reduce the threat. So is there something that we can do to bring that drone down before it does actual property damage? This bill doesn't address that."[37] Senator Wendy Davis Schuler pushed to keep the word "substantially," positing a scenario in which some students are using a drone for a class project, and the drone briefly enters the landowner's airspace here and there. She said they don't want to see those teenagers fined or sent to jail for a few months for something like that. In response, Mr. Moline noted, "I would appreciate having the word 'substantial' taken out, and I will sit by that, ma'am. We can agree to disagree—I will put it that way, ma'am."[38]

Jim Magagna with the Wyoming Stock Growers Association, also testifying on behalf of the Wyoming Wool Growers Association lobbies, did not want the landowner to have to prove that the trespassing interfered with his use and enjoyment of the land: "[I] have to wonder if that wouldn't be a better approach here—that if a drone without authorization flies over private land, that they're guilty of trespass. And don't have to prove impact on use and enjoyment—they trespassed, just like if they physically trespassed on the land." Travis Deti, director of the Wyoming Mining Association, responded, "We share concerns, also especially in the coal industry, about trespassing as well. Coming over and, you know, snapping some photographs and then using them on a website for negative impact for our industry."[39]

On January 16, 2023, Tara Nethercott, Republican Wyoming state senator and a lawyer by training, turned to the trespass questions—new (drones) and old (data trespassing): "What we're doing is we are repealing this particular statutory section." She explained to her fellow state senators that the Joint Judiciary Committee had passed and sponsored some four to five bills about trespass. "You're dealing with one of those today, and I think you've dealt with another drone-trespass bill. Don't let those helicopters on the drone-trespass bill send you sideways. But this bill actually comes—" Chairman Bill Landen interjected, "We'll try to land them in the right spot, Senator." Nethercott stated, "That's right. That's right. We'll shoot them out of the sky. This bill actually comes from work of the legislature back in about 2014, where there was a lot of discussion and concern over scientists and others accessing land for the purposes of gathering data. That created a great concern with private landowners. The Wyoming legislature responded, and, of course, Senator Case was here then and so was Chairman Landen, by creating this Adjacent Land Resource Data Trespass statute, which made it against the law to do that and created a trespass." Representative Nethercott carefully explained that the data-trespass law was challenged as being "unconstitutional in violation of the First Amendment." It was, she said, "a pretty unique legal argument . . . presented by some out-of-state lawyers. Just as a word to the wise, and I always made this caution to my members of the Judiciary Committee, they challenged the law for being unconstitutional. A federal court did find it to be unconstitutional and awarded attorney fees for over $650,000 to those out-of-state lawyers just to demonstrate to the Wyoming legislature that this is what happens when maybe the zealousness concerning an issue results in an unconstitutional passage of law. That money was paid for by the taxpayers and the people of Wyoming." Cautioning her fellow Republicans, she urged "great caution, despite sometimes what appears to be the need for these issues to come forward." Landen quipped, "I think that's the first and only time that the state senate has been wrong on something."[40]

THEORY OF AGNOTOLOGY: ANTI-EPISTEMOLOGY

During World War II, Claude Shannon developed a mathematical theory of cryptography, proved theorems about unbreakable codes, and began to outline how source material and code keys would work. His theory of cryptography and of information developed hand in hand, and, in 1948, he simplified his diagrams to one for information transfer in general. That diagram (redrawn below; fig. 6.4) outlines the functional components of any message.

Conceived as a guide to *blocking* information, Shannon's diagram could

FIGURE 6.4. Anti-epistemology—stopping information—categorized by which block is halted within Shannon's diagram of information flow.

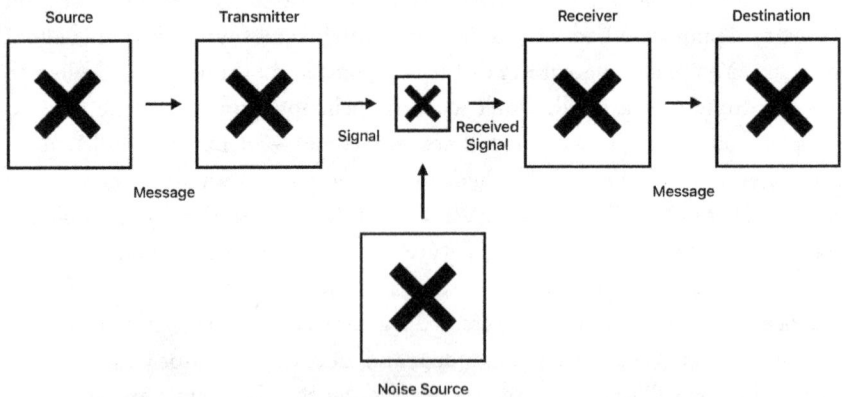

Source: Author's re-creation and modification based on Claude E. Shannon, "A Mathematical Theory of Communication," *Bell System Technical Journal* 27, no. 3 (July 1948): 381, https://doi.org/10.1002/j.1538-7305.1948.tb01338.x.

serve us well as a foundation for an anti-epistemology. Instead of seeing the functional elements of a successful transmission of a message (against noise, for example), one could see it as an apparatus through which one can identify the *failure modes* of communication. In other words, if the source, transmission, noise, receiver, and/or destination is attacked, the message will not be conveyed, understood, or interpreted. Grasping the different roles these blocks play can help articulate the ways by which knowledge is not acquired, how knowledge is not secured.

Start with transmission. Here lies in fact the bulk of the modern secrecy system: knowledge exists but is blocked from moving in certain directions. Classified information functions in this way: the recipient must have the corresponding clearance (e.g., secret, top secret, or confidential, and, in addition, be given access to particular programs). Even unclassified information may be controlled, as, for example, in the United States, when given the designation "Not Releasable to Foreign Nationals" (NOFORN), "an explicit foreign release marking used to indicate intelligence information that may not be released in any form to foreign governments, foreign nationals, foreign organizations, or non-US citizens."[41] *Censorship* is a means of effecting this secrecy by physically excising information. If one of the three-letter agencies—the CIA (Central Intelligence Agency), NSA (National Security Agency), FBI (Federal Bureau of Investigation), DIA (Defense Intelligence Agency), and NRO (National Reconnaissance Office), among others—imposes restrictions on circulation, its classification is designed to reflect the damage that would result to the United States should that information fall into the wrong hands. And on that basis, the potential penalties rise accordingly, all with the aim of containing information that already exists and is to be kept flowing in the right channels behind secrecy walls of varying height.

Next, reading left to right on the anti-information diagram, is "noise." Noise might be physical, acoustic noise used to make overhearing a conversation impossible. It might, for example, be electronic, jamming radio, television, or radar broadcasting white noise. But noise can also be deliberately imposed in ways other than by physically broadcasting interference. Information to be protected might be drowned in competing accounts, frustrating any attempt to extract the target. Robert N. Proctor has, for example, drawn attention to how assiduously tobacco-industry lobbyists worked to produce uncertainty about health damage from smoking. As the now-infamous R. J. Reynolds document of 1969 puts it, "Doubt is our product since it is the best means of competing with the 'body of fact' that exists in the minds of the general public. It is also the means of establishing a controversy."[42] Naomi Oreskes and Erik M. Conway track a similar strategy advanced by overlapping lobbyists and publicity firms to produce an artificial "debate" over climate change, even when a widespread scientific consensus existed.[43] The production of artificial controversy is precisely a form of noise generation—no less than a 1945 vacuum tube that spewed interference.

Importantly, these interventions to block knowledge presuppose that knowledge to exist. Stepping further along the communicative chain gets us to the "receiver." This could be done technically—the cheap and widely distributed Nazi *Volksempfänger* (peoples' receiver) was hard-pressed to pick up anything but local (state-controlled) stations; only German stations, and later Austrian ones, were marked on the dial. Troublesome films and books could be blocked at the border. In more recent times, Jonathan Zittrain and his colleagues have been tracking worldwide internet blocking. They categorize this receiver blockage into three scales: "Pervasive" filtering, blocking a wide content area; "substantial" filtering, covering the moderate stopping of several categories or wide-angle blocking of a single one; and "selective" filtering, which takes on a particular issue, such as any mention of Tiananmen Square or even its date (June 4 [2015]).[44]

At the end of the information (or anti-information) chain is the destination—those who actually get the information. Arresting someone for possession of banned political, cultural, or obscene material hit at the destination of the message. It might be possessing the wrong book or tuning into the wrong music. Once, during the Regime of the Colonels, I saw a fearful beach crowd in Crete pick up their belongings and flee in fear when someone defiantly played a record by Mikis Theodorakis, scourge of the regime, on a portable player.

All these blocking forms, all these modes of agnotology (transmission, noise, reception, destination), have one critical feature in common: each presupposes that the information—the fact, circumstance, image—already exists. Ultimately, the most devastating angle of attack on knowledge is the disruption of fact formation itself, obviating any need to block it downstream. Such a move against data gathering and assembly was *precisely* the aim of the Wyoming Resource Data Trespass laws, with the 2015 criminal and civil laws being the first attempt to prohibit the gathering of data on all "open" (unincorporated) lands, public or private. When that bid failed, the next step was forbidding the crossing of private land to get to public lands with the intention of gathering data for federal or state regulatory purposes (which was summarily shot down as a violation of the First Amendment). Measuring *E. coli* in streams, rivers, and ponds, taking place-specific notes, filming or photographing with GPS metadata? Recording land degradation? Documenting archaeological sites? Forbidden.

When these measures failed, the legislature debated a "drone-trespass bill" to prevent observation of lands with the intent to collect data where pedestrian trespass laws were insufficient. This died (at least in 2022–2023) before being enacted. These laws aimed to forbid the assembly of facts themselves—*this* stream has a geometrical *E. coli* density mean of one thousand colony-forming units per one hundred milliliters, *that* eight-hundred-year old pit house is located at a specified location, etc.

Such legislation joined another set of laws, which have been adopted in more than a dozen states: the so-called ag-gag laws. In one sense, these were more wide-ranging. They targeted the surreptitious documentation of large-scale industrial agriculture, mostly by still images but especially directed toward (time-based) moving images and recorded sound. These were wider because in many instances, the laws did not require trespass. But they were also narrower: they were specifically about the threat such images and sound recordings presented to the big, typically multinational, meat-production companies. By contrast, the Trespass to Unlawfully Collect Resource Data laws covered a full range of scientific fields from archaeology through history to zoology, across air, water, and land and involving minerals, animals and plants.

In this sense, the theory and practice of anti-information has jumped ahead of anything that the positive practice of communicative efficacity had imagined. Surely, Claude Shannon never had in mind the problems of the original message being splintered, formlessly, into nonexistence before anyone touched a telegraph key, a radio microphone, or a computer. This annihilation of the very possibility of knowledge is the asymptote of agnotology.

SEVEN

Why We Wrongly Imagine Adam Smith as a Free Market Fundamentalist

NAOMI ORESKES *and* ERIK M. CONWAY

AGNOTOLOGY SCHOLARS RECOGNIZE THAT ignorance comes in many forms and can be created in many ways. One way to create and sustain ignorance, emphasized particularly by the late sociologist Charles W. Mills, is through selective *attention*: we attend to things we want to know (or want others to know) and ignore, dismiss, disparage, or repress things we don't want to know (or don't want others to know).[2] Mills directed his readers' attention to the cultural construction of whiteness and the way it had often been downplayed by previous scholars. He asked us to look and think again particularly at the history of the Caribbean and of race and inequality in America. In this paper, we call attention to a related matter: the technique of selective *presentation*. By this, we mean the processes by which a set of events or the work of a particular thinker is presented selectively in ways that distort an individual's meanings and/or intent. Our focus here is on one of the most influential thinkers in Euro-American history: Adam Smith.

Classical liberal economists—including Adam Smith—recognized that government serves essential functions, including building infrastructure, raising taxes to pay for public goods, and regulating banks (which, in the pursuit of self-interest, could destroy an economy). But in the mid-twentieth century, a group of self-styled "neoliberals" began to shift economic and political thinking radically. They argued that government action in the marketplace, even if well intentioned, compromises the ability of individuals to do as they please—and therefore threatens freedom. Political and economic freedoms are "indivisible," they insisted: any compromise to the latter—even to address an obvious ill, like child labor, or control a deadly product, like tobacco—is a threat to the former. Moreover, these thinkers held, free markets are a bulwark of democracy, because by distributing decision-making, they prevent the concentration of power in centralized government. Free markets create a firewall that protects citizens from the potential tyrannies of concentrated governmental power. Crucially, they argued, expanding and protecting economic freedom serve to expand and protect political freedom. In the later twentieth century, they used this claim to justify the weakening of workplace and environmental protection, of minimum wage laws, and of antitrust enforcement.

The indivisibility thesis was promoted and developed in a variety of ways but most famously at the University of Chicago in the Department of Economics and the program in Law and Economics. In the past four decades, it has been highly influential in American life, underpinning a good deal of both neoliberal and conservative thinking. It has been invoked by corporate leaders, such as J. P. Morgan's Jamie Dimon; political leaders, including Ronald Reagan, Paul Ryan, and Ted Cruz; culturally influential figures, including Rush Limbaugh and Glenn Beck; and by the long-time head of the U.S. Federal Reserve Alan Greenspan. The idea that government regulation of the marketplace threatens personal freedom underlies much conservative hostility to "Big Government" and has been used to defend a variety of policy positions that limit the reach of government, including federal tax cuts (particularly for the rich), privatization of social security, the promotion of charter schools, and an all-volunteer military.

The problem with the indivisibility thesis is that it is fundamentally flawed, both logically and historically. The flaws in the argument are various, but three are obvious. The first, which economists have long recognized, is that markets can fail in two distinct ways. One way markets can fail is by not providing things we need, such as public goods. These include the economic and cultural benefits of road, bridges, and other forms of infrastructure, as well as the economic, social, health, and psychological benefits of government provision of water supplies, public education, healthcare, and support for the arts. Economists sometimes refer to the latter as "external benefits," or "positive externalities." Markets can also fail when they yield products or activities that do (or threaten) substantial harm. Economists refer to this as the problem of external costs, or "negative externalities." A powerful recent example involves the damage done to stratospheric ozone—popularly known as the "ozone hole"—caused by a class of chemicals known as chlorinated fluorocarbons. While the chemicals at issue were useful for refrigeration and other purposes—and entirely legal to produce—their continued use threatened the future of life on earth—an essentially infinite external cost. This market failure was remedied by international governance: the 1987 Montreal Protocol on Substances that Deplete the Ozone Layer (which came into force in 1989), which bans the use of these dangerous chemicals.

A second flaw in the indivisibility thesis is that it ignores the impact of monopolies and anticompetitive business practices. In fact, the idea of markets protecting democracy by distributing decision-making only works if power is *not* concentrated in the hands of businessmen. But this concentration is precisely what happened in late nineteenth-century America when important parts of the economy were monopolized by industrial giants, such as Standard Oil, Carnegie Steel, and Cornelius Vanderbilt's empire of railroads and steamship lines. The leaders of these companies—John D. Rockefeller, Andrew Carnegie, and Vanderbilt—used their stranglehold on oil, steel, and transport not only to distort the workings of the marketplace but also to accumulate vast fortunes, which they used to corrupt and distort American politics. The Sherman Anti-Trust Act, passed in 1890, was designed to protect both markets and democracy from the distorting impacts of power concentrated not in the hands of centralized government but in centralized industry.[3]

A third flaw in the argument is that history offers substantive refutation of it. Any broad claim about what is true or false in history will inevitably face exceptions and counterexamples, but no attentive historian could easily deny that the relationship between capitalism and freedom is complex and vexed. To be sure, there is a case, as Martin Wolf has recently made, that the individualistic spirit of capitalism—in particular, the decision-making function of individuals in the marketplace—has a resonance with the individualistic impulses of democracy.[4] That said, history offers abundant reasons to question the simplistic argument that capitalism and freedom are indivisible.

Market-based capitalism first began to flourish in the United Kingdom in the sixteenth and seventeenth centuries, but political, civic, and religious freedoms did not quickly follow for the majority of its denizens. England—a nation with an established church, the Church of England—famously suppressed religious dissenters, some of whom fled to the American colonies. Nor was economic freedom necessarily widely available. Well into the eighteenth century, various statutes restricted the rights of laborers to move in search of work; those without work could at times be forced into poor houses or even imprisoned.[5] In practice, workers may sometimes or even frequently have found ways around those statutes. Nevertheless, the fact that they long persisted refutes any simple claim that the ideals and demands of the free market system necessarily led to expanded economic freedoms. When we turn to the question of political freedom in the United Kingdom, we see a similar pattern: the rise of capitalism did not quickly lead to a broad expansion of political freedom. Universal male suffrage was not implemented until 1918, and universal female suffrage, not until 1928.

Then, there is the deeper question of how we understand freedom. Workers who struggled to survive on the tellingly labeled "starvation wages" cannot be said to have been free in any meaningful sense, hence Karl Marx's famous declaration in his 1848 *Communist Manifesto* that the workers of the world had nothing to lose but their chains and Anatole France's wry comment a few decades later that the law, in its majestic equality, forbade rich and poor alike from sleeping under bridges.[6]

While not literally enslaved, many industrial workers lived lives of scant opportunity for substantive individual agency. The same was broadly true in other industrialized capitalist nations, including the United States, where capitalism had not prevented the enslavement of ten million people of African descent, the dispossession of Native Americans, or the denial of legal and political rights to American women.

The indivisibility thesis is based on a logical fallacy—in fact, two logical fallacies. The first involves a critique of Soviet-style centralized planning. Soviet-style centrally planned communism was certainly associated with widespread denial of political, civic, and religious freedom. But it did not follow that if an economy were *unplanned* that freedom would follow, much less be maximized. In effect, the indivisibility thesis is based on the illogical argument that if x imperils y, then $\sim x$ must protect y. It also involves the fallacy of the excluded middle: that the alternative to centrally planned communism is unregulated (or only very weakly regulated) capitalism and that there is no middle ground to be found or held.

Given these difficulties, an obvious question arises: How and why did this view of things come to be so influential? Why did so many people accept a worldview that Adam Smith himself, often considered the father of free market capitalism, would have rejected? Like all historical phenomena, the answer is complex and involves the actions of many groups and individuals, but a key part involves the revisionist reconstruction of Adam Smith as a market fundamentalist by the Chicago economist George Stigler.

In the early twentieth century, key American business leaders tied to famous corporations, such as Dupont and General Electric, and captains of industry, such as Alfred P. Sloan and J. Howard Pew, began a century-long campaign to fight government regulation of the marketplace. They argued that laws to limit child labor, for example, or to establish programs of workers' compensation denied freedom: the freedom of businessmen to run their businesses as they saw fit and the (alleged) freedom of workers to work where they saw fit. During the New Deal, these businessmen dug in deeper, fighting Franklin Roosevelt's programs to establish minimum wages, protect the right to collective bargaining, and regulate banks (among other things). Their core argument was not just that these government programs were inefficient but that they were un-American, socialistic, and threatened *freedom*.

For the most part, these arguments failed, in part because during the Great Depression, the "free market" had so conspicuously failed.[7] But after World War II, with the economy more or less back on track, they tried again—this time by funding research at the University of Chicago intended to shore up the intellectual foundations of their otherwise obviously self-interested claims. A key part of this effort was the reconstruction of Adam Smith as a free market fundamentalist through the selective presentation of his classic work *The Wealth of Nations*.

A classmate of the famous Chicago economist Milton Friedman, George Stigler earned his PhD in 1938 under the direction of Frank Knight—often considered part of the "first generation" of Chicago economists. (Knight was instrumental in getting Friedrich von Hayek's classic neoliberal text *The Road to Serfdom* published in the United States.) After graduate school, Stigler taught at Iowa State, the University of Minnesota (where he was again a colleague of Friedman), and Columbia University before returning to Chicago in 1958. He was a founding member of the Mont Pèlerin Society, serving as its president from 1976 to 1978, and would win the Nobel Memorial Prize in Economics in 1982 for his work on economic regulation.[8] Stigler generally opposed government regulation of economic activity; he coined the term "regulatory capture" not as a critique of corporate corruption but to suggest that many, if not most, regulations would fail to achieve their goals because vested interests would manipulate the system to their advantage.

In the 1950s, Stigler took on the task of producing an edited version of Adam Smith's foundational work *The Wealth of Nations*. At over a thousand pages, the original 1776 text was not something a professor could assign in its entirety to undergraduates (nor for that matter to most graduate students). Stigler had to pick and choose the most salient parts.[9] His treatment does an impressive job encapsulating Smith's central antimercantilist argument that commerce and wealth can arise from individuals' actions without the guidance of a monarch or other central authority. In Smith's words that are so famous as to have become a mantra for conservative economists, "It is not from the benevolence of the butcher, the brewer, or the baker that we expect our dinner, but from the regard to their own self-interest." Yet, at the same time, Stigler expunged the passages where Smith acknowledged the limits of that view. Conspicuous among the topics elided (or omitted entirely) are Smith's arguments for the necessity of regulation when self-interest fails and for taxation to pay for public goods that markets by themselves either do not provide or cannot sustain.

To be sure, Smith advocated open trade and competition, but he also acknowledged the need for restraints on the marketplace to protect public safety. He also identified the problem of wages that were sometimes inadequate and the necessity of taxation to pay for public goods, such as education and infrastructure. That Adam Smith is nowhere to be found in Stigler's Chicago version.

Consider this passage on banking from book 2, chapter 2 of *The Wealth of Nations*:

To restrain private people, it may be said, from receiving in payment the promissory notes of a banker, for any sum whether great or small, when they themselves are willing to receive them, or to restrain a banker from issuing such
notes, when all his neighbours are willing to accept of them, is a manifest
violation of that natural liberty which it is the proper business of law not to
infringe, but to support. Such regulations may, no doubt, be considered as in
some respects a violation of natural liberty. *But those exertions of the natural
liberty of a few individuals, which might endanger the security of the whole
society, are, and ought to be, restrained by the laws of all governments, of the
most free as well as of the most despotical. The obligation of building party
walls, in order to prevent the communication of fire, is a violation of natural
liberty exactly of the same kind with the regulations of the banking trade
which are here proposed.*[10]

Here, Adam Smith insisted on the necessity of regulating banks. (He may also
be fairly read as endorsing the necessity of building codes.) His explanation is
clear and simple: *restraints on liberty are justified when the actions of a few
individuals endanger the rest.*

Smith spent many pages explaining his argument for bank regulation; chapter 2 of *The Wealth of Nations* is entirely dedicated to the issue. The discussion
is not merely theoretical; it is based on problems that had already arisen in the
banking system. In the late eighteenth century, many banks issued promissory
notes as a form of currency; this system depended on the "fortune, probity, and
prudence" of the banker to produce gold and silver when demanded in exchange
for those notes.[11] Anticipating the modern notion of reserves, Smith allowed
that the banker does not have to stockpile gold and silver in exact proportion
to the notes he has issued; he need only stockpile enough to meet demand as it
arises. "By this operation, therefore, twenty thousand pounds in gold and silver
perform all the functions which a hundred thousand could otherwise have performed."[12] This is a good thing, but it only works if the banker is honest and
carries adequate reserves, and this had proved to be something one could not
assume.

In Smith's native Scotland in the previous twenty-five to thirty years, the
number of banks issuing paper money and promissory notes had grown substantially. Many of these banks also offered lines of credit—often on "easy
terms . . . of repayment"—which helped to stimulate economic activity. Then,
as now, credit gave businesses room to maneuver.[13] The merchant in Edinburgh,
where such accounts were available, could do more than a comparable merchant in London, and so the Scottish economy boomed. But many of these
banks did not carry adequate reserves. If people realized this, it could lead to a
run on the banks. In a worst-case scenario, banks would collapse, and deposi-
tors would go bankrupt.

There was another problem. Because paper money was of no use abroad, British merchants with excess paper were returning it to the Bank of England, which was obliged to buy gold and silver to redeem it. This drove up the price of these metals, forcing the bank to pay higher costs for coinage of the same market value. A shilling was still a shilling, but the silver needed to make it cost more, so the bank lost money. The Scottish banks, Smith explained, "paid . . . dearly for their own imprudence and inattention: but the Bank of England paid very dearly not only for its own imprudence, but for the much greater imprudence of almost all the Scotch banks."[14] Through the fault of *others*, British bankers were suffering. An excess of Scottish liberty was doing harm in England.

Smith stressed that he was not antibank. On the contrary, "the judicious operations of banking can increase the industry of the country." The problem was that many bankers did not act judiciously, and their self-interest did not always prove benevolent. The conclusion was inescapable: banks needed to be regulated. Smith proposed that the proliferation of paper money in excess of reserves could be curtailed by regulating the circulation of paper to dealers (as opposed to consumers) or by limiting paper money to large denominations. (He suggested that ten pounds would be an appropriate lower limit and certainly not less than five pounds.)[15]

Smith anticipated the objections that Chicago school economists and diverse latter-day neoliberals and conservatives would make: that regulations are an infringement of liberty. And so they are. But sometimes, Smith insisted, they are justified and *necessary*. Smith—the hero of libertarians, the father of free market economics, the patron saint of self-interest—spent a significant section of his most famous work discussing banks and banking precisely *because* it illustrated an essential and nonnegligible point: that regulations *do* infringe liberty, but they are necessary when the "natural liberty of a few individuals . . . endanger[s] the security of the whole society."[16]

What does Stigler do with Smith's long, detailed, and thoughtful analysis of the need for banking regulation? Nothing. Smith's discussion of the problems of late eighteenth-century banking and the need for banking regulation is entirely omitted from Stigler's version of the volume. *The Wealth of Nations* is admittedly an enormous and complicated book, and any reduction of it to barely more than one hundred pages will inescapably invite critique. But considering how central money and its relation to value are to Smith's argument, and particularly his prominent treatments of money, wealth, and value in books 1 and 4, Stigler's omission of Smith's crisp articulation of the necessity of banking regulation is more than a little lacuna. And considering that liberty is arguably the motivating concept in the neoliberal defense of market economics, it is astonishing that Stigler omits Smith's discussion, for Smith did not merely *acknowledge* the problem, he *solved* it.

Some 168 years later, Friedrich von Hayek penned the work that is often viewed as laying the foundations of neoliberalism: *The Road to Serfdom*.[17] In it, Hayek insisted that he was not antigovernment but merely wanted a clear set of principles by which to distinguish when regulation was justified.[18] In 1776, Smith had given it; in fact, he offered it in a single principle. But in 1957, Stigler expunged it.

Banking is not the only area where Smith discussed the warrant for regulation. In book 1 of *The Wealth of Nations*, Smith took up the topic of wage regulation with a degree of vigor that would surprise many contemporary readers. In a chapter entitled "Of the Wages of Labour," Smith suggested that power imbalances between workers and factory owners might make it appropriate to level the playing field, particularly considering that existing laws were biased in favor of masters. Wages could be understood as a contract between two parties, who "are by no means the same. The workmen desire to get as much, the masters to give as little, as possible. The former are disposed to combine in order to raise, the latter in order to lower, the wages of labour."[19] So far so good.

The problem is that under nearly all circumstances, the contract is not signed on a level playing field. The masters have the advantage because they control the workplace. Moreover, they are more likely to be able to manipulate the situation to their advantage, as "being fewer in number, [they] can combine much more easily: moreover, the law authorizes or at least does not prohibit, their combinations, while it prohibits those of the workmen. We have no acts of Parliament against combining to lower the price of work, but many against combining to raise it." The reason is simple: the ruling classes tend to pass laws that protect their own interests. Moreover, Smith noted, in any conflict, the masters, having more assets, can hold out longer than the workers.[20] So they are multiply advantaged: by their intrinsic position, by their assets, and by the legal protections that they have granted themselves.

Even where owners don't combine formally, Smith noted, they "are always and everywhere in a sort of tacit, but constant and uniform, combination, not to raise the wages of labour," and sometimes they "enter into particular combinations to sink the wages of labour." These activities typically are conducted with "the utmost silence and secrecy." On the rare occasions when workers organize to fight back, their activities "are always abundantly heard of" and often met with demands for legal action to prevent or punish them.[21] Two centuries before American conservatives, businessmen, and the Chicago school decried unionization as a form of restraint on trade in the twentieth century, Smith noted that workers were far less likely than owners to be able to manipulate the marketplace.

Could wages ever be set so low that workers would struggle to survive on them? In theory, the answer was no, because in a rational system, workers would reject starvation wages, but in practice, they have often had no alternative. Wages were sometimes so low that workers' children did starve; the situation was so bad—and so common, Smith observed—that laborers routinely tried to rear at least four children in order that two would survive.[22] Wages also varied from year to year in ways that did not correspond to price variation, so laborers had years of plenty and years of want for reasons beyond their control; in the latter, children died.[23]

Smith believed many workers were better off in 1776 than in earlier decades given the lower cost of "an agreeable and wholesome variety of food."[24] He noted that where there is high demand for labor, wages will increase sufficiently so that most families will be able to rear children to adulthood. In this way, labor, no less than other commodities, would respond to the forces of supply and demand. But it did so in a tragic way. "It is in this manner that the demand for men, like that for any other commodity, necessarily regulates the production of men."[25] This may at first seem cold-blooded, as Smith implied that infant mortality is the unavoidable (and possibly acceptable) result of the interplay of supply and demand, but Smith absolutely affirmed that workers are entitled to a minimum standard of decency: "No society can surely be flourishing or happy, of which the far greater part of the members are poor and miserable. It is but equity, besides, that they who feed, cloath and lodge the whole body of the people, should have such a share of the produce of their own labour as to be themselves tolerably well fed, cloathed and lodged."[26] If one is not persuaded by arguments of equity, Smith offered the practical argument that well-fed workers are better workers.[27] "That men in general should work better when they are ill fed than when they are well fed, when they are disheartened than when they are in good spirits, when they are frequently sick than when they are generally in good health, seems not very probable."[28] Thus, he concluded that when a *regulation, therefore, is in favour of the workmen, it is always just and equitable; but it is sometimes otherwise when in favour of the masters.*"[29]

As a result of these passages, Smith was considered in his day to be a friend of the poor and, in the 1790s, would be invoked by British advocates of minimum wage regulation.[30] Economist Amartya Sen finds that a full reading of Smith reveals him as someone who was "deeply concerned about the inequality and poverty that might survive in an otherwise successful market economy."[31] Smith's rejection of mercantilism and defense of free trade were *not* a rationalization of the impoverishment of laborers nor a rejection of the idea that the government might set appropriate standards for wages. So what does Stigler do with this discussion? Like the discussion of bank regulation, he omits it.

Stigler's edit of *The Wealth of Nations* also undermines a key passage on taxation and the provision of public goods. In book 5, Smith identified four major domains that market forces might not adequately address: defense, justice, public works and institutions, and the "expense of supporting the dignity of the sovereign." Some of these can pay for themselves, as when tolls cover the costs of building and maintaining roads. Law courts can, to some extent, be financed by court fees (although, as Smith noted, that creates a risk of corruption). But there are limits to how much these activities can be made self-financing. The clearest example is defense. Clearly, countries need to be able to defend themselves, and there is no coherent way for a standing army to pay for itself (and militias entail other problems). But these other domains, too, are "beneficial to the whole society, and may therefore, without injustice, be defrayed by the general contribution of the whole society."[32] Nor was this list exhaustive; other activities that broadly benefit society may rightly be undertaken by governments and paid for by the public. Smith discussed various means to cover these costs—including rents from sovereign lands and government functions that raise their own revenue, such as the post office—before working his way around to taxation. He did not oppose it but articulated four principles, or "maxims," regarding how taxes should be applied:

I. The subjects of every state ought to contribute towards the support of the government, as nearly as possible, in proportion to their respective abilities. . . .

II. The tax which each individual is bound to pay, ought to be certain and not arbitrary. The time of payment, the manner of payment, the quantity to be paid, ought all to be clear and plain to the contributor, and to every other person. . . .

III. Every tax ought to be levied at the time, or in the manner, in which is most likely to be convenient for the contributor to pay it. . . .

IV. Every tax ought to be so contrived, as both to take out and to keep out of the pockets of the people as little as possible, over and above which it brings into the public treasury of the state.[33]

Smith offered a detailed examination of different tax policies around Europe and elsewhere. The whole discussion, spanning seventy pages, makes it clear that while Smith was not enamored of taxes, he was by no means offering a general treatise against them, much less a polemic. For Smith, taxes are legitimate and necessary; he just wanted them to be levied in a manner that did not create undue or unfair burdens.[34]

So long as the tax system is fair, transparent, and certain, Smith also had no problem with it being *progressive*. Indeed, he believed that "a very considerable degree of inequality . . . is not near so great an evil as a very small degree of uncertainty." In essence, this means take what you need but no more, and do it in a civilized, consistent way. Don't be arbitrary, don't make the process vexatious, and, as far as possible, don't waste money on the system of taxation itself. Twenty-first-century readers who consider Smith anti-Marx may be surprised to find that Smith suggested that the appropriate way to raise revenues is by "all the different members contributing, as nearly as possible, in proportion to their abilities."[35]

Stigler is fairer to book 5 of *The Wealth of Nations* than to book 2. While the latter vanishes completely, the former's 199-page argument gets twenty-six pages. Stigler follows Smith's outline dutifully—until he gets to taxation. Indeed, here, Stigler absolutely distorts the argument as the sections where Smith described means *other* than taxation to finance public goods (such as tolls on roads and bridges or profits from government enterprises) are kept, but the parts where Smith concluded that taxes may be justified are deleted. And Smith's defense of progressive taxation is entirely gone. A student reading Stigler's condensation might easily get the impression that Smith believed that *all the legitimate governmental expenses could and should be self-supporting* and that *taxation plays no valid role*. That impression would be utterly wrong.

Again, any volume as slim as Stigler's will be full of choices, some more and some less defensible, but it seems fair to summarize the situation this way: Smith was a good scholar and astute thinker who considered potential objections to his theory and granted that some of these objections had merit. He recognized that the market alone would not attend to all of society's needs, he provided well-reasoned frameworks for economic regulation and taxation, and he was deeply sympathetic to the plight of workers. That Smith is missing from Stigler's volume.[36] In short, the reader of Stigler would be ignorant of the real Adam Smith.

This was clearly no accident. Smith's arguments for regulation and taxation had earlier been highlighted by Stigler's mentor, economist Jacob Viner. In a 1927 paper inspired by the 150th anniversary of *The Wealth of Nations*, Viner noted that in numerous instances, "Smith supported government restrictions on private initiative." Smith was "not a doctrinaire advocate of *laissez-faire*," Viner explained. "He saw a wide and elastic range of activity for government, and was prepared to extend it even farther if government, by improving its standards of competency, honesty, and public spirit, showed itself entitled to wider responsibilities." Viner even wondered if Smith had not gone so far as to undermine his own argument "by demonstrating that the natural order, when left to take its own course, in many respects works against, instead of for, the general welfare."[37]

Viner resolved the problem by understanding *The Wealth of Nations* as "a specific attack on certain types of government activity" that operated against national prosperity, such as international-trade restrictions and the apprenticeship systems that limited where and how people could work. "Smith's primary objective was to secure the termination of *these* activities of government."[38] Viner concluded that Smith adopted a *presumption* against government regulation, but it was a presumption that could be challenged by evidence of adverse outcomes, as it was in the cases of banking transgressions and wage inequity.

There is, furthermore, an exception in Stigler's book that proves the rule: Smith's discussion of the regulation of school fees and professors' salaries. Smith clearly carried a grudge against academics because here his otherwise temperate tone and prolix style becomes crisp and dyspeptic.[39] Professors were not only generally paid a fixed salary irrespective of the quality of their teaching, Smith grumbled, they were also typically shielded from competition. "Rivalship and emulation render excellency," but there was little of that in education because schools operated as virtual monopolies in their regions, and professors, as monopolists in their subject.[40]

Most professors were "indolent" and "indifferent," showing up to "attend upon his pupils" only the set numbers of hours to which they were formally obligated. Colleges were set up "not for the benefit of students, but for the interest, or more properly speaking the ease of the masters. Its object is, in all cases, to maintain the authority of the master and whether he neglects or performs his duty, to oblige the students in all cases to behave to him as if he performed it with the greatest diligence and ability." Smith here assumes that most academics are lazy and will necessarily be as "careless and slovenly" as "authority will permit."[41]

Smith's solution to the education problem is to introduce competition, to highlight and condemn the factors that bind students to colleges irrespective of the quality of education being offered, and to condemn scholarships attached to specific colleges. Students should be free to go wherever is best for them, Smith insisted, and to study with whomever they found most inspiring. Pointedly, Smith criticized any regulation that prohibited the members of a college from leaving and going elsewhere without getting permission from the college they intended to leave.[42]

This discussion is scarcely central to the overall thesis of the *Wealth of Nations*, but that is just the point. This ill-tempered portion, peripheral to Smith's major concerns, is one that Stigler chooses to include. Indeed, the *only* appearance of the word "regulation" in Stigler's version of the *Wealth of Nations* is in this hellish context where students are prisoners of the academic institutions they attend, stuck in classrooms with boring and pretentious dons and no means of escape.[43] In short, Stigler preserves a discussion that is peripheral to Smith's main thesis but where regulation is detrimental while excising several much more salient passages where Smith frames regulation as potentially beneficial or even essential.

Stigler's edited volume was published in a series presenting great works with little or no historical context, and it is odd that a scholar known for his interest in the history of economic thought would make scant effort—in this book and

elsewhere—to situate Smith in the intellectual climate of his times.[44] This may
not be an accident, for when one places Smith in context, one must acknowledge that he was no conservative and certainly no libertarian. As historian
Emma Rothschild notes, in his day, Smith was considered a politically seditious
radical, reviled by many conservatives, including Edmund Burke.[45] Many scholars point out that in Smith's "other" book, *The Theory of Moral Sentiments*, he
argued that self-interest is but one part of our instinctual apparatus; equally important is our capacity for empathy.[46] Morality is rooted in concern for others,
which is as much a part of our nature as is self-interest. To understand Smith's
philosophy as a whole is to understand how we counterbalance self-interest in
economic activities with morality in guiding society.[47] Yet this is precisely what
Stigler and his Chicago colleagues failed—or refused—to do. Nor did they account for how it was that Smith went from being reviled by conservatives to
being beatified by them.

Amartya Sen offers perhaps the best evidence that Smith was not the ideo-
logue the Chicago school made him out to be. A few years before Smith died,
Sen remarks, philosopher Jeremy Bentham complained that Smith had failed to
adequately appreciate the *benefits of markets*! (Among other criticisms, Bentham singled out Smith's discussion of banking regulation.)[48] Economist Craig
Freedman dryly concludes that Stigler "reduced Smith's economics to fit a pattern contiguous with his own beliefs."[49]

Admittedly, Stigler was not alone in reconstructing Smith as "a mouthpiece
for the unalloyed virtues of the market." According to Sen, Smith's makeover
as "an uncomplicated champion of pure market-based capitalism" was already
well underway in the nineteenth century. But it is clear that Stigler and the
Chicago school did much to push that view further in the twentieth century.[50]
Indeed, when the chair of the U.S. Federal Reserve testified in the U.S. Congress in the aftermath of the 2008 financial crisis—the most severe since the
Great Depression—he blamed it in part on an overconfident belief in the views
of Adam Smith. He (Greenspan) had "made a mistake," he allowed, "in pre-
suming that the self- interest[s] of organizations, specifically banks and others,
were such that they were best capable of protecting their own shareholders and
their equity in the firms."[51] Had Greenspan read and absorbed the real Adam
Smith—rather than accepting the flattened Chicago school version—he would
have understood that in any market-based economy, you have to regulate the
banks.

In 1935, when arguing against the New Deal, Chicago economist Henry
Simons had said that what his generation needed more than anything, "especially the so-called brain-trust," was an understanding of Adam Smith.[52] Two
decades later, George Stigler would make it easy for anyone to read Smith but
hard to understand him. The readers of Stigler's Smith get little sense of Smith's

concern for those whom markets failed to serve, of the need for regulation when self-interest threatened the common good, or of the wide space Smith found for government activity. What the readers get is a stick figure version of Adam Smith saying everything could be left to self-interest and that government plays no useful role.

Croft Classics published Stigler's edited volume in a small inexpensive edition intended for teaching. Large numbers of American students—perhaps even Greenspan—have likely encountered Smith through Stigler's version of him. And if they read only this version of Smith, they will never know that he offered—again, following Sen—"a balanced argument for supporting a society with multiple institutions in which the market would play its part without being hostile to the important role of other institutions, including those of the state."[53] They also will not know that Smith, avant la lettre, offered a solution to the problem that vexed Hayek and is at the root of all democratic governance: how to decide when infringements on liberty are warranted and when they are not. And the ignorance of this version of Smith may continue into the future, as six decades after it was first published, Stigler's version of *The Wealth of Nations* is still in print.[54]

Stigler's selective presentation of Adam Smith as a free market fundamentalist is an example of deliberate agnotology. Stigler made a series of editorial choices that turned Smith into something that he was not and that led directly to a form of structural ignorance. Generations of students reading only Stigler's text would have had no way to know that Smith rejected the sort of extreme laissez-faire views that Stigler and his Chicago colleagues advanced.[55] The "advocate of laissez-faire who objects to government participation in business on the grounds that it is an encroachment upon a field reserved by nature for private enterprise" (as Jacob Viner put it almost a century ago) typically defends this position by reference to Adam Smith.[56] But that position doesn't come from Adam Smith. It comes from those, like George Stigler, who have built antigovernment arguments on false foundations and continue to do so today.

EIGHT

Gun-Lobby Agnotology

Degrading the Truth about Firearms

JOHN J. DONOHUE

MUCH AS THE TOBACCO INDUSTRY distorted the truth about the health consequences of cigarettes beginning in the 1950s, virtually every comment to emerge from the gun lobby on the topic of American gun policy today is either irrelevant, incorrect, or misleading. Over the last half century, the Republican party and the gun lobby have allied powerfully to promote falsehoods, attack unfavorable research, limit federal funding of impartial scholarship, reverse centuries of law on the scope and meaning of the Second Amendment, and push for expansions of "gun rights." This aggressive progun advocacy—enabled because many Americans are easily swayed by political and industry efforts designed to promote fear of and ignorance about the incidence and causes of crime—has deepened divisions within our country, damaged our democracy and our courts, and thwarted the will of the people, while needlessly increasing the net harm from firearms.

AMERICA LEADS THE AFFLUENT WORLD IN HOMICIDES

While empirical researchers struggle to definitively answer complex questions about guns and crime, the gun lobby has an easier time making eye-catching claims, unfettered by the pursuit of truth. When insisting that regulation is not needed, they call American gun violence a minor problem while seemingly searching for definitions of "mass shooting" that allow them to claim that "the U.S. has much less than its share of public mass shooters."[1] At the same time, they hawk the need to carry weapons at all times in defense of themselves and their loved ones.

John R. Lott's Crime Prevention Research Center (CPRC), for example, has published a plot of 2008 homicide rates across 192 countries, with the apparent goal of minimizing the United States' homicide problem.[2] The figure employs some of the tools described in Darrell Huff's *How to Lie with Statistics*, putting onto the same scale some very high-homicide nations, like Jamaica and Honduras, to make the level of homicides in the United States seem close to zero by comparison.[3] The indiscriminate amalgamation of nations also obscures where the United States ranks among its highly affluent peers. This creates the erroneous impression that homicide is not a serious problem in the United States—and certainly not one caused by guns, given that we have the highest gun-ownership rate and what appears on the chart to be a below-average murder rate, close to zero. This CPRC figure has been used to make these arguments in numerous gun-lobby publications, including those on the websites Firearms Owners against Crime and the Truth About Guns.[4]

Figure 8.1, however, reveals a very different, and more accurate, characterization of homicide in the United States. As the regression equation under this scatter plot of 112 nations shows, a doubling of per capita gross domestic product (GDP) corresponds with a statistically significant 41 percent decline in

FIGURE 8.1. Per capita GDP versus homicide rates in 112 countries with populations greater than 5 million, 2019. All 31 countries richer than Russia (GDP/capita of \$27.2K) are essentially at or below the regression line, except for the United States. These 30 non-U.S. countries range in GDP per capita from \$28.2K (Turkey) to \$98.4K (Singapore), and their population-weighted averages are \$43.3K per capita GDP, 1.08 per 100,000 homicide rate and 0.25 per 100,000 gun-homicide rate. The regression line equation is ln (homicide rate) = -0.41*ln (GDP/capita) + 4.99, with R^2 = 0.18 and p = 0.000.

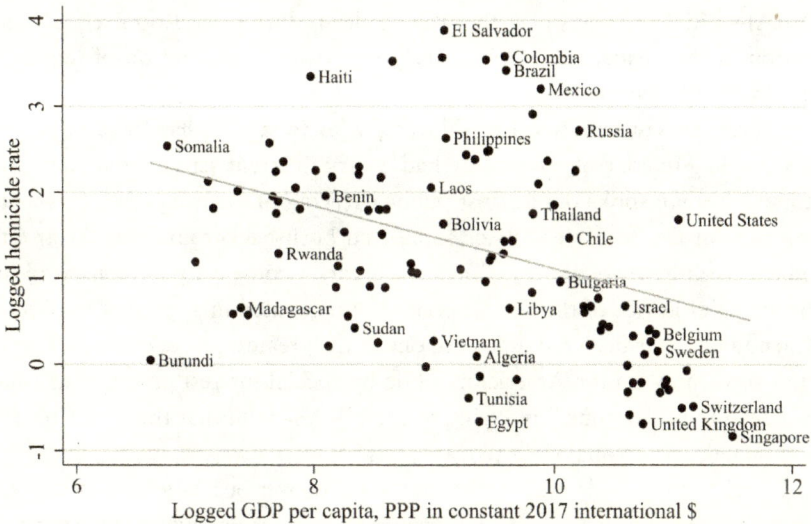

Sources: For the homicide data, see Institute for Health Metrics and Evaluation; for the GDP data, see World Bank.

the murder rate. Specifically, every nation at least as rich as Turkey has a murder rate at or below the regression line except for one: the United States, which sits far above with a murder rate of 5.4 per one hundred thousand, while the predicted murder rate for a country this wealthy would only be 1.64 per one hundred thousand. This figure accentuates the CPRC's failure to acknowledge—let alone control for—the fact that other affluent nations have addressed the problem of homicide far more effectively than the United States has. Note that the thirty countries at least as rich as Turkey have an average GDP per capita of $43,400 and a murder rate of 1.08 per one hundred thousand, compared to the United States GDP per capita of $68,200 and murder rate of 5.4. In other words, the U.S. homicide rate is roughly five times the average in the world's thirty richest countries, overwhelmingly driven by our higher firearm-homicide rate. This is not a record to be proud of.

This record stands out even more as we focus on firearm versus nonfirearm homicides in thirteen affluent nations, as seen in figure 8.2, which reveals that the vastly higher U.S. murder rate in 2020 is due to the higher level of gun homicides. Without the scaling problem of the CPRC report, it becomes clear that the United States is far more homicidal than the other listed countries. Rather than confirming Lott's theory that guns are not a major problem driving U.S. homicide, figure 8.2 conveys exactly the opposite impression.

Thus, the United States faces a dual problem: an enormously high murder rate driven by our firearm-homicide epidemic and an active effort to obscure gun availability as the root cause. Interestingly, the group that is often assumed most culpable in this effort to distort the truth—the National Rifle Association (NRA)—did not always play that role. Indeed, during the Roosevelt administration in the 1930s, the NRA actively supported the adoption of important gun-safety legislation.[5]

Moreover, conservative Republicans, who today are hardcore opponents of sensible gun control, previously had a very different sense about such laws. Ronald Reagan spoke out against public carrying of firearms when he was the governor of California and later supported both background checks and the federal assault weapons ban. Warren Burger, a conservative Republican and former chief justice of the U.S. Supreme Court, stated in 1991 that the Second Amendment "has been the subject of one of the greatest pieces of fraud, I repeat the word fraud, on the American public by special interest groups that I have ever seen in my lifetime," referring to the NRA's claim that the U.S. Constitution includes a personal right to own guns.[6]

So what happened? Not surprisingly, the answer begins with a strong economic motivation in the gun industry, stemming from two trends that were hurting sales: a strong decline in hunting and the enormous drop in crime during the Clinton administration (1993–2001)—one of the most remarkable shifts in

FIGURE 8.2. **Gun- and non-gun-homicide rates by country. Years of observation: Switzerland (2021); Australia, Spain, Sweden, and the United States (2020); Canada and the Netherlands (2018); Italy and Japan (2017); Belgium and France (2016); Germany and the United Kingdom (2015).**

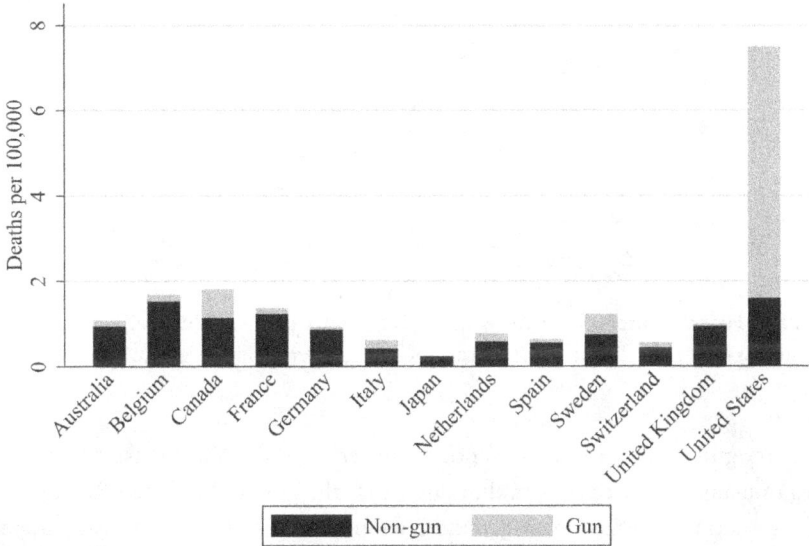

Sources: Author's adaptation based on Sara B. Heller and Max Kapustin, "Gun Violence in the U.S.," EconoFact, Nov. 14, 2022, https://econofact.org/gun-violence-in-the-u-s; underlying data as of Oct. 7, 2022, from https://www.gunpolicy.org/firearms/region/united-states.

American history. Figure 8.3 reveals the stunning magnitude of the reduction in the number of homicides over this period—down from almost twenty-five thousand annual murders in 1993 to nearly fifteen thousand by 2000. While saving ten thousand lives per year was wonderful for the country, it proved a significant threat to the gun industry since fear of crime drives gun sales for personal protection.

Faced with these two troubling trends, the industry expanded the demand for guns by creating two new markets: one for concealable guns to be carried for personal protection and one for assault weapons. Widespread restrictions and even outright bans on the concealed carry of guns curtailed creation of the first market—for example, Texas had prohibited carrying of handguns from 1871 until 1995, when the long-term prohibition was ended under Governor George W. Bush. Circumventing this obstacle required widespread legislative change, leading the gun lobby to ally with the Republican Party on a number of initiatives.

This prompted the effort to adopt right-to-carry (RTC) laws allowing citizens

FIGURE 8.3. Reported homicides, 1990–2004. The enormous decline in homicides during the Clinton administration harmed gun sales.

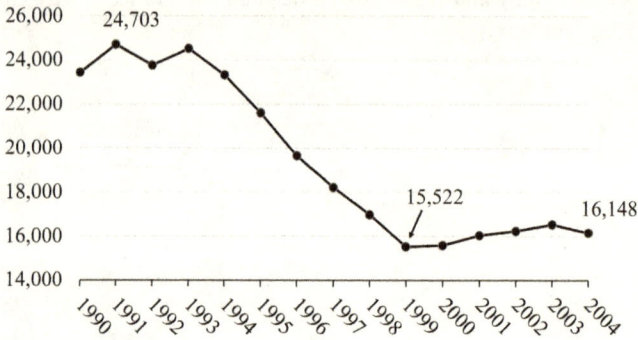

Source: Federal Bureau of Investigation, U.S. Department of Justice, "Crime in the United States, 2005," Sept. 2006, https://ucr.fbi.gov/crime-in-the-u.s/2005.

to carry guns with far fewer exceptions for restrictions. When these gun carriers began using their weapons in dubious ways, the gun lobby joined Republicans in advocating for stand-your-ground laws encouraging more aggressive "defensive" gun use, thereby reducing the legal risks of gun carrying and stimulating demand for such weaponry. Finally, the industry started a three-part campaign designed to (1) fund arguments for the existence of an individual federal constitutional right to firearms (what Chief Justice Burger called a massive fraud), (2) appoint federal judges who embraced this view,[7] and (3) litigate to broaden the scope of the Second Amendment and restrain the ability of government to address gun violence in any way that might diminish sales.

At the same time, the industry began to advertise civilian versions of U.S. military weapons as "assault rifles," opening up an entire line of sales that had never previously existed. This advertising piggybacked on the imagery of American military action during both Bush administrations, targeting (especially) weak and ineffectual men. After convincing President George W. Bush to renege on his campaign promise to maintain the federal assault weapons ban, which lapsed in 2004, the industry, in 2005, obtained a federal-immunity statute—the Protection of Lawful Commerce in Arms Act—that protects its aggressive marketing efforts by barring lawsuits against the gun sellers and manufacturers.

The final element of the sales plan introduced the idea that consumers should constantly be upgrading their guns to keep up with the latest technologies, combined with fearmongering that the looming danger of crime could only be addressed by the possession of weapons. While those planning to commit mass violence were highly attracted to this increasingly lethal weaponry, the truth is that these products did nothing to enhance the safety of Americans.

THE GUN LOBBY INCREASES IGNORANCE
ABOUT FIREARMS AND CRIME

Rational gun policy, like a rational decision about whether one should carry a gun for protection, requires an accurate assessment of costs and benefits, which at the very least would be based on knowing the level of crime. Unfortunately, Americans are strikingly bad at assessing even the most basic level of crime risk. Figure 8.4 illustrates this point by showing two time series: the upper line is the percentage of Americans who state that crime has risen in the last year, and the lower line shows what actually happened to violent crime in the last year. Even as crime fell dramatically in the 1990s, 71 percent of Americans reported that crime was rising. From 1993 to 2014, violent crime fell astonishingly, from eighty crimes per one thousand individuals (over age twelve) to twenty per one thousand individuals. Yet for the last decade of that period, roughly two-thirds of Americans believed that crime had actually risen each year.

This shocking level of ignorance was cultivated by and beneficial to the gun lobby, Republican politicians, and much of the media.[8] To distract Americans from the sharply falling crime rate going into the 1996 presidential election with a Democrat in office, the Republicans and the gun lobby were intent upon fabricating an alleged coming crime wave of "superpredators." This "prediction" was designed to aid Republican electoral prospects, blunt the Clinton administration's efforts at gun-safety regulation, and prop up the declining sales of firearms.

FIGURE 8.4. Perceptions of crime and victimization rates, 1990–2023. High percentages of Americans believe crime is rising even when it is declining. The violent-crime-victimization rate is per 1,000 persons aged 12 or older.

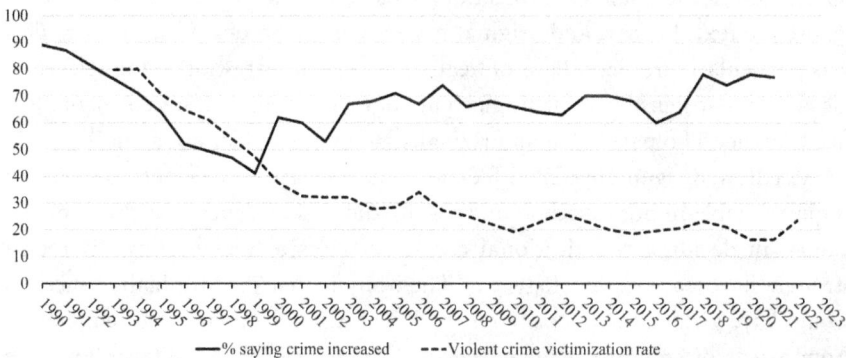

Sources: Crime data from National Crime Victimization Survey; beliefs data from Gallup.

Similarly, Americans failed to assess the risk of violent crime in the National Survey of Economic Expectations from 1994 to 2004, believing that their chance of being robbed in the coming year was 15 percent when the actual rate was 1.2 percent.[9] Being off by an order of magnitude usually means you simply have no idea about reality, but this is not a case of general ignorance: the study found that Americans are actually quite good at estimating other bad outcomes, such as their chance of losing a job or their health insurance. But that is because there is no political or economic interest steadily feeding them false information about these other likelihoods.

Concurrently with this safety misinformation campaign, the gun lobby has attempted to reinterpret the Second Amendment, which states, "A well regulated Militia, being necessary to the security of a free State, the right of the people to keep and bear Arms, shall not be infringed." This first clause, referring to "a well regulated Militia," undermines the view that the amendment establishes an expansive individual right to own guns with little possibility for regulation. If the amendment's goal is to provide for a citizen militia to circumvent the need for a standing army, then it really has nothing to do with personal protection—especially in modern America, where arms for personal protection are not well suited to military use. Moreover, the specific call for wise regulation makes it difficult to argue that sensible limitations are forbidden.

Chief Justice Burger's accusation of fraud referred in part to the gun lobby's incessant denial of the first clause of the Second Amendment. One crude attempt at perpetuating this fraud, a pre-*Heller* NRA poster, featured a colonial-era pistol carefully positioned over the militia clause of the amendment. Seventeen years after Burger's admonition, however, Justice Antonin Scalia wrote the 2008 majority opinion in *Heller* that effectuated the elimination of this first clause. Representing the five-to-four majority, Scalia deemed the introductory text to be irrelevant, allowing the amendment to create an individual right to have a gun in the home for protection. This decision endorsed NRA efforts to ignore the text, history, and original purpose of the Second Amendment. This was particularly ironic in light of Scalia's statement just prior to the release of his opinion—ostensibly to distinguish his views from those of his colleague Justice Clarence Thomas—"I'm an originalist and a textualist, not a nut."[10]

As fellow Reagan-appointed federal judge Richard Posner noted, "[*Heller*] is questionable in both method and result, and it is evidence that the Supreme Court, in deciding constitutional cases, exercises a freewheeling discretion strongly flavored with ideology. . . . The irony is that the 'originalist' method would have yielded the opposite result."[11] The Second Amendment, emanating from a fear of standing armies, reflects a preference that militias should be called up to fight wars that would secure the state. Since we now have a standing army that guarantees American security, the amendment became anachronistic by 2008. Scalia turned the amendment on its head, saying that weapons for personal protection could not be banned, while weapons for military purposes could. Nor did the opinion discuss whether granting an individual the constitutional right to a gun promoted any articulable social benefit or public purpose. *Heller*'s incoherence, its violation of the constitutional text and history, and its absence of any evidentiary support for declaring a new right to firearms more than two hundred years after the amendment appalled many traditional Republicans. Retired justice John Paul Stevens wrote in 2019, for example, that "*Heller*, which recognized an individual right to possess a firearm under the Constitution, is unquestionably the most clearly incorrect decision that the Supreme Court announced during my [thirty-four-year] tenure on the bench."[12]

Scalia's opinion in *Heller* did include some limiting language, apparently needed to secure the fifth vote of Justice Anthony Kennedy. Specifically, the ruling did *not* guarantee a constitutional right to carry a concealed weapon, stating, "Like most rights, the Second Amendment right is not unlimited. It is not a right to keep and carry any weapon whatsoever in any manner whatsoever and for whatever purpose: For example, *concealed weapons prohibitions have been upheld under the Amendment or state analogues.*"[13] Fourteen years later, however, the court backed away from this highlighted language to enormously expand the reach of the Second Amendment through *NY State Pistol and Rifle Ass'n v. Bruen.* This decision established a right to carry outside the home, while striking down a 109-year-old law in New York that had restricted carrying to those with a particularized need (the "proper cause requirement"). While the Court of Appeals for the Second Circuit had sustained New York's law as "substantially related to the achievement of an important governmental interest," *Bruen* was uninterested in whether a gun-safety measure promoted public safety, adopting instead an extreme nonconsequentialist view that empirical evidence on gun policy should have no impact on judicial opinions. The absurd rationale behind this puzzling view was that the founders had essentially sealed the fate of victims of gun violence merely by enacting the Second Amendment, leaving no room for discussion of its consequences.

The progun lobby has characterized this disregard of present-day damage from gun violence as consistent with "the Framers' vision of governance, which emphasizes a judiciary that plays an important but limited role and leaves policymaking to" the legislative branch of the government.[14] But this defense is untenable. One cannot announce that the legislature is more institutionally competent to evaluate the best empirical evidence concerning gun policy while also claiming that this evidence is irrelevant when interpreting the Second Amendment. The Supreme Court in *Bruen* was foreclosing legislative action, not deferring to it. This recent wave of virulent Second Amendment advocacy is the height of antidemocratic judicial activism wedded to antiscientific proclivities, resulting in bad public policy.

Bruen opened the floodgates for a host of gun-lobby cases designed to strike down virtually every beneficial gun-safety measure. Gallup polls reveal the antidemocratic nature of this massive litigation campaign: 66 percent of ordinary Americans in 2022 believed "that the laws covering the sales of firearms should be made more strict," and less than 15 percent wanted to see less strict laws.[15] But *Bruen* and its progeny have ensured that only the 15 percent will be heard.

The magnitude of the support for certain gun-safety measures is now so great that even Fox News reported on an April 2023 poll in which respondents overwhelmingly supported background checks (87 percent), better enforcement of existing laws (81 percent), a minimum purchase age of twenty-one (81 percent), mental health checks (80 percent), and thirty-day waiting periods (77 percent).[16] In today's polarized America, it is difficult to get 80 percent approval for virtually any public policy measure. Even NRA members overwhelmingly support universal background checks, yet they have little influence over NRA policies, which are driven by industry demands for gun sales and the industry's political goals.[17] The Supreme Court's demonstrated hostility to gun policies with such broad public support offers striking proof of the power of the gun lobby to thwart the will of the people and undermine American democracy.[18]

THE SUPREME COURT HAS EMBRACED A CRIME-INDUCING "GUNS-EVERYWHERE" PHILOSOPHY

It is hard to overstate the success of the fifty-year campaign to enable the proliferation of firearms and their use. In the mid-1970s, all but five states had either banned concealed carry or limited the carrying of firearms to trustworthy applicants with permits. A decade later, as figure 8.5 shows, sixteen states still did not allow the carrying of guns, and only one—Vermont—allowed carrying without a permit. Figure 8.6 shows that by 2023, not a single state banned gun carrying, and a staggering twenty-seven allowed carrying without a permit.

A 1997 study claiming that allowing citizens to carry handguns would substantially *reduce* crime provided an early academic boost for the so-called right-to-carry or shall-issue laws.[19] Frank Zimring and Gordon Hawkins expressed their skepticism at the time that "a statistical analysis has been released with the flamboyantly specific claim that relaxing the remaining restrictions on concealed handguns in the United States would prevent 'approximately 1,570 murders, 4,177 rapes, and over 60,000 assaults' each year. According to the study's authors, John R. Lott Jr. and David B. Mustard, the 'estimated annual gain from allowing concealed handguns is at least $6.214 billion.'"[20] This article

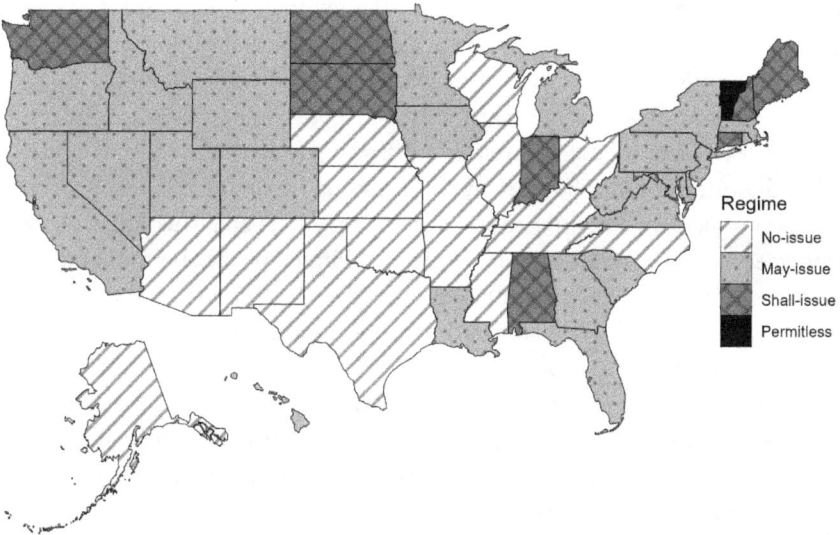

FIGURE 8.5. **Right to carry, 1986.** In 1986, only Vermont allowed gun carrying without a permit.

Source: From each state's legislation active in 1986.

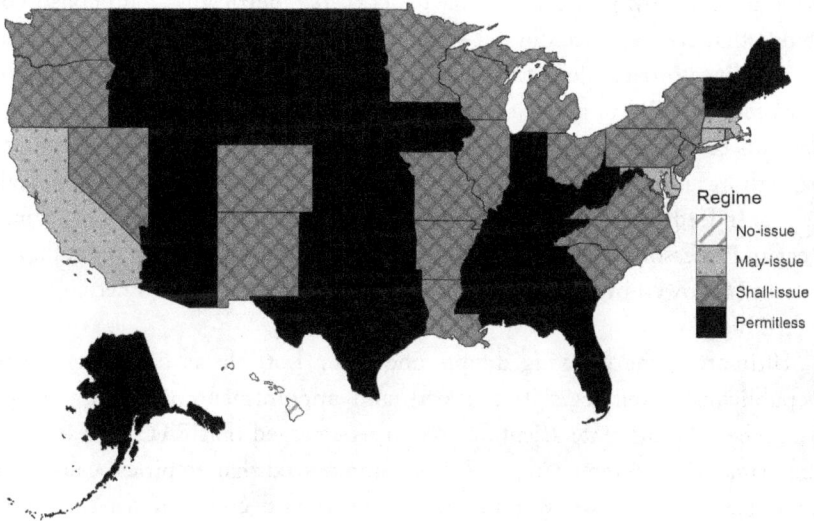

FIGURE 8.6. **Right to carry, 2023.** By 2023, 27 states allowed gun carrying without a permit.

Source: From each state's legislation active in 2023.

launched Lott to rock-star status among the gun lobby, which he rode to an appointment in the first Trump administration Justice Department, eventually writing a paper claiming that Trump had actually won the 2020 election.[21] He defended this same Trump-won position ("highly likely") in federal testimony seeking to overturn California's ban on assault weapons.[22]

When Lott's work—expanded in his best-selling book *More Guns, Less Crime*—started to influence legislators and courts, the National Research Council (NRC) created a committee to investigate the claim that RTC laws reduced violent crime. The state of the academic literature in 2005, however, only permitted the committee to conclude that the Lott and Mustard study was econometrically problematic in many ways.[23] Fifteen of the sixteen committee members emphasized that the findings were highly sensitive to the choice of explanatory variables, concluding that the panel data through 2000 were too fragile to support any conclusion about the laws' true effects. One dissenter—James Q. Wilson—did partially endorse the Lott thesis by agreeing that murders seemed to fall when RTC laws were adopted, although the other fifteen members of the committee pointedly criticized Wilson, saying that "the scientific evidence does not support his position."[24]

Wilson's dissent allowed Lott to claim that the one honest voice on the panel supported his thesis that RTC laws reduced crime. Yet Wilson's presence on the committee was problematic. First, typically one would not be a member of an NRC report if one had already taken a position on the key issue to be addressed—and Wilson had written in support of the Lott and Mustard paper prior to serving. Moreover, while the panel's econometricians unanimously concluded that the evidence did not support the Lott and Mustard work, Wilson was willing to overrule them even as he conceded his comparatively limited econometric skills (admitting in the second sentence of his dissent that he was "not an econometrician").[25] Especially in light of the more recent evidence strongly refuting Wilson's claims, it appears his insistence was ideologically driven. Indeed, Wilson was both a lifetime member of the NRA and the Ronald Reagan Professor at Pepperdine University. The contrast between the standards of proof followed by the advocates of progun positions versus other academics is striking.

Ultimately, the growing doubts about the Lott thesis did little to slow Republican-backed legislative efforts and appointment of progun judges. And when the ultimate scientific consensus emerged that RTC increased violent crime, the Supreme Court in *Bruen* announced that empirical evidence of the impact of gun laws would be irrelevant to its decisions under the Second Amendment.

While the 2005 NRC report deemed the early literature on the impact of RTC laws on crime inconclusive, researchers have since accessed both more complete data and improved empirical methodologies. In 2005, for example, the synthetic control methodology had not yet been adopted, and the critical nature of the parallel-trends assumption was not yet fully appreciated (or even mentioned in the NRC report). Using these and other improved methodologies, diverse researchers in the last seven years have produced eighteen empirical papers finding that RTC laws lead to *higher* rates of violent crime and/or homicide.[26]

A 2019 paper by John J. Donohue, Abhay Aneja, and Kyle D. Weber (hereinafter DAW), for example, examines violent crime after the adoption of state laws granting the right to carry outside the home, either with or without a permit requirement.[27] DAW uses two empirical approaches employing data from all fifty states and the District of Columbia from 1979 to 2014: a panel data analysis with state- and year-fixed effects and a synthetic controls analysis generating estimates for the thirty-three states that adopted RTC laws over that same time period. Both reveal a similar pattern: RTC adoption led to an immediate and enduring increase in violent crime. The panel analysis estimates that RTC laws generated a 9 percent increase in crime on average across the entire postadoption period, and the synthetic control analysis indicates that the average increase in violent crime had grown by 13–15 percent ten years after adoption. Plausible estimates of the elasticity of crime with respect to incarceration were around -0.15, which means that the average RTC state would have to double its prison population to offset the increase in crime caused by RTC adoption. In other words, the social costs of allowing citizens to carry guns outside the home are substantial.

Figure 8.7 depicts the panel data estimates of how RTC laws influence crime in each year after adoption. The DAW econometric model works well: *prior* to adoption, the model suggests the law has no effect—a sensible prediction for a law that is not yet passed.[28] As soon as the law goes into effect, however, there is a clear and distinct increase in the rates of violent crime. This panel data model controls for a variety of criminal justice, socioeconomic, and demographic factors that could also influence violent crime, such as the lagged incarceration rate, the lagged police-employee rate, real per capita personal income, unemployment rate, poverty rate, beer consumption, the percentage of the population living in metropolitan areas, and six age-sex-race categories. The overall estimated 9 percent increase in violent crime was highly significant ($p = .002$).

The last of the eighteen RTC studies referenced above examines data from sixty-five cities with an average population of over 250,000 between 1979 and 2019,[29] concluding that RTC laws increase violent crime by 20 percent, gun theft by 50 percent (since guns taken outside the home are more vulnerable to theft), and lower violent-crime clearance rates by 9 percent. This last point suggests that carrying guns increases violent crime partly by degrading police effec-

FIGURE 8.7. **Violent crime rises with the passage of RTC laws.**

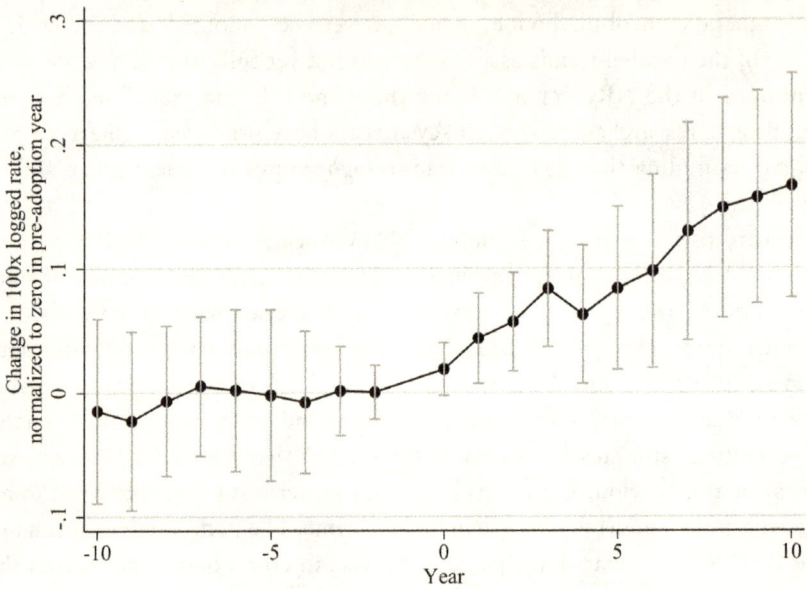

Note: Ninety-five percent confidence intervals using cluster-robust standard errors and the number of states contributing to each estimate displayed.

Source: Author's adaptation based on John J. Donohue, Abhay Aneja, and Kyle D. Weber, "Right-to-Carry Laws and Violent Crime: A Comprehensive Assessment Using Panel Data and a State-Level Synthetic Control Analysis," *Journal of Empirical Legal Studies* 16, no. 2 (2019): 198–247, fig. 2.

tiveness in identifying and apprehending criminals. Furthermore, another paper in the online listing finds that total shootings of citizens by police—justified and unjustified—rise by 13 percent with the adoption of RTC.[30] Law enforcement officers seem to change their behavior in ways that enhance violent crime and decrease public safety. The dramatic video-captured shooting of Philando Castile by Minneapolis Police highlights the fact that police shoot more quickly when they are confronted by an armed individual—even a law-abiding citizen with an RTC permit, like Castile. Such changes in policing undoubtedly anger the community and undermine the delivery of justice.

Road rage also becomes more dangerous when citizens are armed, and criminals will increase their own gun carrying once the public becomes more highly armed. Of course, there are some anecdotes of "good guys with guns" that appear to support the decision in *Bruen,* but on balance, the creation of a right to carry guns outside the home will increase violent crime and impair public safety.

REPUBLICAN-APPOINTED JUDGES EMBRACE FALSE CLAIMS

Perhaps not surprisingly, the NRA hired Paul Clement, one of the most talented oral advocates in the country, to argue that New York's 109-year-old law restricting gun carrying was unconstitutional under the Second Amendment. Referencing unpublished work by a supportive researcher,[31] Clement told the Supreme Court in the *Bruen* oral argument that allowing more citizens to carry guns would have no impact on crime while ironically conceding that even the most pro-NRA assessment found no benefit from more carrying. Ignoring all the recent evidence and relying solely on the single unpublished paper, Clement added, "So, Justice Breyer, I would . . . point you to two things that maybe would give you some comfort. . . . [O]ne is the experience of the 43 states, and there really isn't a case that those 43 states that include very large cities like Phoenix, like Houston, like Chicago . . . have . . . had demonstrably worse problems with this than the five or six states that have the regime that New York has."[32]

Of course, we have just seen that the evidence strongly shows that RTC laws have made violent crime demonstrably worse. Clement continued to mislead the justices of the Supreme Court, stating, "Experience does tell you a lot. . . . [S]even of the 10 largest cities in America, measured by population, are in shall issue jurisdictions. And I've mentioned them, cities like Phoenix, Chicago, Houston. These are large cities where it hasn't been a problem."[33] But the aforementioned study of sixty-five large cities in fact found that violent crime was seriously *worsened* by RTC adoption. Indeed, Chicago, Houston, and Phoenix have substantially higher homicide rates than New York City, as seen in figure 8.8 below. One of the crowning ironies of the *Bruen* oral argument was the NRA lawyer's brazenly inaccurate description of his favored policy's impact on violent crime, which the less talented attorney arguing the case for the state of New York was not able to parry.

While the dissenting judges in *Bruen* did try to correct the record on the best empirical evidence, Justice Samuel Alito responded that "the dissent's presentation of such studies is one-sided." To support this claim, he cited an early review by the RAND Corporation, *Effects of Concealed-Carry Laws on Violent Crime*, that preceded the eighteen aforementioned studies. In fact, a few months after *Bruen*, RAND updated its stance to a much clearer statement that RTC laws impose substantial crime burdens. Specifically, the latest RAND report found—at its highest evidentiary standard—that RTC laws would increase "total homicides, firearm homicides, and violent crime."[34]

Alito also cited the unpublished work by William English that Clement had commended to the justices. English, of course, had access to all the same studies as RAND, yet he nonetheless declared, "The overwhelming weight of statistical

FIGURE 8.8. Homicides per 100,000, 1990–2019. NYC is unquestionably safer than these other cities that allow concealed carry permissively.

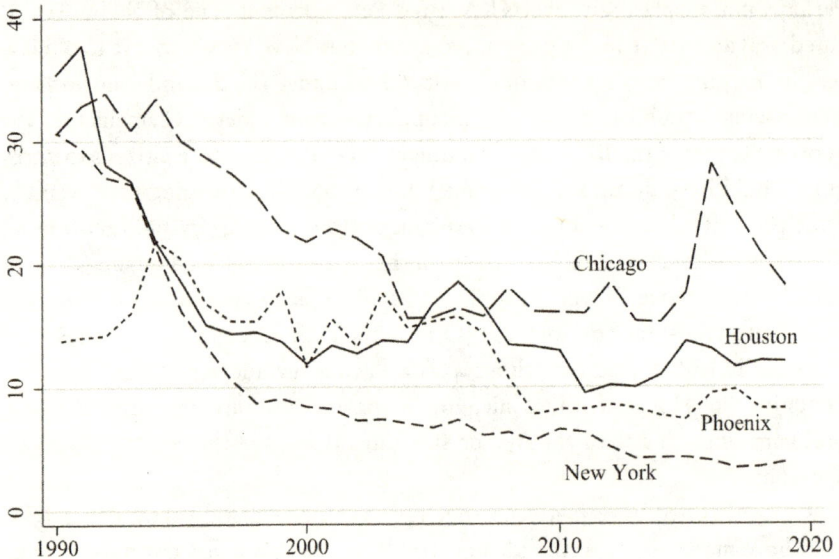

Source: Uniform Crime Reporting, Federal Bureau of Investigation, accessed through the Inter-university Consortium for Political and Social Research (ICPSR).

analysis on the effects of [RTC] laws on violent crime concludes that RTC laws do not result in any statistically significant increase in violent crime rates."[35] So while the RAND report concludes that the empirical literature firmly establishes harm from RTC laws, Clement and Justice Alito cite a single scholar's unpublished work stating the complete opposite. This is a hallmark of the gun-lobby playbook on "interpreting" evidence. First, they worked to defund and thwart empirical research linking guns and crime. Second, they incessantly advertised the Lott and Mustard study and when that study was largely discredited, championed the unpublished work by English—all while trying to discredit what we now know is more accurate research reaching the opposite conclusion.[36]

For his part, Justice Alito embraced these gun-lobby talking points in his concurring opinion in *Bruen*:

In 1791, when the Second Amendment was adopted, there were no police departments, and many families lived alone on isolated farms or on the frontiers. If these people were attacked, they were on their own. It is hard to imagine the furor that would have erupted if the Federal Government and the States had tried to take away the guns that these people needed for protection. Today, unfortunately, many Americans have good reason to fear that they will be victimized if they are unable to protect themselves. And today, no less than in 1791, the Second Amendment guarantees their right to do so.

Alito champions the theory that the Constitution—or at least the Second Amendment—is frozen in time. Amazingly, he evokes the circumstances of 1791 to justify why the United States of today—a large wealthy urban nation with substantial police resources—should have a permissive gun regime. Beyond this jarring disconnect, Alito seemed oblivious that his support of greater gun carrying falsely assumed that more guns will lead to *fewer* victimizations.

Beyond standing in opposition to sensible public policy, the *Bruen* majority's claims about constitutional law also conflict with the history of American gun regulation and longstanding patterns in state-to-state crime differences. Remarkably—from today's vantage point, at least—Texas banned carrying guns outside the home for protection from 1871 to 1995.[37] Under *Bruen*, this would now be deemed unconstitutional; apparently, no one in Texas noticed this Second Amendment violation for 125 years. But in 1995, prior to adopting a "guns-everywhere" public policy, Texas had almost a 20 percent lower murder rate than California and an only slightly higher murder rate than New York. After twenty-five years of tightening gun restrictions in California and New York, Texas now has a 27 percent higher murder rate than California and a 75 percent higher murder rate than New York. *Bruen* may help close those gaps—by making California and New York as deadly as Texas.

Decades of evidence attest that firearm access *increases*, rather than *decreases*, the risk of homicide victimization and suicide.[38] A new meticulous study with extraordinarily detailed individual data from 2004 to 2016 finds that almost six hundred thousand Californians who did not own guns but lived with others who did were at a considerably higher risk of dying by homicide.[39] The authors used extended Cox proportional hazards models adjusted for cohort members' gender, age, race and ethnicity, and, partially, for the presence of a long gun in the home to isolate the impact of gun carrying. The models allowed the baseline hazard to vary by census tract, thus comparing people who resided with handgun owners (exposed) only with people living in gun-free homes (unexposed) in the same small neighborhood. This adjusts for local factors, such as crime rates and economic conditions, that may have confounded the relationship of interest. The bottom line from this study is that having a gun in your home elevates your risk of death, contrary to the NRA's view that private gun ownership on balance protects household members from homicidal victimization. The truth is that the public needs protection *from* guns, not a feckless hope that guns will protect them.

Since *Bruen*, gun-lobby lawyers have flooded federal courts with attempts to eliminate effective measures to reduce gun violence on the grounds that these are not consistent with the history and tradition of the Second Amendment.[40] This onslaught has imperiled longstanding bans on assault weapons and high-capacity magazines, age restrictions to purchase or carry firearms, and prohibitions on possession by felons.

Atrociously inaccurate claims circulated in Far Right circles now routinely make their way into federal court decisions interpreting the Second Amendment. In 2017, U.S. district court judge Roger Benitez asserted, for example, that "nationally, the first study to assess the prevalence of defensive gun use estimated that there are 2.2 to 2.5 million defensive gun uses by civilians each year. Of those, 340,000 to 400,000 . . . were situations where defenders believed that they had almost certainly saved a life by using the gun."[41] Since by 2017, the highest number of murders ever in the United States had been about twenty-five thousand in 1992, the idea that private gun ownership was saving four hundred thousand lives a year is wildly exaggerated. While progun advocates and supportive federal judges do not hesitate to parrot these figures, they are widely acknowledged to be among the most inaccurate referenced in U.S. policy debates.[42]

Consider another example of misinformation at the highest judicial level in Judge Benitez's ruling that California's assault weapon ban violates the Second Amendment in part because militias could be forced to settle for "less-than-ideal" weapons rather than the "ideal" AR-15 rifle. He explained, "That may not be a severe burden today when the need for the militia is improbable. One could say the same thing about the improbable need for insurance policies." Benitez then went on to assert that the use of an assault weapon in a mass shooting is an "infinitesimally rare event," adding, with no citation, that "more people have died from the COVID-19 vaccine than mass shootings in California."[43] Of course, since assault weapons bans effectively *reduce* mass-shooting deaths, it would be ludicrous to eliminate California's assault weapons ban *because* of its low number of mass-shooting deaths.[44]

On top of minimizing the number of deaths caused by assault weapons and firearms equipped with high-capacity magazines, Benitez's argument is misguided for three additional reasons. First, the deaths and injuries caused by mass shootings around the country are increasing at an alarming pace. Second, the social harm from these traumatic events is far larger than the mere casualty counts. Third, the constant increases in firearms' lethality in order to increase sales means that ever-larger mass shootings, possibly resulting in episodes with hundreds of deaths, may well be our fate. Only concerted and effective governmental action, including bans on assault weapons and high-capacity magazines, can address this menace. Recall that a twenty-year-old with an AR-15 came within millimeters of damaging our democracy in the attempted assassination of Donald Trump.

Armed with absurd empirical claims and an untenable constitutional theory, the gun-lobby attack on sensible firearm regulations continues to generate troubling legal decisions. In March 2023, the U.S. Court of Appeals for the Fifth Circuit invoked *Bruen* to strike down a federal law prohibiting subjects of domestic violence restraining orders from possessing firearms.[45] In this case, a drug dealer in Texas with a history of armed violence (Zackey Rahimi) assaulted his girlfriend and threatened to shoot her if she told anyone, leading her to obtain a restraining order. The order suspended Mr. Rahimi's handgun license and barred him from possession. He subsequently threatened another woman with a gun, and then in a two-month spell, he opened fire in public five times—including shooting an AR-15 rifle into a former client's home and firing several bullets into the air when a fast-food restaurant declined a friend's credit card. When Rahimi was charged with violating the federal law, his Second Amendment challenge was initially rejected; he pled guilty and received a six-year prison sentence. While the Fifth Circuit initially affirmed his conviction, the appellate court reversed it after *Bruen*, deeming the federal law unconstitutional because it had no historical analogue in 1791.[46] Thankfully, the Supreme Court overturned this bizarre decision in June 2024, but there is no question that *Bruen* will further expand the presence of firearms—in their most lethal forms—in the hands of Americans while generating unwanted consequences.

GUN-LOBBY AGNOGENESIS

In pursuit of narrow political and economic goals, the gun lobby has distorted and thwarted empirical research, degraded constitutional decision-making, violated historical precedent, advanced false narratives that undermine the intelligence and knowledge of the electorate, and coalesced around the baleful encouragement of revenge violence. As one op-ed columnist for the *Washington Post* observes, "[A] violent revenge theme is common in conservative rhetoric, whether it's gun advocates talking about how they need AR-15s to fend off home invaders or wage war on the government or Trump telling cheering crowds about his desire to inflict violence on liberals. ('I'd like to punch him in the face,' he said about a protester.) It's why Kyle Rittenhouse is seen by conservatives not as a dumb kid who put himself in a dangerous situation with tragic results, but as a hero and a righteous avenger."[47] At the very moment I was presenting this sad saga about America's self-inflicted gun wounds at the Agnotology: The New Science of Creating and Preventing Ignorance conference at Stanford on May 6, 2023, eight people (including three children) were fatally shot and seven others were injured at a mall in Allen, Texas. The AR-15-toting killer, armed with two other guns plus five more in his car, was a Nazi-sympathizing lunatic wearing a patch with the acronym RWDS—standing for "Right Wing Death Squad."

The horrible tragedy and some of the reactions it prompted reflect some of the lamentable harm and folly around the issue of guns in America. An "active shooting instructor" advised Fox News viewers to "have a plan to kill everyone you meet because you never know."[48] At the same time, House Republican Marjorie Taylor Greene called for an investigation into antidepressants "and other factors that cause mass shootings."[49] The local Texas congressman, Republican Keith Self, pushed back on accusations that thoughts and prayers were not enough to prevent mass shootings, telling CNN that "those are people that don't believe in an almighty God who is absolutely in control of our lives. I'm a Christian, I believe that He is."[50] While the public supports greater restrictions on guns, our damaged democracy is paralyzed by special interests and shackled by our misguided Supreme Court, leaving it unable, at this point, to effectively address the growing problem of increasingly lethal weaponry.[51]

NINE

Gluttony and Sloth?

Personal Responsibility versus the True Cause of Obesity

ROBERT H. LUSTIG

"PERSONAL RESPONSIBILITY"—the ideological construct that states that each individual is in charge of their own actions and therefore must suffer the consequences of those actions—is championed as a general solution for many societal ills. In reality, it is the public expression of what is known as "alexithymia"—a kind of emotional ignorance. Indeed, it is much easier to blame the individual and their behavior than society or the larger environment, which are much harder to influence.

When it comes to health and healthcare, the problem is that every so-called personal responsibility issue eventually morphs into a public health crisis, which ultimately requires a public health solution. Lead poisoning, vitamin deficiencies, and asthma were all initially blamed on personal responsibility, but science made it clear that each of these resulted from exposures. Teen pregnancy, smoking, and HIV were also initially attributed to personal responsibility until the sheer weight of the costs necessitated public health responses. But consider also our current obesity and diabetes pandemics. Diet and exercise, gluttony and sloth—"it's your fault"—the standard mantra is that each individual exerts their own "freedom to choose" as to what they put in their mouths and whether they exercise appropriately. But what if you don't really have a choice? Or what if your choice is compromised by your biochemistry? What if the root cause of obesity is biochemical and that the biochemistry drives the behavior? What if your food is toxic and addictive, and the food industry likes it that way? And what if society cannot afford the health consequences of the global industrial diet? "Personal responsibility" did not originate in law; its origin was the tobacco industry rationalizing yet another reason for people to smoke, constructing the epidemic as externalized, personalized, and privatized. The same playbook is currently used to stigmatize those suffering from obesity, resulting in "freedom to blame." Personal responsibility is a form of cultural agnotology that disavows biology. The truth is that each of these public health crises cannot

133

be resolved one person at a time; they require coordinated individual and societal intervention.

The global crisis of chronic diseases continues to chew through lives and money in the United States and around the world. Despite amazing strides in the advancement of medicine, virtually no headway has been made in stemming the tide of this tsunami. In 2000, for example, the global burden of diabetes was estimated at 151 million individuals. But that number would grow to 285 million in 2010 and 539 million in 2021. And it is now expected to reach 563 million by 2030 and a whopping 785 million by 2050.[1] Likewise, cancer and dementia continue to grow with no prospects of slowing down.[2] The age at which they occur is also getting earlier, increasing the cost of healthcare even as fewer people are able to pay into healthcare-delivery systems. The cost is bankrupting healthcare budgets and placing burdens on healthcare delivery.[3] Chronic diseases also increase the prevalence of both absenteeism (can't go to work) and presenteeism (can go to work but can't do one's job), reducing corporate productivity and increasing corporate payouts for health insurance. It's very clear: you can't fix healthcare until you fix health.

EMOTIONAL IGNORANCE AND FREE WILL

While health is a complex construct, the consequences of good health depend on both individual and societal actions—actions that sometimes yield unintended results. Health insurance was ostensibly designed to mitigate catastrophic costs for healthcare, for example, yet insurance can also cause people to exercise poor judgment, thinking, "Well, I'm covered"—a disincentive known as "moral hazard." More perniciously, the population can be driven by dark forces designed to result in poor judgment. For instance, it is hard to imagine good health in a population where a lot of people smoke. But are the chronic diseases of smoking due to individual failures, failures of society, or the malfeasance of an industry dedicated to keeping people smoking? Cigarette makers love to blame individuals for their own diseases—a kind of "immoral hazard." Corporate connivance to keep people from acting in their own best interests is a key example of agnotology.

Many examples of ignorance fit within the taxonomy of the three forms of "cognitive ignorance" described by Nikolaj Nottelman: (a) *factual ignorance*—not wanting to believe that something happened; (b) *object ignorance*—not wanting to believe that something exists; or (c) *technical ignorance*—not wanting to learn how to do something.[4] There is also a fourth version of ignorance, however—*emotional ignorance*, also known as "alexithymia"—characterized by not wanting to empathize with others.[5] Alexithymia does not meet the criteria for personality disorder in the *Diagnostic and Statistical Manual-5 (DSM-*

5), but countless examples of alexithymia have been elucidated. The version of emotional ignorance that I find manifesting collectively, and increasing rapidly, having overtaken medicine and public health in the twentieth and twenty-first centuries, is the ideology of "personal responsibility."

Our modern notion of personal responsibility is an outgrowth of a libertarian view of the self as an independent actor. It is a scaffolding based on three interlocking premises, which when laid upon one another can be used to assign blame. Each of these three premises may appear to be undisputable or even common sense, but each is in fact more nuanced:

(1) *Role responsibility*: Your body belongs to yourself. Your body is your own property, and you are in charge of that property. While this may appear self-evident, it is not supported by either the Declaration of Independence or the U.S. Constitution, and recent U.S. Supreme Court rulings—for example, *Dobbs v. Jackson Women's Health Organization*—cast doubt on this premise entirely.

(2) *Causal responsibility*: Your health is in large part determined by your personal choices. This also appears self-evident given how we tend to view smoking, alcohol, and opioid consumption. I will debunk this premise in the context of the science of the current obesity epidemic (see below).

(3) *Liability responsibility*: You should bear the costs of undesirable consequences—a view also known as the Pottery Barn principle: "You break it, you buy it." There are also some confounding issues here, such as cost and negligence. Liability responsibility is certainly not universally accepted.[6] Depending upon which principles girding personal responsibility are adopted, various ethical and judicial notions of paternalism, utility, fairness, and/or compensation can be manufactured to abrogate empathy and blame the individual. The health insurance industry was supposed to mitigate this principle, but through moral hazard, has only perpetuated the assignment of individual blame.

Free will is a prerequisite that must be met in order to assign personal responsibility. After all, you can't be responsible if you are not in control. The debate surrounding free will is at least as old as Plato, with many scholars denying its unqualified existence. On the philosophical front, John Calvin (1509–1564) espoused the theory of predestination—that God had already chosen your fate, which was immutable. Pierre-Simon Laplace (1749–1827) proposed the idea of causal determinism—that events are determined by preexisting causes. Thomas Hobbes (1588–1679) was a proponent of a kind of "soft determinism" that allowed for moral agents to act upon their desires. And Arthur Schopenhauer (1788–1860) maintained that free will is an illusion because "you are free to do what you want, but not to want what you want." More recently, Anthony R. Cashmore has argued that free will is a construct developed to undergird the criminal justice system in order to identify and punish transgressors. Cashmore claims that free will is a combination of DNA, environment, and some random processes,[7] which would suggest that altering the environment can alter behavior. This is clear in the science behind the current obesity pandemic.

OBESITY AND THE MYTH "A CALORIE IS A CALORIE"

Obesity has been around for a long time (fig. 9.1), although in ancient times, it was a rare occurrence. In the Middle Ages and the Renaissance, obesity was considered a symbol of affluence (think of the Rubenesque body type). Today, by contrast, there are more obese people in the world than people who are underweight.[8] Our current explosion of adult and childhood obesity only came to broad public attention some fifty years ago, when a fairly stable childhood-obesity-prevalence curve went parabolic. Today, one in five American children

FIGURE 9.1. *The Venus of Willendorf,* a 13-inch sculpture now in the National History Museum of Vienna, carbon dated to 24,000 BCE.

Source: Matthias Kabel, photograph, CC BY-SA 3.0, Wikimedia Commons.

and 6 percent of European children are obese.[9] Estimates for adult obesity vary by country, but obesity rates in the United States have tripled since the 1970s, increasing from 30 percent in 2000 to 42 percent in 2018—a huge and rapid increase. But what is the cause?

Discussions of obesity almost invariably invoke the first law of thermodynamics, which states that "the total energy inside a closed system remains constant." The first law can be profoundly misleading, however, when applied to human nutrition.

The most common interpretation of the first law as applied to obesity is *"If you eat it, you had better burn it, or you're going to store it."* The prevailing wisdom (i.e., dogma) can be summed up as *"A calorie is a calorie."*[10] That is, in order to maintain body weight, a calorie eaten must be offset by a calorie burned. The calorie eaten can come from anywhere—meat or dairy, vegetables or soda. And when burned, it can go to anything: sleeping or watching TV or vigorous exercise. In this simplistic interpretation, excess energy intake over energy expenditure is primary, and weight gain is the (secondary) result. This is commonly attached to the presumption that what you eat is your own choice, much like whether you exercise or not. So if you are fat, it must be because you either chose to eat too much or to exercise too little or both. Obesity, by this logic, is the natural consequence of two "aberrant behaviors," *gluttony* and *sloth*—both of which are supposed to be under individual control. Thus, obesity results from failures of our own personal choices and defective moral character and therefore is an archetype of personal responsibility.

Multiple stakeholders hold fast to this interpretation—really, an ideology—including the *processed food industry* (which profits from the current food environment), the *health insurance industry* (which raises rates on people with obesity), the *medical profession* (which provides prescriptions and procedures like liposuction), the *obesity profiteers* (who make money by promising personal cures), the *fat activists* (who claim that obesity is a medical myth), and the *federal government* (which derives income from taxes on processed foods). Examples of this misapplication of the first law of thermodynamics to buttress personal responsibility includes the statement by former governor Jesse Ventura (Independent, Minnesota): *"What happened to willpower? I love fat people. Every fat person says it's not their fault, that they have gland trouble. You know which gland? The saliva gland. They can't push away from the table."*[11] Or consider the admonition of former U.S. representative James Sensenbrenner (Republican, Wisconsin), author of the Cheeseburger Bill: "This bill says, 'Don't run off and file a lawsuit if you are fat.' It says, 'Look in the mirror because you're the one to blame.' "[12] To this day, most framing of obesity in the media blames individual behavior and recommends personal behavioral change rather than social or environmental causes and cures.[13]

This interpretation of the first law cannot explain many aspects of our obesity pandemic. It cannot explain why body temperature has declined over the last 150 years in the United States,[14] suggesting a metabolic defect in mitochondrial function and an involuntary reduction in energy expenditure. It cannot explain why animals bred in captivity have been gaining weight over the last twenty-five years,[15] suggesting the presence of dietary obesogens (chemicals that cause weight gain in excess of their inherent calories) to which animals, as well as humans, are exposed. It cannot explain the increase in obesity in six-month-olds,[16] and it cannot explain the increase in neonatal adiposity (newborn fat) in many parts of the world.[17]

The prenatal and neonatal environment has long been considered a critical period for the origins of obesity and cardiometabolic disease in humans.[18] The *in utero* environment can contribute to the risk of obesity and metabolic disease in childhood and adulthood by increasing the sensitivity or susceptibility to gain weight even before birth.[19] This paradigm applies to altered nutrition, stress, and environmental chemical exposures prior to birth, thus questioning the role of personal responsibility.[20]

Those who fixate on the first law of thermodynamics tend to assume a metric of calories as the driver of weight gain. However, our group and others have documented the central role of high insulin levels, unrelated to calories, as a primary driver of weight gain.[21] For example, our group at St. Jude Children's Research Hospital demonstrated that children who became obese following therapy for brain tumors did so because their brains signaled their pancreases to release more insulin, and that when we suppressed insulin release using a medication, both their food intake decreased and their physical activity increased, resulting in weight loss.[22] Furthermore, we identified a subcohort of obese adults who exhibited increased insulin release in response to glucose administration and who lost weight and increased physical activity in response to pharmacologic insulin reduction.[23]

More recently, Kevin D. Hall and his colleagues at the National Institutes of Health documented that the same dietary composition consumed as ultraprocessed foods would lead to weight gain, while consumption of whole unprocessed foods would lead to weight loss.[24] Thus, the *quality* of the food is more important than the *quantity* as it relates to weight gain and adiposity. The standard American diet (SAD; also known as the Western diet), replete with ultraprocessed foods, acts like endocrine disruptors that drive adiposity and alter mitochondrial ATP production.[25] The recent advent, validation, and utilization of the NOVA classification of food processing divides foods into four categories:[26] (1) unprocessed (an apple, for example), (2) minimally processed (apple slices), (3) moderately processed (such as apple sauce), and (4) ultraprocessed (a McDonald's apple pie). And it has been demonstrated that only the ultraprocessed food category causes obesity,[27] diabetes,[28] heart disease,[29] cancer,[30] dementia,[31] and other mental health disorders,[32] even after controlling for total calories.

Thus, a better interpretation of the first law can be stated: "If you're going to store it (i.e., an obligate weight gain driven by biochemical factors out of your control, such as high insulin), and you expect to burn it (i.e., normal energy expenditure for normal quality of life), then you'll have to eat it." In this interpretation, the behaviors associated with gluttony and sloth are secondary to biochemical stimuli, which themselves are secondary to environmental changes—in this case, the advent of ultraprocessed foods. The biochemical drivers of behavior are, therefore, beyond one's control and not consistent with free will.

ADDICTION VERSUS "PERSONAL RESPONSIBILITY"

For decades, we heard a lot about addiction to things like heroin, cocaine, and alcohol being traced to "personal responsibility,"[33] with the root cause being some kind of personal character defect. Over the past forty years, however, as animal experiments, psychopharmacology, and neuroimaging have revealed the biochemical nature of the reward-addiction response, scholars and the courts have become more willing to recognize drug addiction as more like a disease than a moral failing or lack of willpower.[34] Personal responsibility in the context of addiction is therefore misplaced given that rehabilitation and societal intervention (e.g., legislation) are typically required to reduce the social burden of addiction. The same applies to obesity.

Can foods or food components be considered addictive? The two on which there is most agreement are sugar and caffeine.[35] Indeed, other than caffeine, the component with the highest score on the Yale Food Addiction Scale is sugar.[36] Soft drinks are the most addictive of all "foods," given that they often involve combinations of the two.

While the concept of sugar addiction continues to engender scientific controversy, the criteria for addiction are clearly met in rodents.[37] Oral sucrose (table sugar, composed of one glucose molecule and one fructose molecule bound together) administration activates the reward pathway similar to morphine, and functional magnetic resonance imaging (fMRI) studies demonstrate establishment of hardwired pathways for craving.[38] In one frequently quoted rat study, sweetness surpassed cocaine as a reward.[39] Anecdotal reports in humans describe the feeling of sugar withdrawal using terms such as "irritable," "shaky," "anxious," and "depressed," similar to withdrawal from opiates.[40] Sugar's addictive nature is also evidenced in the fact that fast food, soft drinks, and juice are very price inelastic, meaning that consumers will keep on buying the product even when the price increases—one of the hallmarks of addiction.[41]

To summarize, added sugar (and specifically the fructose moiety) activates the reward centers of the brain and dopamine receptors, as one finds in other drugs of abuse. Sugar also demonstrates the effects of tolerance and dependence, as one expects from an addicting substance.[42]

Here, it is important to recall that many substances once blamed on personal failings are now recognized as problems requiring a public health solution. Table 9.1 is a compendium of health crises that originally fell under the rubric of "personal responsibility." Each of these has since been reclassified as a "disease," however, and/or a public health target. Some of these (e.g., lead poisoning) are exposures, but others are addictions. Addictions to heroin, cocaine, nicotine, and alcohol are now recognized as diseases and therefore worthy of insurance coverage and rehabilitation rather than incarceration. Addictions to gambling and sex are also now considered diseases and listed in the *DSM-5*, and internet gaming and social media addictions are being considered for inclusion. And although addiction to ultraprocessed food is not yet part of the *DSM-5*, it is part of the World Health Organization's *ICD-11* manual, where it is considered a substance of abuse. Addictions are biochemically driven, and insofar as addiction compromises "free will," then obesity, too, cannot be entirely blamed on "personal responsibility."

TABLE 9.1. Types of chronic diseases that require both individual and societal interventions for mitigation.

Exposures	Chemical addictions	Behavioral addictions
Lead poisoning	Tobacco/Vaping	Gambling
Asbestos	Cocaine	Internet gaming / social media
TB	Opioids	Shopping
Guns	Crystal meth	Teen pregnancy
COVID/Vaccines	Ethanol	Pornography

BIG TOBACCO'S INVENTION OF "PERSONAL RESPONSIBILITY"

"Personal responsibility" is an ideological rubric for discounting or ignoring the misfortune of others. But where does this peculiar form of neglect—or redirection of blame—come from? An early expression can be found in Homer's *The Odyssey*, where Zeus implores humans to stop blaming their misfortunes on the gods: "Look you now, how ready mortals are to blame the gods. It is from us, they say, that evils come, but they even of themselves, through their own blind folly, have sorrows beyond that which is ordained." But where does our modern obsession with personal responsibility come from? A review of the tobacco-industry literature reveals some surprising origins.

The association between smoking and lung cancer was first made by German and Argentine scholars in the 1930s.[43] As the medical evidence became stronger, newspapers and magazines began to report on the relationship,[44] prompting cigarette makers to devise a response: the so-called denial campaign (actually a conspiracy) launched at the Plaza Hotel on December 14, 1953.[45] As lawsuits began to be filed, accusing the industry of causing cancer and fraudulently denying harms, cigarette makers developed a coordinated legal strategy that focused on "assumption of risk," implying that people had the knowledge and the right to make their own choices.[46] It was not until 1982 and the onset of class action lawsuits (notably, *Cipollone v. Liggett Group*) that the industry began to market the concept of personal responsibility as a means to deflect blame away from themselves and onto the individual consumer, responsible for their own choices and all consequences of those choices.[47]

Cigarette makers were not alone in attempting to push personal responsibility. The automobile industry attempted a similar campaign in the 1970s to block seat belt requirements, which continued until Mothers against Drunk Driving was able to petition state legislatures to impose "click-it-or-ticket" laws.[48] They and the alcohol industry also waged a short-lived campaign citing "personal responsibility" to battle designated driver laws.[49] And the opioid industry successfully staved off regulation for many years, both by denying that opioids were addictive and by insisting that individuals must exercise "personal responsibility"[50]—until multiple state attorneys general sued Purdue Pharma under the legal doctrine of *parens patriae* (the state as the parent). The net effect is that many public health catastrophes once blamed on personal responsibility have been shown to be public health crises, demanding public health solutions. The one important exception has been our ongoing (and worsening) obesity pandemic, which, despite a robust scientific literature, continues to be viewed by the public as a moral failing and a quintessential failure of personal responsibility. Obesity is a perfect example of alexithymia because even though it has become the norm in American society and the science now points to environmental triggers,[51] public opinion has changed very little. No doubt this is in part driven by corporate propaganda stoking the rubric of cultural agnotology.

CORPORATE-DRIVEN AGNOTOLOGY

Food industry advocates try to keep the public ignorant of the larger causes of obesity so that they can continue to promote alexithymia while minimizing their own corporate culpability.[52] This is performed in myriad overt and covert ways by public relations campaigns. For a more complete treatise on this subject, the reader is referred to the book *Dark PR* by Grant Ennis.[53] But here is a sampling of their techniques:

(1) *Weaponizing words*: Words are important and can be twisted to obfuscate the truth. Here are just two examples. "Sugar" can refer to either "blood sugar" (blood glucose) or "dietary sugar," consisting of any of the following additives in foods: the monosaccharides glucose, fructose, or galactose, or the disaccharides maltose (glucose-glucose), sucrose (glucose-fructose), or lactose (glucose-galactose). The processed food industry intentionally obfuscates the meaning of the word "sugar," leading to the common misimpression that dietary sugar is the same as blood sugar. This can be true, but more often, it is not. Another example is "fat," which can refer to either "body fat" or "dietary fat." The food industry has promulgated the notion that dietary fat leads to body fat. This is one reason the food industry reduced the fat in its products for several decades: the plan was to sell low-fat high-sugar products because fat was more expensive to produce than sugar, and also because sugar is addictive, leading to increased consumption.

(2) *Supporting shoddy science*: The food industry has pushed the idea that sugar is less energy dense (four calories per gram) than fat (nine calories per gram) in order to suggest that eating a low-fat cookie, such as that offered by the brand SnackWells, would be a good way to lose weight. The problem is that the sugar in a cookie is not displacing the fat; it's displacing the water. The recipe for SnackWells is two grams less fat (eighteen calories), for example, but thirteen grams more carbohydrates (fifty-two calories), four of which are sucrose. And while one might expect that the detrimental effects of sugar to be reflected in systematic reviews or meta-analyses,[54] many of these studies are funded by the food industry in order to dilute the available data or to obfuscate any significant effects.[55] Lenard Lesser and his colleagues found, for example, that studies funded by the industry are nearly eight times more likely to show a conclusion favorable to the industry.[56] The industry's influence in distorting public health scholarship even extends to institutions responsible for ensuring scientific integrity, such

as the University of Sydney, who protected scientists who falsified data by claiming there is an "Australian paradox"[57]—that sugar consumption in Australia has declined while obesity has risen—in order to exonerate sugar as a cause of obesity.[58]

(3) *Ignoring diabetes*: Food industry operatives attempt to divert the public health conversation away from diabetes and onto obesity.[59] One reason this is possible is that the data correlating sugar consumption with obesity are relatively weak, accounting for only about 10 percent of the observed effect.[60] And this has become one of the food industry's main talking points. If sugar is only one of many causes of obesity, this allows them to claim once again, "A calorie is a calorie"; it's all about gluttony and sloth, diet and exercise, and energy balance. So if you're fat, it's your fault. Yet when weight and calories are factored out, the correlation between sugar consumption and diabetes becomes much stronger.[61] There are countries where diabetes rates are astronomical, while obesity rates are low, such as India, Pakistan, and China, where sugar consumption has increased by 15 percent in the past six years alone.[62]

(4) *Buying scientists*: The Sugar Research Foundation (SRF) paid off two Harvard scientists to write two articles in the *New England Journal of Medicine* in 1967 to divert attention away from sugar and onto saturated fat as a cause of cardiovascular disease.[63] Similarly, the SRF bought their way onto the National Institute of Dental Research study section in 1971 to divert attention away from sugar as a cause of dental caries.[64] Since then, those with either sugar, high-fructose corn syrup, beverage, or processed food industry concerns have paid scientists to be complicit in marketing sugar as healthy.[65] For instance, Coca-Cola handsomely paid the short-lived Global Energy Balance Network, composed of three scientists, to promulgate the idea that obesity was the result of insufficient exercise.[66] The former head of the Centers for Disease Control (CDC), Dr. Brenda Fitzgerald, also took money from Coca-Cola to say that obesity can be treated with movement.[67] More recently, an analysis of Web of Science citations from 2008 to 2016 searching for "Coca-Cola funding" identified 779 articles[68]; a subsequent comparison with Coca-Cola's own Transparency Website identified 128 articles and 471 authors not disclosed by Coca-Cola and nineteen academic investigators who had email contact with the company.

(5) *Marketing hyperpalatable foods*: In an ostensible attempt at diversification, cigarette makers owned several food companies during the years 1988–2001, during which time Philip Morris, makers of Marlboro cigarettes, was the nation's leading food producer. A recent analysis demonstrates that

those companies were significantly more likely to produce foods high in fat and sugar or salt and sugar than their competitors.[69]

(6) *Co-opting public health organizations*: The U.S. Academy of Nutrition and Dietetics, the British Dietetic Association, and the Dieticians' Association of Australia all receive annual contributions from food-industry concerns.[70] The British Dietetic Association, which once claimed to be "delighted to work with the sugar bureau," has ignored scientific evidence by stating on its website that sugar does not cause diabetes. The scale of this problem can be seen from the fact that Coca-Cola and PepsiCo have provided funding for ninety-six different public health entities.[71]

(7) *Creating astroturf groups*: So-called citizen nonprofit groups mask their sponsors to appear as though they are grassroots organizations. In the United States, for example, the Center for Organizational Research and Education (CORE; formerly the Center for Consumer Freedom) is funded by fast-food, meat, alcohol, and tobacco corporations.[72] Its founder encouraged industry executives to attack those opposed to industry interests, with the admonition they could either "win ugly or lose pretty."[73] Similarly, in the United Kingdom, the Institute of Economic Affairs (IEA) claims to be a "free market think tank" independent of any political party or organization. The IEA produced a report arguing that lack of physical activity was driving the obesity epidemic,[74] with funders including cigarette makers but also Coca-Cola and Tate & Lyle.

FREEDOM TO CHOOSE VERSUS FREEDOM TO BLAME

To qualify under the rubric of "personal responsibility" for diseases like obesity, diabetes, and metabolic syndrome more generally, four prerequisites must first be met:[75]

(1) *Knowledge*: We must be aware of the problem, the causes, and the consequences in order to be able to make good choices. The public is kept in the dark about the pernicious role of sugar and ultraprocessed food in causing obesity and chronic metabolic disease.

(2) *Access*: Many poor communities don't have immediate access to healthy foods and products. These are often called "food deserts," but they are more like "food swamps."[76]

(3) *Affordability*: We have to be able to afford good foods. The consumer price index (CPI) is currently 50 percent lower for ultraprocessed foods thanks

to federal subsidies for corn, wheat, soybeans, and sugar. Conversely, the CPI is 100 percent higher for fresh produce. A 2002 British study shows that healthy food is twice as expensive as processed food and growing more expensive (relatively) over time.[77] Many poor people simply cannot afford to eat healthfully and are coerced into the poor choice. The problem is that Medicare will be broke by the year 2026 due to the chronic disease burden associated with that poor choice.

(4) *Externalities*: Your choice shouldn't hurt other people. Smoking morphed from a personal responsibility issue into a matter of public health once the harms of secondhand smoke became known. Alcohol morphed from a personal responsibility issue to a public health issue thanks to drunk driving. Obesity and diabetes have enormous social costs—costs that threaten Social Security and Medicare and clog emergency rooms as ever-younger people suffer from heart attacks, cancer, and kidney failure. These and other maladies reduce national productivity due to both absenteeism and presenteeism.

From a philosophical point of view, the problem is that personal responsibility is too often perceived as a natural (and positive) consequence of our freedom to choose—a perception predicated on the presumption that "choice" is synonymous with "liberty." As we've learned from tobacco litigation, however, cigarette lawyers have weaponized personal responsibility, using it as a kind of "license to blame." A "poor choice" in this scheme becomes essentially synonymous with "fault." Cigarette makers often win in court by claiming that all liability must be placed on the smoker (the consumer), with none going to the producer.[78]

Invoking personal responsibility has become one of the chief ways cigarette makers and other corporate malefactors win in court. But the notion of personal responsibility also makes it harder for people to obtain health insurance and to access medical care, as insurers can jack up the rates on current and former smokers and may not provide obesity care delivery unless the patient has co-existing diabetes.[79] It also makes it harder for the medical establishment to empathize with victims: a recent analysis of orthopedic nurses demonstrates reduced motivation to care for patients they consider opioid addicts, citing perceived failures in personal responsibility.[80]

As for insurance companies, they are less consistent in applying personal responsibility to eligibility for medical coverage. Roughly three-quarters of all U.S. healthcare dollars are currently spent on diseases caused by cigarettes, alcohol, HIV, opioids, and other maladies said to stem from "self-destructive behavior." Insurance companies will pay for lung cancer, cirrhosis, and protease inhibitors but only rarely for obesity therapy. Why not? Because it would probably break the bank. The ideology of free will, a myopic interpretation of the first law of thermodynamics ("A calorie is a calorie"), and our desire to blame only "gluttony" and "sloth"—all of this gives the insurance companies a license to deny.

This narrative may eventually change. The success of the new weight-loss drugs semaglutide (Ozempic, Wegovy) and tirzepatide (Mounjaro, Zepbound) has caused physicians and patients to rethink obesity as a "disease" and has insurers reexamining the cost-benefit balance between treating and not treating.[81] Currently, these drugs cost about $1,300 per month, but obesity-related medical costs are estimated at $1,800 per year per person, counting the obese and nonobese.[82] Here, we will find a distinct profit motive for the pharmaceutical industry and possibly for the food industry as well. It will be interesting over the next decade to see whether market forces will alter their narrative of obesity as "personal responsibility" or whether our current epidemic of alexithymia will continue to prevail—and to kill.

TEN

How Big Meat Has Created and Legitimized Ignorance

JENNIFER JACQUET

ALTHOUGH CLIMATE CHANGE IS the result of the "slow buildup of insensible gases in the atmosphere,"[1] one group of experts—climate scientists—have shown that the buildup is human caused; that it was not, by geologic standards, slow; and that if left unchecked, it will cause an existential threat to humanity and other forms of life. Scientists proved these things long ago even as the effects of climate change were only predicted rather than observed. Despite this now longstanding and vast knowledge, there are no binding global regulations to mitigate climate change, and the concentration of atmospheric greenhouse gases continues to increase. One cause of the absence of policy is the extensive "climate change countermovement," funded mainly by oil and gas producers, which has worked to create ignorance about climate change and thereby stall regulation and allow proliferation of pseudo-solutions.[2]

Many of the scientific experts involved in producing the science and the public understanding of climate change were—and are—employed as professors at universities. Universities are committed to a broad suite of goals. Stanford University's founding grant refers to "the studies and exercises directed to the cultivation and enlargement of the mind."[3] The University of Miami's mission includes "the freedom to think, to question, to criticize, and to dissent." Harvard University's seal is a shield with the Latin motto "VERITAS" (meaning "verity," or "truth") written across three open books.[4] Universities have been essential to the creation of knowledge, including the scientific and popular understanding of climate change. But universities and their faculty and researchers have also played a role in the creation of ignorance, including undermining the public understanding of climate science and public policy. Some of the most prominent climate contrarians have been legitimized by their positions at reputable universities.[5]

Corporate polluters can benefit from having university-affiliated researchers as their spokespeople because universities enhance those spokespeople's credibility.[6] In the twentieth century, cigarette makers used the Council for Tobacco Research to fund more than a thousand scholars, all of whom—knowingly or not—helped the industry to prevent government action and maintain the industry's power and legitimacy.[7] The tobacco industry would announce new grants to Harvard scientists in press releases and used the Harvard name whenever it could in PR announcements.[8] The industry recruited Dr. Carl Seltzer, for example, an anthropologist and part-time staffer for the Peabody Museum at Harvard, and regularly referred to him as a "Harvard scientist." An R. J. Reynolds Tobacco Company newsletter headline from March 25, 1971, reads, "Harvard Scientist Challenges Heart-Smoking Relationship" (and cites one of Seltzer's letters to the editor of the *New England Journal of Medicine*).[9] One headline in the *Boston Globe* reads, "Harvard Scientist Says Elderly Smokers Needn't Quit." The article begins by claiming that "a Harvard research scientist said yesterday people over 65 who like to smoke won't live longer if they stop."[10]

Universities engage in pernicious contradiction when they work to create knowledge in one domain and take industry money to create ignorance on that very same topic. Consider Stanford University, which employs a number of prominent climate scientists as professors, including the now-deceased Stephen Schneider, who codirected several important climate-related initiatives at Stanford in the early 2000s. During that same time, Stanford accepted funds from Exxon to spread climate disinformation; the Union of Concerned Scientists found that Exxon had donated $295,000 to the Hoover Institution between 1998 and 2005, including $30,000 in 2003 for "global climate change projects."[11] The Hoover Institution made climate contrarians Sallie Baliunas and S. Fred Singer fellows and published articles by Singer calling global warming a "hoax" and a "charade,"[12] as well as an article titled "Sure, the North Pole Is Melting: So What?," in which Singer claimed that "there is no sound scientific evidence that the globe is warming."[13]

Here, I examine industry influence on one particular individual researcher and institution to create ignorance and therefore stall regulatory action on an important current issue—the contribution of livestock to climate change. Before I turn to the role of Professor Frank Mitloehner and the University of California, Davis (UC Davis), I provide a short account of attribution research that singles out the livestock sector as a source of anthropogenic climate change because it was this knowledge that provided the impetus for the subsequent creation of ignorance. I also quickly detail how the industry responded to this research.

EARLY ESTIMATES OF LIVESTOCK EMISSIONS
AND HOW THE INDUSTRY RESPONDED

There are many ways to assign attribution (or responsibility) for anthropogenic change, and different approaches have different utilities.[14] A series of articles in the 1980s brought attention to methane as an important source of global warming,[15] for example, and at least one points out that one of methane's major sources is intestinal fermentation from ruminant livestock (mainly cattle, but also sheep and goats, which chew and regurgitate their food, mainly grasses).[16] The U.S. Environmental Protection Agency (EPA) published a report in November 1989 estimating that globally ruminant animals "account for nearly one fourth of the total anthropogenic [methane] emissions, or about 15 percent of total emissions."[17] The *New York Times* covered the report, emphasizing the role of livestock while highlighting, "Of all the 'greenhouse' gases methane may be the easiest to control." Notably, they did not include an industry response.[18]

Six months later, the *New York Times* published another article—this one focused on the impacts of livestock—that mentioned again the EPA's methane calculations. This article included the industry's perspective: "This figure [referring to the EPA estimate of 15 percent of methane emissions] is fiercely contested. In a study commissioned by the National Cattlemen's Association (which, in 1996, became the National Cattlemen's Beef Association, or NCBA), Professor F. M. Byers of Texas A&M University concluded: 'Cattle are not a significant source of greenhouse gases. The U.S.A. amounts to less than 0.5 percent.'"[19] A week or so beforehand, Byers had published an editorial in *USA Today* defending meat eating and claiming, "Cattle production is an efficient use of energy. U.S. agriculture uses less than 3% of all fossil energy in the USA; beef production uses a small fraction of that."[20] Byers also reportedly wrote a twenty-three-page research paper on beef, methane, and global warming for the National Cattlemen's Association. Byers emphasized how small the U.S. contribution is relative to global emissions and focused on fossil-energy use rather than methane emissions—two common arguments that will repeat themselves in the meat industry's creation of ignorance about its role climate change.[21]

In 1992, Jeremy Rifkin published a trade book titled *Beyond Beef*, drawing attention to some of the many problems with beef production, including a chapter focused on its contribution to global warming.[22] Soon thereafter, Rifkin's team started a campaign in North America to encourage people to cut their meat consumption in half. The book and campaign received a lot of press coverage. In response, the beef and dairy industry embarked on a countercampaign titled Don't Blame It on Bossie.[23] The National Cattlemen's Association (later NCBA) and the Grocery Manufacturers Association called a number of reporters to tell them that Rifkin's book was "misleading," and the public relations firms

Edelman and Burson Marsteller sent press packets that were positive about beef and offered reporters free trips to Wyoming.[24] Rifkin would eventually cancel the remainder of his book tour due to threats.[25]

Meanwhile, the science exploring the contribution of livestock to global warming continued to develop. In 2006, the Food and Agriculture Organization of the United Nations (FAO) published a landmark report titled *Livestock's Long Shadow*, which included all forms of greenhouse gas emissions across the full supply chain (including feed-related emissions) as well as manure. The FAO attributed fully 18 percent of global emissions to the livestock sector, amounting to 7.1 gigatons of CO_2 equivalent per year, with beef and dairy production accounting for the majority.[26] This was the first major international report to highlight the animal-agriculture sector's responsibility for climate change, and it is difficult to overstate its importance. One member of the beef industry referred to it, with some understatement, as a "public relations challenge."[27]

In response, NCBA funded Frank Mitloehner, a professor of animal science at UC Davis, to assess the FAO report. He would put forward many of the same points Byers had earlier, only Mitloehner would also become essentially a mouthpiece for the industry, challenging research that highlighted the role of livestock in climate change and opposing efforts to reduce meat consumption or production. The following is a brief summary of what followed.[28]

AGNOGENIC ACCOUNTING: FRANK MITLOEHNER IS FUNDED TO ADDRESS A PUBLIC RELATIONS CHALLENGE

In the early 2000s, NCBA provided $26,000 to Frank Mitloehner, then an associate professor in the Department of Animal Sciences, to conduct an assessment of the FAO report. Mitloehner had a degree in animal science and up to that point, had never published on the topic of climate change. His UC Davis teaching included an undergraduate course in domestic livestock production and a graduate course titled Grant Procurement and Administration.[29]

Mitloehner led an article published (that was presumably peer-reviewed) in 2009 in *Advances in Agronomy*, titled "Clearing the Air: Livestock's Contribution to Climate Change."[30] His team did not present a new global estimate of emissions but instead emphasized the facile point that the relative *percentage contribution* of livestock in the United States is lower than the global average (global statistics are global, so this point also cuts the other way: in New Zealand, for instance, livestock emissions account for 37.5 percent of total emissions, not 18 percent). Mitloehner also repeatedly pointed out that the report underestimated transportation emissions in order to claim that media reports that livestock accounted for more emissions than the transportation sector were incorrect. The article did not disclose any funding.

With the paper's publication, UC Davis published a press release—reminiscent of the Don't Blame It on Bossie frame—titled "Don't Blame Cows for Climate Change" (this press release included a mention of the industry funding).[31] Mitloehner also presented his work at an American Chemical Society conference in 2010, which published a press release titled "Eating Less Meat and Dairy Products Won't Have Major Impact on Global Warming" (this press release did not disclose the study's funding).[32] Note that neither press release title can be justified by the actual contents of the *Advances in Agronomy* article.

Again, Mitloehner's published research article does not challenge the absolute emissions presented in *Livestock's Long Shadow*, nor does it present a different global estimate. However, the media framed the study as a debunking of the FAO results.[33] For instance, a *Daily Mail* article begins, "Dr Mitloehner says meat and milk production generates less greenhouse gas than most environmentalists claim—and highlights the source of confusion as a report from the UN."[34]

NPR's (National Public Radio) Ira Flatow asked Mitloehner about his talk at the American Chemical Society: "So you're saying that it's better to reduce your driving if you want to do something than to stop eating meat and cheese and milk?" Mitloehner responded, "Well, according to EPA, the Environmental Protection Agency, they have done an emission inventory last year in 2009, and they claimed that transportation contributes 26 percent of all greenhouse gasses versus the 5.8 percent for all of agriculture, and about half of that 5.6 or 5.8 percent is due to livestock. So in the United States, approximately 3 percent of all greenhouse gas is associated to livestock, and about half of that three is associated to beef and dairy. So your relative contribution to climate change is not really a major factor when eating these products."[35] Here Mitloehner also ignored the role of imported products—a mistake difficult to imagine from someone who claims to study the climate impacts of the food system—and made a leap from U.S. production of livestock to the livestock consumed by U.S. consumers, which are not the same thing.

Flatow then asked Mitloehner, "Does any of your funding come from the cattle industry or the beef industry?" Mitloehner did not mention his study's direct funding from the beef industry but instead emphasized, once again, relative percentages: "About 5 percent of my funding comes from agriculture as a whole, and approximately 3 percent from the beef industry." The similarities are striking in how Mitloehner discussed livestock's emissions ("5.8 percent for all of agriculture, and about half of that 5.6 or 5.8 percent is due to livestock") and his own funding from the livestock industry—both good examples of agnogenic accounting.[36]

Mitloehner's first foray into the question of climate change was funded by the industry, and it would not be his last. Just as it is difficult to overstate the importance of the FAO report *Livestock's Long Shadow* to the scientific and popular understanding of livestock emissions, it is difficult to overstate the role of Mitloehner in downplaying the problem of livestock-related emissions. Mitloehner has been prolific in shaping both public policy and public understanding through social media and the mainstream press.[37]

Cattlemen appear to have been pleased with the outcome of their investment because Mitloehner has become a prominent university researcher on the topic of livestock and climate change—more prominent, in some cases, than actual climate scientists who work on livestock and climate change. Since 2009, he has received at least $5 million in industry funding,[38] although the full extent of his funding is uncertain, and Mitloehner's speaking and consulting fees are not currently available (it is possible they could be made available through public records requests, though many universities do not require the disclosure of dollar amounts for consultancies, and all university disclosure rests on the good faith reporting of individual faculty).

Just like Byers (who was also supported by the beef industry) in 1990, Mitloehner, in 2009, spoke about the industry's impact in relative, rather than absolute, terms and focused on the United States rather than on global contributions. And Mitloehner has continued to emphasize that the United States is "vastly different from the global picture."[39] After noting the relatively small fraction of U.S. emissions from beef cattle (as opposed to absolute emissions from livestock), Mitloehner has said, "If you consider that the official number that was used by the FAO globally is 18%, then these are huge differences [between the United States and the rest of the world] that certainly do not warrant discussions of 'Meatless Mondays,' and so on."[40]

One might wonder, then, whether discussions of "Meatless Mondays" are warranted in a country like New Zealand, where livestock emissions represent more than a third of all emissions from that country. In 2022, Mitloehner gave a talk to a conference of three hundred "delegates" in Christchurch, reported by the *Otago Daily Times* under the headline "Methane's Impact on Warming Misrepresented US Scientist Tells Red Meat Sector Conference."[41] In 2023, Beef + Lamb New Zealand sponsored two lectures from Mitloehner, whom they billed as an "international GHG [greenhouse gas] communicator." Mitloehner's talk was entitled "How Managing Methane from Livestock Can Be a Climate Solution."[42]

Mitloehner is active on social media, where he has issued countless tweets challenging livestock-related climate research, policy proposals, and even individual researchers.[43] He uses the handle @GHGGuru—meaning "greenhouse gas guru." At least one of Mitloehner's tweets—claiming that the ingredients of plant-based burgers from Impossible Foods and Beyond Meat are indistinguishable from dog food—was investigated by the fact-checking site Snopes and found to be false.[44] Documents obtained by a Greenpeace Freedom of Information Act (FOIA) request revealed a confidential memo from an executive at a feed-industry group (and funder of Mitloehner) that discussed his Twitter (now X) account and noted that "Dr. Mitloehner, UC Davis, and IFEEDER [Institute for Feed Education and Research, an arm of a feed-industry trade association] have made significant progress on both increasing Dr. Mitloehner's messaging and exposure."[45] It is difficult to imagine the National Science Foundation issuing a memo about a principal investigator's number of Twitter followers or progress in increasing them.

Mitloehner's value is not just in shaping public understanding. Alongside the livestock industry,[46] Mitloehner has also helped shape emissions-accounting efforts. After the 2006 release of *Livestock's Long Shadow*, FAO formed a partnership with private stakeholders, known as the Livestock Assessment and Performance Partnership (LEAP), to create FAO's "global livestock environmental assessment model" (GLEAM) for estimating livestock emissions. This LEAP-FAO collaboration also included input from industry groups, such as the International Feed Industry Federation and the International Meat Secretariat, with Mitloehner listed as a LEAP partner and its chair in 2013. A 2013 FAO report estimated total emissions from global livestock again at 7.1 gigatons of CO_2 equivalent per year but noted that this represented only 14.5 percent of all anthropogenic GHG emissions, down from its earlier estimate of 18 percent.[47] In October 2022, GLEAM released a lower estimate of 6.19 gigatons of CO_2 equivalent in livestock emissions—despite an increase in livestock production—and calculated livestock as only 11.2 percent of all anthropogenic GHG emissions.[48]

Mitloehner has also been politically active. He has testified before the U.S. Congress, as well as before governments in other countries. Mitloehner has said he has had "a lot of interaction with the EPA and others" and that "it probably has an impact."[49] To a degree, he has even gotten politicians to drop meat from their climate-related efforts. *Politico* reports on how a tweet by Mitloehner helped fuel a backlash to the Green New Deal plan's mention of cows and how even Alexandria Ocasio-Cortez had reached out to Mitloehner "to set up a call to discuss the potential for climate mitigation efforts in agriculture," which, according to Mitloehner, "lasted more than an hour." *Politico* refers to Mitloehner as "a leading scientist on agricultural emissions at the University of California, Davis."[50]

Mitloehner has repeatedly claimed cows should not be blamed for climate change. He says their impact is minimal and that the focus of remediation should be on aviation, food waste, or some other carbon source. He has reportedly said that "the real problem the livestock sector faces is convincing consumers and policy makers that animals aren't the bad guys of the global warming challenge."[51] Besides saying cows should not be blamed for climate change, Mitloehner has also insisted that cows can be a climate solution. This is a paradox reminiscent of the tobacco industry's argument that while everybody has always known that smoking causes cancer, nobody has ever really been able to prove it or their rejected claim that some tinker will solve the problem—like filters, low tar, or "lights."[52] Funded by numerous feed companies, Mitloehner has regularly promoted feed additives as a solution to climate change.[53] He has said that California is well on its way to achieving the goal of "climate neutrality" by 2027 and that its dairy industry should be able to sell carbon credits to other sectors by then.[54]

Mitloehner's efforts to undermine *Livestock's Long Shadow* have been well documented, as have his ties to the industry and the long history of his efforts on behalf of the livestock industry and climate change.[55] The website DeSmog, a clearinghouse dedicated to exposing global warming misinformation campaigns, has recently added a long entry for Mitloehner.[56] Mitloehner has now produced some peer-reviewed articles on livestock and climate change, but the vast majority of his activities have been aimed at influencing public understanding and greenhouse gas accounting, while also engaging in political lobbying.[57]

THE ROLE OF UC DAVIS

UC Davis disclosed Mitloehner's meat-industry funding in its press release about his challenge to the *Livestock's Long Shadow* report,[58] but since then, the university has been far less transparent about the millions of dollars in industry funding Mitloehner's has received, as well as the overhead rates and funds the university has taken (which is normal for universities). When a *New York Times* reporter sought comments from UC Davis about Mitloehner's funding, the university declined to comment.[59] Only after Greenpeace's FOIA request did Mitloehner's university center add a description to its website that included a mention of its industry funding.[60]

Along with a veneer of impartiality, UC Davis has given Mitloehner legitimacy. The university amplifies Mitloehner's industry-friendly talking points. The university hosts the website for Mitloehner's industry-funded center, which does public outreach and self-promotion. UC Davis also regularly promotes Mitloehner with its own press releases, which repeat the talking point that U.S. cattle make up "just 2 percent of direct emissions" and cite a popular article Mitloehner wrote stating that "forgoing meat is not the environmental panacea many would have us believe" (for which he was required to include a disclosure statement; his statement did not mention any industry funding).[61] When the UC Davis press office reported on his receipt of an industry-sponsored communications award, they repeated his claim that by "overstating the emissions," the

FAO had "set people off on the wrong solutions—such as cutting back on meat and milk consumption—that would only stymie efforts to feed a growing world population."[62]

Although DeSmog, focused on climate misinformation, has deemed Mitloehner worthy of his own (dishonorable) entry, UC Davis features Mitloehner on its website of climate experts.[63] The university claims that it "leads the cutting edge in climate science" and that "the extension of knowledge to serve the public good" is part of its land-grant mission.[64] But the university has clearly allowed private interests to use its reputation to propagate ignorance. The university has accepted money to downplay the contribution of livestock to climate change while at the same time employing (other) researchers to create honest knowledge about climate change, its causes, and its social implications.

While perhaps the most visible, Mitloehner is certainly not the first or the only UC Davis professor to front for industrial animal agriculture. UC Davis has a longstanding relationship with the industry through its Department of Animal Science, dedicated to a field that is impossible to disentangle from Big Animal Ag because the industry funds much of the research, sponsors and attends its academic conferences, and is either financially or editorially responsible for many of its research journals. In 1992, UC Davis professor and dean of the College of Agricultural and Environmental Sciences from 1975 to 1989 Charles Hess helped to defend the meat industry after the publication of Rifkin's book *Beyond Beef.*[65] The *San Francisco Examiner* coverage states that Hess was "responding to the book for the beef industry"[66]—a description the media has rarely used for Mitloehner.

WHAT AN INDUSTRY GAINS FROM
FUNDING UNIVERSITY EXPERTS

Industries are aware that there is more public trust for academic scientists than for industry spokespeople, and many corporations have relied on university scientists and scholars to help maintain their social license to operate.[67] The main way industries have succeeded in getting university professors to speak on their behalf has been by providing financial resources. Like other enterprises, the animal-agriculture industry sees university funding as an investment that will ultimately help maintain profits. This is what separates industry money from other kinds of money and why it is false to claim that "any funding is a conflict of interest, even when the funder could not benefit in any way from the findings of the research."[68]

Imagine a civil society group, such as the People for the Ethical Treatment of Animals (PETA), donating millions of dollars to a university professor to represent their interests. The end goal of such a donation, while clearly ideologically motivated, would not be so that PETA could promote a product to make money. Again, that is what makes donations from industry (and industry-related foundations) distinct. Robert N. Proctor points out, "A grant from Reynolds or Philip Morris is not like a grant from, say, the Ford Foundation. An application submitted to the Ford Foundation is not run by the legal department of the Ford Motor Company—which is more like what happens in the tobacco context. Tobacco grants are approved by damage-control experts at the companies, which means that tobacco money is more like development research—or a kind of advertising."[69] Funding from a polluting industry or its affiliates, such as industry-sponsored foundations or nonprofits, usually comes with strings attached.

THE MOTIVATIONS OF INDIVIDUAL RESEARCHERS

Each individual has their own values, motivations, and priorities. Some people may want to produce knowledge, answer to a higher calling, make money, or become influential. Ideological motives can have powerful outcomes on scientific understanding,[70] and some scientific journals have gone so far as to ask researchers to declare them. But many researchers may not even be fully aware of their ideological motives, let alone capable of or willing to declare them to a scientific journal.

A financial interest (which some journals have delineated as greater than $10,000, and others, at $5,000, while many leave it to the individual's discretion) is distinct from, and should not be confused with, ideology. Studies show that financial interests influence decisions directionally in favor of the funder at every point in the scientific process, including problem selection, hypothesis formulation, experimental design, sample choice, data collection and analysis, and framing the results.[71] Others point out that it is "a mistake to object to the constraints on financial gain by complaining that there are other kinds of influence,"[72] and conflating "conflicts of interest" with "interests" in general "serves to muddy the waters about how to manage conflicts of interest, generating confusion as to the nature and definition of the problem and doubt as to whether conflicts of interest can be addressed at all."[73]

Not all conflicts of interest involve financial interests; an editor for a scholarly journal handling a paper that includes a relative among its list of authors, for instance, is also a conflict of interest. But a conflict of interest often does involve a financial interest. This has long been clear in the field of medical research, where the primary interest of researchers—which, according to the Hippocratic oath, is the welfare of the patient—comes in conflict with secondary interests, such as financial gains from pharmaceutical companies.[74]

To some extent, the notion of "conflict of interest" has been misunderstood in purely financial terms, and with regard to that common misunderstanding,

Mitloehner does have a conflict of interest. However, if we understand conflicts of interest to be something about which the individual researcher has an internal struggle (i.e., a primary and secondary interest coming into conflict), then there may be no "conflict of interest." Mitloehner has significant financial interests, for example, but those are perfectly aligned with his primary (research and personal) interest, which, as an animal scientist, is to maintain or improve the production of livestock.

Mitloehner's condition is a more serious issue than a conflict of interest: he is a shill—someone who gives the appearance of an impartial endorsement of something in which they themselves have an important stake (table 10.1). He has downplayed and even hidden his ties to the animal-agriculture industry.[75] The media coverage has repeatedly failed to note his industry ties. At this point, if Mitloehner were to offer a full disclosure of his funding, it would be unlikely

TABLE 10.1. A proposed typology of university researchers working for the industry.

Type	Characteristics
Independent	Receives no industry money, declares the funding received for research, does not agree to any arrangement in which funder has input into study design or final say about publication, and discloses paid board positions on their curriculum vitae (CV).
Hard up	Receives some industry money at some points to do research but discloses it in publications, press releases, on his or her CV, and in media appearances.
Sellout	Receives large amounts of industry money over the course of his or her career and/or serves on industry boards and/or takes large speaking fees from the industry but discloses all the funding publicly (i.e., not only through a given university's required disclosure efforts).
Industry shill	Receives large amounts of industry money, is involved in a lot of public outreach and advocacy, and spends a lot of time challenging other existing scientific work or scientists. Remarks are always in favor of the industry, and talking points are notably similar to those of the industry. Funding is rarely or not fully publicly disclosed (even if it might be disclosed with the individual's university), and in existing disclosures, industry ties are downplayed.
Merchant of doubt	Works across several scientific issues in which they have limited expertise, is ideologically motivated in that he or she shares the industry's desire for zero regulations or a desire simply to occupy the contrarian space in the press, and may or may not take large amounts of industry funding, which is rarely or not fully publicly disclosed (even if it might be disclosed with the individual's university).

to compensate for the fifteen-plus years of communications work he has done for the industry under the guise of impartiality. In press and talks he has given, it should have been stated that he, like Charles Hess, was "responding . . . for the beef industry."

When called out for failing to disclose their industry funding, university experts have often responded by claiming that they are completely independent, that they are not required to disclose, that they are being attacked personally when the focus should be on the research, or that there is nothing wrong with working for the industry—or that industry funding should be part of the solution.[76] The same day Greenpeace and the *New York Times* published articles exposing Mitloehner's deep ties to the industry, Mitloehner published a blog post (using his center / UC Davis site) in which he facetiously responded,

> There's a shocking revelation out there, and I am at the heart of it. Are you prepared for this?
> Animal scientists work with animal agriculture. That's it. That's the exposé, the conspiracy that so many activists and journalist want to share with you.[77]

CAN UNIVERSITIES CUT TIES WITH IGNORANCE?

UC Davis helped Mitloehner achieve his reputation as an independent expert. The university frames him as a climate expert and has not required disclosure of his center's funding on its website (hosted by UC Davis) or in communications. Why do universities help keep industry funding in the shadows?

Universities themselves have a conflict of interest. Universities are supposed to be sites of knowledge creation, but they frequently receive money from industry to create ignorance. Big Meat and UC Davis are just one example. Big Tobacco, Big Pharma, Big Oil, and the agrochemical conglomerate have also funded universities and their experts to help create ignorance. Many universities have administrative units devoted to doting on sponsors, including corporate ones. In some cases, universities may even help an industry use back channels that allow them to avoid public scrutiny (in one case, a University of Florida researcher suggested that the Monsanto Company could avoid being in a conflicts-of-interest account by directing funds through a program within the nonprofit University of Florida Foundation).[78]

But universities are not public relations firms—they do not exist merely to do work on behalf of their paying clients. Some faculty have been motivated to try to improve systems of disclosure among faculty. Universities could require transparency from their faculty, requiring an electronic long-form disclosure made available on the university website and in email signatures. Some universities have refused to take certain kinds of money altogether, such as tobacco-industry money. A similar ban has been suggested for fossil fuel money for climate-related research.[79] Brown University has recently discussed a screening process that would prevent funding from any group that supports disinformation, and the same university's Climate Social Science Network has launched an ambitious program to study "climate obstruction" both in the United States and abroad.[80]

One could go further and suggest that universities could consider prohibiting any donations from industry altogether, knowing what we do about its influence. Some civil society groups, of which the largest and most vocal about it is Greenpeace, have committed to not taking corporate money. Greenpeace screens "all large private donations to identify if there is anything about them which could compromise our independence, our integrity or deflect from our campaign priorities. If we find something then we will refuse or return the donation."[81] Perhaps the institutions that have dedicated themselves to the creation of knowledge deserve similar protections.

ELEVEN

On the Burial of the Palestinian Nakba[1]

ROSEMARY SAYIGH

Introduction

In his seminal work on 'white ignorance,' the philosopher Charles Mills writes the following:

> In the heyday of white racism and formal European domination of the planet, global white ignorance took the form of the acceptance of the inferiority . . . of people of colour, the normative legitimacy of white rule, [colonialism and imperialism, indigenous expropriation, displacement, and killing; racial slavery], and the corollary of racialized assumptions and frameworks, blindnesses and indifferences, necessary to render such domination consistent with both asserted fact and proclaimed moral principle.
>
> Mills, 2015: 219

Mills' analysis of the production of ignorance is reminiscent of Trouillot on the importance of political power in creating silences:

> The play of power in the production of alternative narratives begins with the joint creation of facts and sources for at least two reasons. First, facts are never meaningless: indeed, they become facts only because they matter in some sense, however minimal. Second, facts are not created equal: the production of traces is always also the creation of silences.
>
> Trouillot 1995: 29

I begin by presenting myself as an exemplar of 'white ignorance.' I am a woman born, brought up and educated in England, who married a Palestinian in 1953 and went to live in Lebanon. When I arrived in my new home, I had no idea why Palestinian 'refugees' were there, and knew nothing about Palestine. That the history textbooks I read for A-levels in Britain did not mention Palestine is not surprising, since what came to be known as the 'Third World' was entirely missing from these books, except as parts of the British or other European empires (Bhambra et al. 2018).

During the first year of our marriage we went to my husband's family home

for lunch every Sunday. These occasions were dominated by silence. Because of
my ignorance I didn't know how to interpret this silence, far less how to inter-
rupt it. I learnt later that silence was characteristic of Palestinian families in the
post-Nakba years. Salim Tamari explains,

> people were so traumatized that they did not actually want to talk about it
> . . . Of course, you know, people were talking within their own families, but
> in a very cryptic way. I know that families on the most intimate terms would
> hesitate in speaking about what happened to them in '48. And when they do – I
> tried because I ran the diary section in the Jaffa work – it was very hard to get
> old people to talk about leaving Jaffa. And until today I am having difficulty
> with people who have critical information.[2]

I sought to reduce my ignorance by visiting the United Nations Refugee and Works
Agency [UNRWA]'s Beirut office.[3] UNRWA brochures told what UNRWA was
doing for the 'refugees' in terms of health, education, and welfare, but nothing about where they came from, nor why they were 'refugees.' In this silence,
UNRWA replicated the policies of the UK and the US, states that dominated the
Middle East for most of the twentieth century. These powers supported Israel
as an obstacle to Arab unification and to guarantee the supply of oil to the West
(Gendzier 2015). They blocked UN resolution 194 on the right of Palestinians to
return to their country, establishing UNRWA instead. UNRWA was the main
instrument for re-settling Palestinians in the 'host' countries as a semi-skilled
labour force. UNRWA welfare subsidized the cost of their labour to host country
employers, and UNRWA's education programme supplied rural Palestinians with
the basic skills needed for economic integration in the host countries.

Exploring UNRWA further, I discovered that the history component of its
schooling programme made no mention of Palestine. This was because when
UNRWA was established it was decided that host country rather than Palestinian curricula should be adopted [Irfan 2019: 8].[4] This decision clearly aimed
to erase memories of Palestine and promote resettlement in the 'host' countries
where UNRWA operates – Jordan, Syria, and Lebanon. The absence of Palestinian history and geography from the UNRWA curriculum was one of many
reasons for the frequent protests of UNRWA teachers, who, besides calling for
better salaries, demanded texts on Palestinian history and geography. In spite
of teacher and student protests, and though it contravenes article 13 of the Convention on the Rights of the Child, this policy remains in place today.

This policy is simply one example in which ignorance of Palestine and the
Nakba of 1948 has been produced through different political and disciplinary
mechanisms, including colonial appropriation, landscape transformation, censorship, memoricide, schooling, or fear of antisemitic labeling. In this chapter,
I analyse similar instances of suppression of knowledge about the Nakba/Pales-

tine, drawing on oral and written testimonies from scholars teaching in North
American universities.

DEFORMING PRISMS

Travellers visited the 'Holy Land' in ever larger numbers during the nineteenth
century, and many of them wrote books describing the country. One of the most
popular was Mark Twain's *Innocents Abroad* [1869]. Twain satirized pious
travellers by exaggerating Palestine's aridity and poverty, a description contradicted by travelers such as William Thomson, whose *Land and the Book* [1859]
describes Palestine as green, fertile, and productive.

A more dangerous discursive deformation was the Balfour Declaration's designation of the Palestinians as 'existing non-Jewish communities' contrasted
with 'the Jewish people' [Cronin 2017]. The political implications of this distinction are evident: a 'people' was qualified for nation/statehood, whereas dispa-
rate 'communities' were not. Framed by the Declaration, the British occupation
of Palestine from 1919 to 1948 was characterized by often brutal suppression of
the indigenous Palestinians in favor of Zionist settlers [Hughes 2010].

Another deforming prism characterizes the books of Europeans who visited
the 'Holy Land.' They viewed Palestine primarily through the Bible, absenting
the physical presence of living Palestinians. Issam Nassar writes of this literature:

> The attitude of nineteenth century European travellers to Palestine can be de-
> scribed as 'textual' because it stemmed primarily from a body of already exist-
> ing European texts. For most travellers at the time, the real Palestine was the
> one described in the books rather than the one they saw before them.
>
> 2003: 8

'One aspect of this focus on biblical Jerusalem was the lack of interest in the history of the people who lived there. Jerusalem was presented as almost an empty
place' [13]. Such descriptions reinforced long-held Christian Zionist beliefs that
Jerusalem was a city to be 'reclaimed' [Sharif 1985]. The popularity of these
books in Protestant countries did much to detach Palestine from the Arab east,
projecting it as integral to Europe, a perspective that favoured Zionism.

1948–9: 'ISRAEL' ERASES 'PALESTINE'

In May 1948 the state of Israel, supported by the US, UK, and USSR, superim-
posed itself by force on what had been Palestine, expelling around 80 percent
of the original population, and depopulating around 560 towns and villages.[5]
Palestine ceased to exist as a recognised territory. Salman Abu Sitta describes
not finding Palestine on any map when visiting libraries in Europe and the US

[Abu Sitta 2016: 269–270], a discovery that inspired his monumental *Atlas of Palestine 1917–1966* [2010].

Once in control, Israel set about transforming the country through: i) demolishing around 500 villages; ii) legalizing the appropriation of Palestinian land and property [Foreman and Kedar 2004]; iii) looting Palestinian archives and libraries [Mermelstein 2011]; iv) planting non-indigenous forests to hide village ruins and transform the landscape [Pappe 2006: 225–234]; v) Hebraizing place names [Masalha 2015]; vi) destroying olive orchards.[6] To quote Ilan Pappé: 'The archaeological zeal to reproduce the map of "Ancient" Israel was in essence none other than a systematic, scholarly, political and military attempt to de-Arabise the terrain – its names and geography, but above all its history' (Pappé 2006: 226). After the 1967 war these measures were extended to the occupied West Bank and Gaza, in addition to: vii) building Jewish settlements across the occupied regions; viii) constructing segregated roads to join settlements to Jerusalem and bypass Palestinian villages. Though at first Arabic was retained as an official language, this was revoked by the Nation-State Law of July 19, 2018.

While such moves are characteristic of colonialist appropriation, they point to a more important aim: memoricide. To quote Piterberg,

> From the start, Israeli officials were well aware of the significance of memory and the need to erase it. Repression of what had been done to create the state was essential among the Jews themselves. It was still more important to eradicate remembrance among Palestinians.
>
> 2001:39–40

Among Israeli measures to erase the Nakba are decrees banning use of this term in school books.[7] Police have been used to remove Nakba day demonstrators. Israeli aims to destroy Palestinian memory is demonstrated by the IDF looting of the PLO Research Center during the 1982 invasion of Lebanon [Sleiman 2016].

The siting of Israel's World Holocaust Remembrance Center, Yad va-Shem on the lands of the ruined village of Deir Yassin, site of the first massacre of 1948, is yet another act of burial. According to its website, Yad Va-Shem is

> a vast, sprawling complex of tree-studded walkways leading to museums, exhibits, archives, monuments, sculptures and memorials . . . 62 million pages of documents and nearly 267,5000 photos, along with thousands of films and videotaped testimonies of survivors . . . and 3.2 million names of Holocaust victims.
>
> Masalha 2005: 6, 7

The graves of those who died in Deir Yassin are unknown and unmarked.

MAINTAINING ERASURE/CONCEALING AGGRESSION

To maintain its good standing, Israel needed to conceal evidence of the violence of its take-over of Palestine. Given that the expulsions were witnessed by journalists, victims, and members of the Zionist forces, a first necessity was centralising all written testimonies, removing those deemed damaging to Israel's reputation, and limiting access to the rest. A *Haaretz* journalist notes,

> Since the start of the last decade, Defense Ministry teams have been scouring Israel's archives and removing historic documents. But it's not just papers relating to Israel's nuclear project or to the country's foreign relations that are being transferred to vaults: Hundreds of documents have been concealed as part of a systematic effort to hide evidence of the Nakba.
>
> Four years ago, historian Tamar Novick was jolted by a document she found in the file of Yosef Waschitz, from the Arab Department of the left-wing Mapam Party, in the Yad Yaari archive at Givat Haviva. The document, which seemed to describe events that took place during the 1948 war, began:
>
> Safsaf [former Palestinian village near Safed] – 52 men were caught, tied them to one another, dug a pit and shot them. 10 were still twitching. Women came, begged for mercy. Found bodies of 6 elderly men. There were 61 bodies. 3 cases of rape, one east of from Safed, girl of 14, 4 men shot and killed. From one they cut off his fingers with a knife to take the ring.
>
> Shezaf 2019

Seth Anziska notes that the Israeli military has gained increasing control over archives,

> in an Orwellian act of self-censorship that began in the early 2000s, the Defense Ministry's secretive security department, Malmab, spearheaded efforts to reclassify documents and methodically remove files from various archives across Israel to hide evidence of Israeli responsibility for the Nakba. Alongside the censoring of interviews with military veterans describing war crimes in 1948, and the sealing of documents that provide evidence of the extent to which the military government controlled the lives of Palestinian citizens of Israel in the first decades of the state's existence, Malmab officials have entered unannounced into the reading room of various archives since 2002 and pressured professional archivists to hand over documents about 1948 without legal authority.
>
> Anziska 2019: 67

ERASURE OF PALESTINE THROUGH EDUCATION

Education is the domain where memoricide comes most effectively into play. As already noted, children in UNRWA schools learn host country history rather than that of Palestine. Until 1967 this meant Jordanian or Egyptian history for children in camps in Jordan and Gaza: 'These textbooks were severely censored by the Israeli occupation authorities until 1994: The word "Palestine" was removed, maps were deleted, and anything Israeli censors deemed nationalist was excised' [Moughrabi 2001: 6]. The 1967 war gave Israel control of the West Bank and Gaza, increasing its power of censorship over curricula and textbooks [Naser-Najjab 2020]. In 1994, after the Oslo Accords, a radically new curriculum for Palestinian schools was produced under the charge of the eminent scholar Ibrahim Abu-Lughod, in coordination with UNESCO. Though the new curriculum was commended by outside observers for its objectivity, yet an Israeli group based in a West Bank settlement wrote a biased report that reached the highest levels of the United States, prompting President Clinton to denounce 'the hateful anti-Israel rhetoric in official Palestinian school books" '. As a result, several governments and the World Bank withdrew promised funding [Moughrabi 2001: 9].

Fear of Israeli and international criticism can be detected in the history textbooks produced by the Palestinian National Authority [NA] created by the Oslo Accords of 1993. In a study of the NA's curricula and textbooks, Nubar Hovsepian writes, 'rarely, if ever do the texts identify the conditions that Palestinians encounter daily – checkpoints, identification cards, curfews, detention, land confiscation, home demolitions, uprooted trees, and closures' [2009: 143]. There is little mention of the Palestinian refugees; 'the word "resistance" does not appear anywhere' [148]. As for the Nakba, it is only 'mentioned obliquely . . . explained by the PA's desire to downplay the contentious issue of the "right of return" '.[8]

In Israel, the textbooks used in Jewish schools have been criticised for their racist depictions of Palestinians. In an authoritative study, Nurit Peled-Elhanan writes,

> I will only mention at this point that none of the textbooks studied here includes, whether verbally or visually, any positive cultural or social aspect of Palestinian lifeworld: neither literature nor poetry, neither history nor agriculture, neither art nor architecture, neither customs nor traditions are ever mentioned. None of the books contains photographs of Palestinian human beings, and all represent them in racist icons or demeaning classificatory images such as terrorists, refugees and primitive farmers.
>
> 2012: 48–9

As for the US, teaching on Palestine was recently excluded in California high schools under pro-Israeli pressure, in spite of a state decision to empower children in subaltern communities [Sokolower 2020]. In the UK, pro-Israeli groups intervened to radically alter the text of school textbooks on the Middle East, removing references to Israeli violence while intensifying references to Arab or Palestinian aggression.[9] The Prevent Duty clause of the UK Counter Terrorism and Border Security Act [2019] has further limited freedom of speech on Palestine [Allen 2019].

SILENCING THE NAKBA

Even in the Arab world, the Nakba has not been widely commemorated [Masalha 2005: 7]. The influence of the global north over education systems worldwide suppresses coverage of the Nakba even in Arab school textbooks. Ilan Pappé points to the dominance of Holocaust commemoration as burying the Nakba through concealing the connection between the two:

> The Zionist movement had the military power to both ethnically cleanse Palestine of its original population and to face a military confrontation with troops from various Arab armies sent to try and prevent the creation of a Jewish state. However, it needed the Holocaust memory to silence any criticism of its ethnic cleansing operation and to prevent any international pressure on it to allow the return of all those expelled from the land after the 1948 war. Europe's guilt at allowing Nazi Germany to exterminate the Jews of Europe was to be cured by the dispossession of the Palestinians.
>
> 2008

> The argument for a Jewish state as compensation for the Holocaust was a powerful argument, so powerful that nobody listened to the outright rejection of the UN solution by the overwhelming majority of the people of Palestine. What comes out clearly is a European wish to atone. The basic and natural rights of the Palestinians should be sidelined, dwarfed and forgotten altogether for the sake of the forgiveness that Europe was seeking from the newly formed Jewish state.
>
> 2008

It was not until almost fifty years after the 1948 erasure of Palestine that the first book about the Nakba was published by a major US university press: *Nakba, 1948, and the Claims of Memory*, edited by Ahmad H. Sa'di and Lila Abu-Lughod, Columbia University Press. Until then, scholars writing about Palestine faced difficulty in finding publishers. Sarni Hadawi, first author to publish on the Nakba in English after 1948, writes:

The experience of publishing my first book, and the difficulties we are encountering at present in finding a publisher for a manuscript of some size of which I am coauthor, have not been encouraging. My Publishers, the Naylor Company of San Antonio, Texas, has been unable to place *Palestine: Loss of a Heritage* in any bookstore in the United States.

1967: xix

Even Edward Said, best known of Palestinian scholars, experienced difficulty in getting published on Palestine. Miriam Said, Edward's widow, writes:

> Edward encountered many difficulties before *Orientalism* was published. . . . The worst incident of censoring . . . took place before *Orientalism* hit the bookstands. Beacon Press . . . asked him to write a book on Palestine. He wrote *The Question of Palestine*. When Beacon received the Ms, his editor wrote back to say that this is not the book they asked him to write . . . Edward then asked the publisher of *Orientalism*, Pantheon Books, if they would be interested in publishing it. Andre Schiffren, the owner of the house, declined. It was obvious that he did not want to publish anything by Edward regarding the Israeli/Palestinian conflict.[10]

In a study published in 2016, Seyyed Hadi Borhany used historical narrative analysis to examine six books on the Middle East most used by Anglophone universities.[11] All but one of these textbooks reproduces the Israeli narrative concerning its presence in the Middle East [Borhani 2016]. Further examination shows that none includes the Nakba in the index; all but one minimize Palestinian history before British occupation; coverage of the 1948 war does not include the massacres, or UNGA Resolution 193 on the right of return. One of the six is a history of Zionism.

EXCLUSION OF PALESTINE FROM THE 'TRAUMA GENRE'

Between 1997 and 2001, three volumes on 'social suffering' edited by Arthur Kleinman and Veena Das initiated the 'trauma genre'. It rapidly proliferated, branching out into studies of memory, mourning and melancholy, the politics of witness, sexual abuse, and racism; specific cases included the Holocaust, Hiroshima, the Armenian genocide, South African apartheid, lynchings in the United States, and death through AIDS. But one searches in vain through this vast literature for the Palestinian Nakba. Some have asserted that the Nakba cannot be compared with other world catastrophes because the Palestinian death toll in 1948 was not remarkable.[12] For example, Gilbert Achcar claims that 'the Palestinians cannot . . . legitimately apply to their own case the superlatives appropriate to the Jewish genocide' [Achcar 2010: 31–32]. But suffering cannot be measured in terms of casualties alone. The Nakba entailed a *continuing* state of rightlessness, with all the varieties of violence that rightlessness exposes people to.[13]

TEACHING PALESTINE IN NORTH AMERICAN UNIVERSITIES

Given that at least 84 American-type universities exist outside the US [of which 18 are in the Middle East/Arab Gulf], it is reasonable to see this type of university as a global model. This gives international significance to the experiences of Palestinian scholars and scholars who teach Palestine in the U.S. and Canada. A threat that affects all such teachers in the US is the Canary Mission:

> Canary Mission is a secretive but clearly non-academic political organization that uses its website to engage in defamatory attacks against college students who advocate for Palestinian rights, against student and other organizations engaged with this issue, and against faculty ho teach, or speak publicly, about the Israeli-Palestinian conflict. Canary Mission's tactics threaten students' right to pursue an education without harassment as well as the employment prospects of those whom the website targets by name, including undergraduate and graduate students.[14]

In what follows I present testimonies by four scholars who teach Palestine/the Middle East in North American universities. First, Julie Peteet, professor of anthropology at the University of Louisville, and author of a study of women and the Palestinian Resistance movement [1991]:

> About *Gender in Crisis* – I did not face problems getting it published; Columbia took it quickly. Issues arose later when a professor at UC Berkeley assigned it as an ethnography for an Introduction to an Anthropology class of 700 students. Parents were up in arms and calling upon the University to dis-allow its usage as a text. With my 2017 book, *Space and Mobility in Palestine*, I faced censorship from University of Pennsylvania Press. I had a contract for the book. It was at the copy-editing stage, when the Board voided the contract. They met without my editor's presence and voted to cancel the contract. They claimed the book was 'biased'.[15]

Ted Swedenburg, author of a history of the Great Revolt of 1936–1939 [1995], and professor of anthropology at the university of Arkansas, writes:

> I had no trouble finding a publisher . . . The problem, at the time (late eighties, nineties) was getting a job. I had interviews at good places but couldn't make it through the final hurdle. Working on Palestine also limited the number of places where one could actually land interviews. These problems led me to start working on other topics . . . I moved away from writing on Palestine because writing on Palestine seemed to be a problem on the job front.[16]

Randa Farah, teacher of anthropology at Western University, London, Ontario, writes:

> If teaching about the Middle East in general is not easy due to deep-seated Orientalist assumptions, teaching about Palestine and the Palestinians is often an 'extraordinary event' and experience. It is Orwellian indeed because merely teaching about Palestinians as a nation and a people with a history of unrelenting land expropriation, colonial settlement, and expulsion is, at best, regarded as a topic that might create 'tension,' implying that it is safer to avoid it . . . I draw not only on my own experiences but those of other colleagues and friends who introduce the Palestinian case from a non-Zionist perspective.
>
> 2018: 13

Randa Serhan, Palestinian, ex-faculty member of City University, Washington, writes:

> In 2010, I came across an ad for Director of Arab Studies and a tenure-line position in Sociology at American University, Washington, DC. I was interviewed and offered the position. On my first day, the director of Israel Studies (non-faculty) showed up in my office, wanting to know where I grew up, and where I had studied. . . . An Ottomanist told me to build Arab Studies, and then hand it over to an Israeli professor in another department. When I co-sponsored an event on Palestinian and Israeli youth activists, the director of Israel Studies asked why I held such an event? There were several attempts to quash or co-opt the Arab Studies program, and hostility towards the program did not wane except when enrollments were so large that it benefitted the department.
>
> I asked to teach a course on my research on Palestinian Americans because I was the only faculty member never to have taught on my own research. My request was denied. My chair told me to 'get over it'. I was denied tenure, and when I appealed the denial one of the things that came up was that I had marginalized the Center for Israel Studies.

DR SERHAN RAISES A CRITICAL ISSUE

This all begs the question, where does the silencing happen? When the attack is so large-scale and so pervasive, the silencing appears as self-censorship. The 'beauty' of such silencing is that is done in ways that can't be called out . . . I wonder how many others are facing more subtle silencing where their case is not 'high profile' enough, where a Zionist group has not placed their name on a website, or where they are too junior, who are self-silenced under the radar. I do not believe my case is unique.[17]

The case of Stephen Salaita calls for attention here. A Palestinian specialist in American Indian studies, Salaita was offered a post at the university of Illinois, but later 'unhired' because he wrote tweets criticizing the 2014 Israeli attack

on Gaza. Nadia Abu El Haj, author of a study critiquing Israel's use of archeological research for political purposes, was subjected to a hostile campaign
against her tenure at Barnard and Columbia university [Abu El Haj 2017].
Based on multiple interviews, Deeb and Winegar write on teaching Palestine in
US universities:

> since the 1970s, academics who research or teach topics against the grain of
> dominant US national narratives about and interest in the region have faced the
> prospects of not having their research funded, not being hired, being accused
> – by parents, students, administrators, and people unassociated with academia
> or their campus – of bias and even treason in their teaching and public lec-
> tures, being targeted by blacklists and hate mail, and even losing their job. This
> is especially the cases for those who teach about Palestinians and/or critical
> perspectives on Israeli state actions, and for scholars of Arab and/or Muslim
> descent, who face additional discrimination.
>
> 2015:2

It is surely hard to find another foreign policy issue with such a pervasive effect
in the academic domain.

CONCLUSION

In this chapter I have presented evidence that the consequences for Palestinians of the expulsions of 1948, through which Israel was created and Palestine erased, have been insufficiently covered in Euro-American histories of the
region. Though justified in terms of the Geneva Convention and international
humanitarian law, Palestinian resistance has been branded by Israel and mainstream media as 'terrorism', while any form of solidarity with the Palestinians
risks being labelled anti-semitism. Space limitation has forced me to exclude
social media, though both Facebook and Zoom have deleted Palestinian messaging. More seriously, I have omitted racism, and the degree to which fear of
Jewish immigration influenced Balfour's Declaration. Racism was integral to
the Declaration, and undoubtedly plays a role in silencing the Palestinian story,
as European guilt over past persecution of Jews is assuaged through support for
Israel, while the racism involved in anti-semitism is re-channeled against the
Palestinians. This is the broader ideological and political context that frames
the burial of the Nakba.

TWELVE

Euphemism in the Architecture and Language of Treblinka

DANIEL AKSELRAD

IN DECEMBER 1942, OFFICERS of the Nazi SS constructed a fake passenger railway station at Treblinka to conceal the true purpose of the extermination camp. Here, at the end of a spur route twenty railcars long, the deputy commandant Kurt Franz ordered imprisoned Jewish craftsmen to paint signs in German reading "First-Class Waiting Room,"[1] along with arrows pointing "to Bialystok and Wolkowysk,"[2] a fake ticketing counter, a painted clock, and other visual deceptions. Treblinka survivor Samuel Rajzman testified at the Nuremberg trials that "there were even train schedules for the departure and the arrival of trains to and from Grodno, Suwalki, Vienna, and Berlin"[3]—schedules for train routes that did not exist.

These elaborate efforts to facilitate the murder of nearly a million people— mostly Jewish men, women, and children of the occupied East—were primarily undertaken to reassure deportees along the Reichsbahn railway. Nazi bureaucrats had already deceived their victims into believing that they were being transported to Ukraine to work on farms or in a "preserve factory."[4] In many cases, wrote Soviet reporter Vassili Grossman (in 1946), the Germans even forced Jewish families to purchase railway tickets to the station of Obermajdan—a name SS officers contrived to conceal Treblinka.[5] Historians have argued that this fake railway station served to keep victims compliant upon arrival at the camp, and certainly, that is true to a degree.[6] But this explanation does not account for the railway facade's role in a wide array of hastily improvised workplace accommodations—infrastructures that enabled German and Ukrainian guards to psychologically repress aspects of their experience in the Nazi exterminationist project.[7]

How did the SS maintain morale inside a Nazi killing center? And, more broadly, how do bureaucratic organizations condition human beings to commit mass murder? My goal here is to explore how the SS developed a vocabulary of visual-rhetorical deceptions to maintain commitment, compliance, and control

171

at Treblinka. As a case study in linguistic and visual agnotology, this banal monstrosity provides an example of how stagecraft enables domination both by manufacturing ignorance and by providing the ideological resources for on-going and systemic subjugation. As survivors' accounts show, Treblinka's false front was just one of many contrivances the deputy commandant installed to reframe the death camp as an ordinary civic center. I argue that euphemism was a rhetorical weapon the Nazis used to carry out atrocities but also that SS officers organized the camp in several ways that physicalized the euphemisms they honed to displace their day-to-day reality at the Reich's peripheries. I want to shed light on how simple yet powerful means of persuasion—euphemism as a means of distorting reality—enabled the violence of the camps. Doing so can deepen our understanding of how language works in environments of mass de-ception, where words, as infrastructure, enable the construction of full-fledged surrounds for the social production of ignorance.

EUPHEMISTICS AND AGNOTOLOGY

Along its vast and dendritic bureaucratic chains of action, Nazism relied on the everyday recodification of violence using language designed to conceal and re-frame Nazi administrative activities. The SS fashioned a lexicon of euphemisms for renaming atrocities—a vocabulary that facilitated the organization and ra-tionalization of Hitler's so-called Final Solution (itself a euphemism). The Nazis named the T4 euthanasia program in Berlin, tasked with gassing thousands of Jewish men, women, and children, the Charitable Foundation for Institutional and Curative Care.[8] The Blood Protection Law (1935) declared German-Jewish intermarriage "racial pollution" and "sexual traffic," and branded Jewish women a "health risk" to the German "body."[9] Police raids and mass executions in Bialystok and Lvov in July 1941 were carried out under the guise of "repri-sals" against Jewish "looters,"[10] and in Kiev, posters beckoned Jewish families to present themselves for "resettlement."[11] Perhaps most notoriously, *Sonderbe-handlung*, or "special treatment," provided an extralegal justification for killing under the pretense of medical therapy.[12] The Reichsbahn railway even extended half-off discount rates and special credit lines to the SS for the transportation of *Sonderzüge*, or "special trains," bringing Jews from Western Europe to Ausch-witz for "evacuation," "resettlement," and "labor utilization."[13]

Scholars have long sought to make sense of this obfuscatory language. In 1947, the Polish-Jewish historian Nachman Blumental published an abbreviated dictionary of these Nazi euphemisms he called "innocent words." He reached only up to the letter *I*, so the volume spans *"Abbruch"* (meaning the eviction and subsequent destruction of Jews and their homes) to "Israel" (an additional first name that Jewish men were required to assume under the Nuremberg Laws and, in 1940, to post at the entrance of their homes).[14] The philologist Victor Klemperer wrote of "cover-up words" in what he called the *Lingua Tertii Imperii*, contrasting, for example, *Rückzug* (retreat) with the Reich's preferred *Rückschlag* (setback).[15] In 1963, the historian and Auschwitz survivor Joseph Wulf indicted the Nazis' "lexicon of murder,"[16] and Hannah Arendt identified the Nazi code for this language system: *Sprachregelung*, or "language rules."[17] The German sociologist Klaus Theweleit later wrote of the "reality-destroying character" of the pre-Nazi *Freikorps*' "secret language,"[18] and Primo Levi spoke of the "Lager jargon" he learned at Auschwitz, which differed from one camp to the next, albeit with a "treasury of words" in common.[19] Discussing the administration of the so-called Jewish question, in which the activities attendant to mass murder were each given a nickname, cryptonym, epithet, metaphor, acronym, or euphemism, Holocaust historian Raul Hilberg offers that paradoxically, "in a sense, nonlabeling became the ultimate camouflage."[20] That is, in a bureaucracy for which every activity must be hidden with a conspicuous name devised to conceal it, ironically, it is the records, timetables, and correspondences for which these efforts at secrecy were abandoned that remained best hidden from prying eyes.

The ability of euphemisms to place reality at a distance, to undescribe atrocities occurring in plain sight, was especially instrumental for SS leadership looking to maintain a strong esprit de corps among the Reich's improbable executioners. This is because for many perpetrators, the act of killing was traumatizing. Heinrich Himmler and his chief SS physician, Ernst Grawitz, were aware of the psychological toll of turning policemen into executioners, with perhaps the most extreme of these views expressed in the words of the *Einsatzkommando* leader (and convicted war criminal) Paul Blobel. It was Blobel who opined after the war, "The nervous strain was far heavier in the case of our men who carried out the executions than in that of their victims. From the psychological point of view they had a terrible time."[21] Berlin even legitimized this position during the war, awarding Treblinka's SS *Obersturmführer* Franz Stangl and other camp commanders a military decoration *für seelische Belastung* (for mental hardship)—which, as Nazi hunter Simon Wiesenthal explains, in Nazi terms meant "for special merit in the technique of mass extermination."[22] Where the trauma of executing families and small children might have been unassimilable, euphemisms by which to rename mass murder in the streets, in forests, aboard boxcars, and in killing centers offered the Reich's executioners a means of undertaking evasive thinking surrounding their own action and inaction.

In all manner of institutional contexts, language plays a crucial role in supporting the deliberate production of ignorance. Orwellian "Newspeak," impenetrable jargon, technical "shop talk,"[23] and other rhetorical patterns make

it possible for bureaucratic functionaries to insulate themselves from the consequences of their own action and inaction but also to excuse or ignore the wrong-
doing of those around them. Organizations invent micro-rhetorics, leverage the
phonesthetic force of speech, coin words that induce semantic overlap and con-
jure phonemic resonance, and assign new meanings that evoke legitimacy to
manufacture consent, exact commitment, and maintain morale. This is not only
to facilitate concealment or efficiency: internal rhetoric as a means of achieving
normative control incentivizes organizational activity by reframing its meanings in favorable terms that increase employee or member buy-in.[24] It is George
Orwell's insight that such strings of words supply the ideological resources of
bureaucratic power—that their service in large organizations is "not so much to
express meanings as to destroy them."[25]

But what about camouflaging acts of violence from one's own self? Such
rhetorical strategies, involving as they do the destruction of meaning, fal-
sify the perception of those outside the institution but also alter the experience of those *inside* the institution in ways that enable them to carry out their
work. This inward turn of persuasion influences individuals' sensemaking in
organizations—their ability to make sense of themselves and their role within a
larger hierarchical institution.[26] Carol Cohn has shown, for example, how the
language of nuclear war—with its acronyms, euphemisms, and sexual double
entendre—allows the U.S. military to avoid speaking candidly about the bloody
realities of nuclear annihilation. In these decision trees, the president's "shopping list" is full of "cookie cutters," "clean bombs," and "penetration aids"
that can take out cities in "sub-holocaust engagements."[27] The linguist Dwight
Bolinger elaborates that "a bombing *raid* is renamed a *mission.... Dead bodies*
are *casualties* or *fatalities*. An explosion does not have *power* or *force*, but
yield [and] the accepted euphemism is *tragedy*," which suggests no blame.[28]
By the same token, modern militaries construct and invade mock cities with
cryptonyms, like "Atropia" and "Donovia,"[29] metonymize compliant soldiers
as "boots on the ground," and brand aberrant defectors "bad apples" or "lone
wolves."[30] This is the sympathetic magic by which the Israeli Defense Forces'
bombardment of Gaza has been dismissed internally as merely "mowing the
grass."[31] This is how the invasion, rape, and torture of a sovereign nation becomes a "special military operation" in Ukraine or how Philip Morris making
cigarettes at Auschwitz becomes the "Oswiecim issue."[32] Whether we are fluent
in this language or not, its use precludes a meaningful understanding of truths
unfolding in plain sight. Where violence is encoded but nonetheless happens,
bureaucratic agnotology makes it possible to maintain such an ignorance mi-
croenvironment between superiors and subordinates.[33]

Understanding the mechanisms through which organizations manufacture
ignorance to induce compliance and conformity is critical toward any proper

understanding of bureaucratic power. This is because any study of knowledge, knowledge practices, and power-knowledge apparatuses must also address the creation of non-knowledge, as well as that which is hidden or unknowable.[34] To this end, Lorraine Daston and Peter Galison define epistemology as the study of "obstacles to knowledge."[35] Nancy Tuana, in the first agnotology volume, theorized entire "epistemologies of ignorance" on the grounds that to fully understand the complex practices of knowledge production, "we must also understand the practices that account for *not* knowing, that is, for our *lack* of knowledge [and for] *unlearning*."[36] By this logic, any epistemology of bureaucracy that fails to account for modes of concealment, deception, and conspiracy, as well as the generation of ambiguity, uncertainty, and doubt, therefore erects a Potemkin village—a false front inadequate to explain the organizational imperative for producing both knowledge and ignoration. False, misleading, purposefully equivocal, and obfuscatory languages, illusions, and deceptions are instrumental for organizations aspiring toward total control of their subjects. In prisons, labor camps, authoritarian regimes, and other "total institutions,"[37] these all-but-innocent words are effective (perhaps inevitable, even) because they capture by antithesis the realities posed by the harms taking place.

For what reasons did Kurt Franz, Franz Stangl's deputy commandant of Treblinka, issue the order to disguise the incoming platform as a passenger railway station, complete with signs reading "Restaurant," "Ticket Office," "Telegraph," and "Telephone"?[38] Imagine the reassurance of arriving at Treblinka and standing underneath the Red Cross flag waving above the doorway to the camp's *Lazarett* (field hospital)—which, in reality, concealed an execution site above a burning pit for the elderly and the disabled (see fig. 12.1). For the two men in the *Lazarett* who "wore white aprons and had red crosses on their sleeves" as they carried the sick and the elderly to their executioners, what was the meaning of their medical garb?[39]

Consider the stagecraft of Nazi medicine, wherein it was of utmost importance to disguise medical executions as the legitimate work of licensed physicians. Auschwitz doctors developed a method of forced sterilization by which they irradiated the genitals of unwitting victims using invisible X-rays all while these "patients" were made to believe they were simply filling in forms in a doctor's office chair.[40] At Buchenwald, a special commando shot to death no less than eight thousand Soviet prisoners of war under the pretense of measuring their height using a specially adapted medical stadiometer.[41] At Treblinka, just beyond Kurt Franz's railway farce, signs reading *Zur Badeanstalt* ("To the Bathhouse") directed victims down the path to a gas chamber fitted with unplumbed sinks and shower heads—a path SS officers called the *Himmelfahrt Strasse* (literally "Ascension Road," or Street to Heaven) and which Jewish inmates euphemized as "the Tube" (*Schlauch*).[42] (And here we can see that there

FIGURE 12.1. Lazarett. At Treblinka, the Red Cross flag disguised an execution pit used for deportees who were wounded, sick, elderly, or children. Two men in medical uniforms carried these victims to the so-called field hospital where, beyond a sight screen, SS *Scharführer* August Miete would shoot them in the neck with a 9 mm pistol.

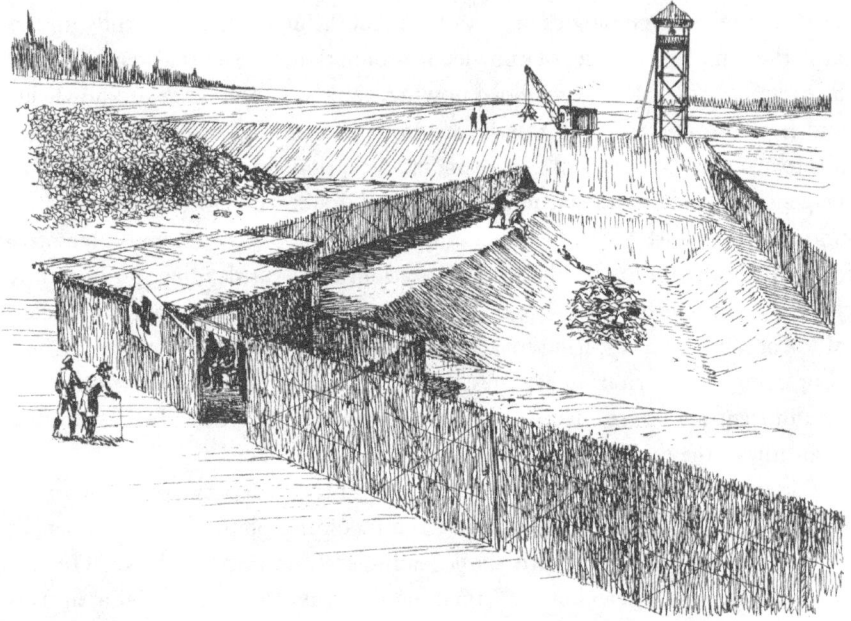

Source: Samuel Willenberg, *Lazarett*, 1981, pencil drawing. Reprinted with permission of Ada Willenberg.

are euphemisms from above and euphemisms from below.) Where such euphemisms provided the rhetorical support for elaborate architectural theatrics, fictions took on material form as infrastructures and sites of action. Euphemizing a gas chamber as a bathhouse soon gave way to signs, sinks, and showerheads. Calling an execution site a "hospital" gave way to red crosses and medical garb. Rhetoric that displaces reasoning is one malice, but when euphemisms harden into "stuff you can kick,"[43] these words become tools through which deadly culture and ideology can be administered at scale.[44]

The decision to euphemize the steps along a killing routine was neither just malice nor simply camouflage—it was an act of psychological repression undertaken by and for the Treblinka staff themselves. These were not acts of self-deception by which SS officers denied the reality of the camp and its purpose in a genocide; to be sure, the perpetrators of these crimes knew exactly what they were up to. Rather, these efforts are an example of what the sociologist Stanley

Cohen called "interpretive denial"[45]—a way of managing one's emotions by performing ideological and category work meant to reframe the Nazi genocide as acceptable, moral, or even "mercy death" (as Hitler himself declared).[46] Franz Stangl would admit twenty-five years later to the investigative reporter Gitta Sereny, "I repressed it all by trying to create a special place: gardens, new barracks, new kitchens, new everything; barbers, tailors, shoemakers, carpenters. There were hundreds of ways to take one's mind off it; I used them all."[47] ("It all" here refers to organizing a mass genocide; note also his use of the word "special.") The tailor shop, the barber shop, the dentist's office, the zoo, the flower beds, and the fake railway station, taken together, were not the result of officers' efforts to undertake "deliberate ignorance" of their own crimes but rather, in the words of Václav Havel, a kind of "evasive thinking" meant to "liquidate reality" through its recontextualization.[48] In this sense, these structures spatialized the functions intended by Adolf Eichmann's "language rules": the signage, the zoo, and all the rest beautified, papered over, and displaced the true meanings of words, structures, and killing routines, even death itself.

To understand these infrastructures of atrocity is to see that they were not the result of a specific order from Hitler's command but rather the product of field-expedient experimentation by local units stationed at the Reich's peripheries. If Nazi Party officials developed a vague and powerful ideology laced with ways of talking around atrocities, then operationalizing their deadly imperative relied on forward commanders in the East to develop their own innovations for performing, escalating, normalizing, and streamlining acts of violence.[49] Rigidity at the Reich's ideological center relied on flexibility at death factories scattered along the Reichsbahn, where deputies like Kurt Franz, who had started as an SS cook just a few years earlier and risen through the ranks, could improvise their own creative means of killing Jews. Much like the gas chambers, which began as improvised mobile gassing trucks, the fake passenger railway platform was an ingenuity contrived in the absence of an instructional manual. As we shall see, beyond the facade, a host of constructions and performances adorned Treblinka so that a broader euphemization of the death camp as an ordinary company town could help officers rationalize their role in a genocide.

OBERMAJDAN STATION

In early 1942, Treblinka's "conveyor-belt" killing routine was still a work in progress.[50] Franz Suchomel, an SS *Unterscharführer* at the camp, explained years later that in this early "phase one," guards on the platform would unleash violence upon arriving deportees at the moment of each train's arrival. Before the camp's new commander, Stangl, arrived, explains Suchomel, there were "thousands of people [i.e., the dead] piled one on top of another at the

train station. . . . Everywhere. They were stacked up like wood."[51] Detainees
who survived the "special" train journeys (and the lesser-known boat journeys)
from all over Europe, most often without food or water, were forced from trains
on arrival at the camp and pushed through a portico inscribed "Women left,
Men right."[52] Treblinka survivor Israel Cymlich describes this process in similar
terms: guards marched victims into the woods and executed them with axes.[53]
As sixty-car transports arrived, Reichsbahn employees transferred twenty railcars at a time to the SS, who found that many of those on board were often
already dead from asphyxiation, starvation, or shooting by the SS en route. The
chaos continued even later when the gas chambers often did not work as intended, slowly asphyxiating victims for up to eight hours. In the winter, corpses
froze by the thousands, making it extremely difficult for Jewish slave laborers
at the camp—the so-called *Dentisten* (dentists)—to extract gold teeth from
mouths frozen shut.

These largely uncoordinated and often improvised acts of violence challenge
the common misconception that Himmler's extermination camps ran as well-
oiled machines with robot-like precision. The German sociologist Wolfgang
Sofsky argued that with an average of three trains arriving per day from August
through December 1942, each sixty cars long and carrying between two and
twelve thousand people, the SS counted on the deportees' cooperation because
"brutality costs time and delays smooth operations."[54] Guards needed families
to prepare their own bodies for execution—to undress, to hand over their valuables to Suchomel's *Goldjuden* (gold Jews), and to move obediently through the
stations of the slaughterhouse built to receive them.[55]

Treblinka's commanders closed the original (pre-camp) Treblinka railway
station to the public on August 27, 1942, to conceal the camp's presence along
a single-track secondary line. In December of that same year, the deputy commandant Kurt Franz ordered several Jewish prisoners to construct an elaborate
fiction as part of their effort to streamline the processing of people from ghetto
to gas chamber. One prisoner-mechanic named Berl Kot stone paved the railway ramp "as smooth as a table top."[56] Others cleaned the platform, covering
it with cinders and laying fresh branches along the perimeter fence.[57] A professional painter from Warsaw, "a man of medium height with a hawk's nose and
a brimmed black hat, and a narrow black bow-tie around his neck," carried out
Franz's order to paint a wooden sign reading "To Bialystok and Wolkowysk" in
black letters against a white background with a directional arrow.[58] He painted
one sign (three meters by eighty centimeters) bearing the word Obermajdan in
the same style and an array of little white signs reading, in German, "First-Class
Waiting Room, Second-Class, Third-Class,"[59] and "Cashier."[60] Carpenters af-
fixed each of these signs above the windows of the repurposed barracks that ran
along the train platform—in actuality a storeroom for warehousing plundered

belongings.[61] Survivor Oskar Strawczynski recalls the ruse: "Over one of the larger windows is a sign that reads 'Tickets.' On the side [of the barracks] appear signs such as 'Information' and 'Station Manager.' On the walls big arrows point: 'To the Washrooms'; 'To the Parking Garage.' A false door is hammered to the wall, and on the door is a sign indicating 'Station Master.' "[62]

On Franz's orders, the prisoners mounted a large clockface high above these fake signs, window frames, and doors along the railway platform. This "time-piece" was not one of the four hundred thousand plundered gold watches and clocks that Herr Stangl delivered from Treblinka to Berlin during the months between October 1942 and August 1943—real ticking clocks of all kinds capable of keeping an empire running on schedule.[63] Nor was it an alarm clock, such as the one Treblinka's SS forced a prisoner named Julian to wear around his neck, along with a cantor's garments, to ensure that no prisoner took longer than three minutes in the latrine; they called him the *Scheissmeister*, or "privy-pit boss."[64] Instead, the railway station clock was a painted clock with numerals and hands that never moved—not a clock per se but the hollow suggestion of

FIGURE 12.2. **An artist inmate preparing misleading signs at Treblinka. In December 1942, Deputy Commandant Kurt Franz ordered an imprisoned painter from Warsaw to create false signs, symbols, and a clockface designed to lend a veneer of order and normalcy to the extermination camp. The name of this painter is lost to history (thus far).**

Source: Samuel Willenberg, *An Artist Inmate Preparing Misleading Signs at Treblinka*, 2001, bronze sculpture. Photo courtesy of Sławomir Kasper, Institute of National Remembrance, Warsaw.

one.[65] Memories of this painted clock splinter across the traumatic accounts of survivors and subsequent reporting. Strawczynski, for example, recalls it being seventy centimeters in diameter.[66] In Samuel Willenberg's sculpture (see fig. 12.2), the hands are frozen at four o'clock. Alfred Spiess, a prosecutor at the Treblinka trials, described a painted clock that "always pointed to twelve o'clock."[67] The novelist Jean-François Steiner tells of "a clock face with hands, painted on a wooden cylinder twenty-eight inches in diameter and eight inches thick," reading three o'clock.[68] Still others recall that it "was permanently set at six o'clock."[69] Some claim that Kurt Franz's painted clock "stopped time in Treblinka" or that the camp was a "place without time,"[70] but such aphorisms either misunderstand or obscure for sake of poetics the purpose of this set piece within a larger scenic deception—one undertaken to "beautify" the killing center with the superficial comforts of a home away from home.

HOW TO "BEAUTIFY" A GENOCIDE

It might be tempting to dismiss these visual euphemisms as mere camouflage, to lump them in a group of other Nazi deceptions used to confuse, misdirect, or mollify those not in on the ruse. After all, the SS would later employ similar tactics to make it appear that Theresienstadt, a concentration camp for Czech Jews en route to Auschwitz, Sobibór, and Treblinka, was in fact a thriving city run by and for Jewish people. The SS called this their *Verschönerungsaktion*, or "Beautification Operation."[71] In December 1943, a cadre of Nazi filmmakers filmed the prisoners there, forcing them to act in a propaganda film they called *Theresienstadt: A Documentary Film from the Jewish Settlement Area.*[72] When Dr. Maurice Rossel, a Swiss physician from the International Committee of the Red Cross, visited Theresienstadt in July 1944, he reported that "the nurseries, children's homes, pavilions and kindergartens are very well built, while a specialist kitchen prepares food for the little ones."[73] Rossel, clearly duped, reported seeing theaters performing Shakespeare, orchestras, an opera, sporting events, and "all the professions we can imagine in a small town . . . the police force, detectives, a corps of firefighters, . . . butchers, bakers."[74] Later, in an interview with Claude Lanzmann, Rossel recalled how thoroughly he had been played, adding that they had put up colored signs with arrows "to the bank," "to the post office," and "to the café."[75]

But the "beautification" of Treblinka appears to have served more than one purpose. Rather than providing a stage set for propaganda films or hoodwinking visiting officials, these structures aided and abetted a kind of inward-facing propaganda for both the guards and the arriving prisoners in radically different ways. Signs and symbols can, after all, be read in different ways by different people, and their meanings can change over time.[76] For the Nazis, there was a powerful ideological motivation tied to the Reich's efforts to euphemize ghettos, labor camps, and extermination camps as "cities." Christian Wirth (also known as Christian the Terrible, who oversaw the death camps under SS *Brigadeführer* Odilo Globocnik) traveled from Lublin to Treblinka to announce to the guards and prisoners that the camp would be one of five cities "built for Jews . . . one of those cities where Jews will live"—knowing, of course, that it was where they would die.[77] And indeed, once the Jewish construction group had finished placing a Star of David atop one of the gas chambers at Treblinka, as survivor Yankiel Wiernik recalls in his memoir, the *Hauptsturmführer* said to his subordinates, "Jew town has at last been completed."[78] Stanisław Kon, a Jewish civil engineer imprisoned at Treblinka, testified that on the gas chamber's outer wall, below the star, officers ordered the inscription of yet another euphemism: *Judenstaat* (the Jewish state).[79] The SS even mounted a curtain atop one gas chamber entrance, pillaged from a nearby synagogue, which read in Hebrew, "This is the gate through which the righteous may enter."[80] Giving material form to the camp-as-city euphemism enabled the commanders to dress up their own genocidal activities for themselves in a way that was comforting and made sense within the Nazi ideological frame. This is particularly clear in Kurt Franz's efforts to beautify the camp's administrative living area (often referred to as the "lower camp" for its lower elevation).

In the weeks following the construction of the fake railway station, Franz—whom the prisoners nicknamed Lalka (the Doll) on account of his soft facial features—fixated on the fabrication of normalcy. He ordered imprisoned craftsmen to erect numerous additional signs, as well as storefronts and attractions throughout the lower camp: a bakery, a zoo, and a tailor shop for SS guards. Oskar Strawczynski characterizes this as Franz's personal obsession—that, "for Lalka, all this is still not enough. He is constantly searching among the new arrivals for engravers" to carve more signage. First, a signpost directed arriving deportees "To the Station," engraved with a campy illustration of bearded and bespectacled Jews dragging their belongings to the train station. Another sign pointed "To the Livestock," engraved with cows, hens, and a shepherd. And yet another depicted Jews carrying shovels and pickaxes, directing arrivals "To the Ghetto." The SS bakery even featured a decorative crescent roll hanging at its entrance.[81] If Franz and the other SS officers were, according to the survivor-historian Eugen Kogon, *gescheiterte Existenzen*, or "failed existences,"[82] then they were at least successful internal propagandists, highly motivated to make the camp-as-city euphemism come alive.

Survivor Isadore Helfing describes Franz's renovated lower camp as a veritable town square that featured—all inside the camp's barbwire perimeter—a laundry for Ukrainian and German guards, "a shoemaker, a tailor, a little quarter with all the shops, and a blacksmith, you know to put shoes on the horses."[83] These included a functioning German barber shop adorned with "three beautifully polished copper plates," a dentist's office (for guards) with a large sculptured molar above the doorway, and a poultry enclosure with what Strawczynski identifies as a "beautifully carved wooden rooster." Franz even ordered one of the condemned to sculpt a stone frog fountain for the modest zoo,[84] which was completed in early 1943 and featured rabbits, foxes, squirrels, a pond for ducks, and a garden as entertainment for guards.[85] Steps away, Franz saw to the installation of what he called a "relaxation area" with straw sunshades for his SS comrades. All these comforts were only a couple hundred meters from the barracks where enslaved Jewish laborers, so-called hair cutters (*Friseurs*), were ordered under penalty of death to shave off women's hair before they were taken to the gas chambers.[86] At Treblinka, Jewish hair—shaved, then packed in sacks and shipped by the trainload—would ostensibly be used to insulate German submarines and, at Auschwitz, to fill German mattresses and palliasses.[87] This was just across the roll call square from the burial pits, where the so-called *Dentisten* were knocking and pulling gold teeth and dental bridges from Jewish corpses using pincers or a small hammer and a bowl of water.[88] It was steps from where arriving prisoners were made to hand over their documents and belongings as a "deposit" and across the camp from where bodies would burn by the thousands on ladder-shaped litters. When they wouldn't burn, the *Knochen-Kolonne*, or "Bone Brigade," would crush them with "iron mallets used to pound gravel on motorways."[89]

Keep in mind that many of the banal structures in the living area had their twisted counterparts in the extermination area, where the gas chambers and burning pits were located. The SS dentist was in the living area (the lower camp), and the *Dentisten* (tooth pullers) were in the extermination area (the upper camp).[90] The SS barber shop was in the lower camp, and the so-called *Friseurs*, or "hair cutters," were in the upper camp. A large SS bathhouse was built in the lower camp; the gas chambers fitted with unplumbed sinks and shower heads were in the upper camp. The SS zoo was in the lower camp; the barracks for male prisoners were halfway from there to the extermination area (see fig. 12.3). And we can see euphemism and dysphemism deployed to do ideological or category work at the level of spoken language too. Kurt Franz—a man who titled his private photo album of Treblinka *Schöne Zeiten* (Good times)—repeatedly reversed animal and human, literally as well as figuratively: he was known to sic his Saint Bernard on Jewish prisoners for entertainment, shouting "Man, grab the dog!"[91] Other officers and uncommissioned Ukrainian guards followed suit: "If not, you will be whipped, you accursed dogs!" "You dog, the bundle is too small!" Such humiliations, according to Stangl himself after the war, served to condition those charged with implementing Nazi policies "to make it possible for them to do what they did."[92]

FIGURE 12.3. General plan of Treblinka extermination camp. The layout reveals how structures in the lower camp (living area) often had corresponding counterparts in the upper camp (extermination area), reflecting a systematic organization that paralleled the ideological distortions embedded in Nazi language.

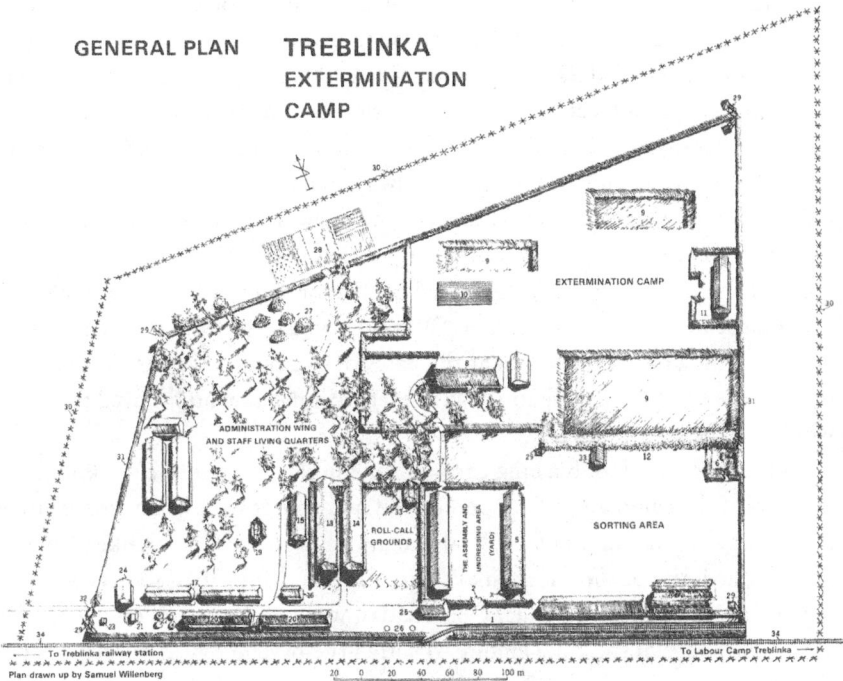

Source: Samuel Willenberg, *General Plan: Treblinka Extermination Camp*, 1981, pencil drawing. Reprinted with permission of Ada Willenberg.

If Jews were to be dogs at Treblinka, if extermination camps were cities, if gas chambers were showers but also a Jewish state, if a cantor was a *Scheiss-meister*, and if the entrance to all of this was a passenger railway station, then euphemisms ceased simply being words to soften reality—they became instruments of terror to disguise and to rationalize atrocities occurring in plain sight. So how, then, should we understand the relationship between words, signs, symbols, structures, and sight screens in this veritable hell? In a top-down power structure, those at the top can develop euphemisms to name a concept, as when Treblinka's SS would refer to Jewish children as "trinkets" or when the SS at Mauthausen named their rubber whip *der Dolmetscher*, "the Interpreter."[93] Violence, it would seem, is a language that cuts across all dialects.

But in that very same power structure, euphemisms can also surface from below, as when prisoners named the path to the gas chambers "the Tube" or called exhausted and emaciated prisoners living at the edge of death *der Musel-mann*, or "the Muslim."[94] From above or below, euphemisms can be brought to bear on a concept to rename and reframe it as though it were a different object—as when Auschwitz prisoners metonymized their nicotinic overseer Ludwig Plagge *Fajeczka* (the Pipe) or when Treblinka prisoners called one SS *Unterscharführer* "the Whip," on account of his gift for brutal beatings.[95] When euphemisms fall like blows from above, those in power are able to recon-struct an idea, an object, or a system in ways that fit their new labels: objects can even begin to take on some of the qualities implied by the euphemisms applied to them. In this formulation, railway signage helps fulfill the fantasy of a death camp as a train station; faucets and shower heads feed the fiction that gas chambers are bathhouses. Calling a gas chamber a "Jewish state" affords a Star of David, much as calling Auschwitz a "labor camp" afforded a sign on the gate that read *Arbeit macht frei*, or "Work sets you free." We can even question whether this wrought iron sign at the gates of Auschwitz was an accommoda-tion for the SS officers there too—Were there SS officers who believed that work would set them free?

Where the stench of burning corpses hung in the air twenty-eight kilometers away from the camp and where bodies and body parts were often strewn about the railway siding upon arrival,[96] false signs and symbols, much like alcohol,[97] anesthetized atrocity for Treblinka's perpetrators. More than to deceive arriv-ing deportees (and them alone), Obermajdan was an improvisation undertaken at the periphery of the Nazi empire to soften the psychological burden of com-mitting murder every day. The barber shop, the zoo, the bakery, the relaxation area, and the painted clock facilitated the kind of evasive thinking that could insulate these men and women from the full force of the conditions—and impli-cations—of their dirty work.[98]

PERPETRATOR TERMINOLOGIES AND
THE EUPHEMIZING OF ATROCITY

Euphemism also infiltrates public memory. When historians speak of mass executions inside Nazi killing centers, harmful softening, displacement, and misdescription concerning atrocities abound. We too often hear of the Nazis' "liquidation" of the camps—an abuse of language that perpetuates the Nazi conflation of human beings with cargo, cash, or debt. We hear of "concentra-tion" and of "entrainment" as the means by which Hitler's "Final Solution" was carried out—euphemisms that, through understatement and misdirection, dis-tance us from the slaughter of millions of people.[99] We too often hear of time-tabled freight cars—the twenty-ton *Güterwagen* (goods wagon)—dysphemized as "cattle cars" and choose to ignore that many of the trains arriving from

Western Europe were luxury trains complete with sleeping cars and dining cars that were—in the words of Yankiel Wiernik—"well supplied with fats" (food, candy, coffee, tea).[100] We hear such trains described as "Jewish transports," as though the boxcars were a Jewish creation, and we still see their inhabitants referred to in earnest as "human cargo"—a euphemism that Franz Stangl himself used even after the war.[101]

Auschwitz tour guides today still refer to the warehouses of plundered belongings as "Canada I" and the "Sauna," and use the perpetrators' slang, "Mexico," to refer to the thirty-two barracks where the SS confined thousands of Hungarian Jewish women.[102] We hear of the "liberation" of the camps—not prisoners—as if those killed by Nazi horrors were freed from the trauma, despair, enslavement, and death. We read of the "euthanasia" carried out under the T4 program, repeating the Nazis' abuse of a medical term that is supposed to mean "a good death."[103] We hear the spectrum of these atrocities summarized as a "holocaust" (literally a "total burning")—if not a euphemism, then at least an oversimplification for what in fact entailed beatings, hangings, drownings, quasi-medical tortures, extrusions and extractions of all life from the men, women, and children who suffered in these "camps." A *Lager*, after all, is not strictly speaking a camp but a storage facility, a warehouse, or a depot.

This is not a proposal to police the language of atrocity. Would anything be more fascist? My point here is that these euphemisms, many of which are inherited from the perpetrators themselves, highlight the risk of rhetorical capture common in the historiography of all atrocity. It is difficult to find a language by which to speak of atrocity in a way that does not repropagate the distancing that these figures of speech were deployed to perform in the first place. Referring to the thirty-year struggle for Irish civil rights as "the Troubles" or to the hundreds of thousands of women forced into sexual slavery by the Imperial Japanese Army as "comfort women" is to drain significance away from such traumas. The problem these euphemisms pose for our public memory of atrocities is that sacrificing specificity for poetics, cryptonymity, brevity, or respect makes it easier to ignore reality: the unremarkable presence of these terms in discourse keeps their meanings camouflaged, beyond our reach, where perhaps it is believed they cannot inflict further harm.[104]

So what is the agnotology of atrocity? It is the playful decoration of the drudgeries of violence: veneers of normalcy, Tangos of Death,[105] and other such theaters of the absurd. Much as Nazi euphemisms serve to undescribe reality, the agnotology of atrocity is more than a gloss or ideological smokescreen—it is the mixed bag of strategies for the creation of ignorance both inside and outside the bureaucratic chains of command that make organized harm possible at scale. What other atrocities lie hidden beneath layers of gravel, soil, and lupine? What false fronts lured hope along the paths of "Indian removal" in the United

States or even still reassure the families of Australia's "Stolen Generations"? Given all we know about such atrocities (and all we don't know), is it rational, or even possible, to use perpetrators' terminologies to dismantle—or at least to better understand—the very houses these euphemisms helped construct?[106] How can we find ways of discussing the atrocities we now face without building barbwire in the brain? I think it is too much to say that such ways of thinking, feeling, and speaking erect concentration camps in the mind.[107] But perhaps we find ourselves inside a fake railway station en route to just such a place—a waiting room filled with all the balmy reassurances of the banal. To speak plainly about atrocity, we will need first to tear down the fake signs that have been erected before us.

THIRTEEN

Hiram Powers, Black Agnotology, and Segregated Art History

CAROLINE A. JONES

ART HISTORIANS WHO STUDY the nineteenth century acknowledge the disproportionate representation of powerful interests in the art of that epoch. Landscapes of the colonizers and portraits of the powerful can simply seem like "the way it is," but that's why we need the concept of agnotology. This essay aims to bring the study of ignorance to art history as a discipline, addressing its cultivated unknowing about the reception of a nineteenth-century sculpture, famous in its time: Hiram Powers's 1844 marble *The Greek Slave*.

The New England transcendentalists offered a beguiling 1841 thought that "sculpture is the pause of art in the swift current of the life of nations."[1] But if we pause in this way with Hiram Powers's *The Greek Slave*, it must be to avoid the actual system of slavery of the time (a *very* swift current in the nation in the 1840s). Nor can any work of static visual art capture the lived reality of enslaved people in our U.S. republic or the continuing production of "diabolical" doubt about what happened.[2] Vermont sculptor Hiram Powers's *The Greek Slave* (fig. 13.1) is the object at the center of this essay because it exemplifies how the edges of actual slavery could be fudged and fuzzed, veiled with White ignorance.[3]

FIGURE 13.1. Dramatically lit photograph of Hiram Powers, *The Greek Slave*, 1844. This first version in marble came to the private collection of the twelfth Lord Barnard of Raby Castle, where it now resides.

Source: Hiram Powers, *The Greek Slave*, 1844, marble. Courtesy Raby Castle, County Durham, United Kingdom © Raby Estates 2023.

This story of art history's agnotological blindness comes via *The Greek Slave* at London's exemplary Great Exhibition of 1851, where the sculpture was the star of the U.S. section. Under the name of one nationality—"Greek"— Powers's sculpture was exhibited as exemplary of another—"American." This took place within Britain's imperial theater of the first world's fair. *The Greek Slave*, once the most famous work of American art, has since been consigned to the shadows of art history. Modern art enthusiasts are embarrassed by its old-fashioned, figurative, sentimental, perhaps even kitschy, aspects. As one such modernist, I wasn't tracking it through the fairs out of admiration but from a sense of duty to the history of art's global circulations. What had once made this "Greek" slave so compelling in the mid-nineteenth century? And, as my twenty-first-century questions sharpened, why was its Black reception so unaccountably obscured?

The particular *Greek Slave* installed at the Great Exhibition was displayed in a velvet-lined niche on a pedestal about four feet high that could be rotated for a fuller view. She had been carved from flawless Seravezza marble to Powers's specification (and with his patented finishing rasp by which he wanted to avoid slick "polish" in favor of skin-like pores). Skilled Italian craftsmen did much of the work roughing out the stone, in faithful obeisance to the artist's plaster model (then housed in his Florence studio), which had metal "pointing" indicators drilled into the plaster surface to guide their work.[4]

Who is this Greek slave? Neoclassical in style but "modern" in subject, she holds a crucifix intended to indicate she is a Christian maiden, enslaved after being taken captive during the Greek War of Independence (1821–1829).[5] Stripped naked, she was put on the block by her nefarious Turkish captors for the titillating commerce in so-called "white" slaves (a color but readily denoting White as a racial category). The racialized framing beckons agnotology already since the sculpture ignores the far more dominant trade in *Black* slaves at the time.[6] Echoing the "modest Venus" (*Venus Pudica*) of classical Greece, Powers's white slave modestly endures the gaze of her purchasers (as well as our own prurient peeking). Our gaze is cleansed by virtue of her virginal religiosity. Our ignorance is sealed with a pious sigh.

This is highly cringe worthy more than a century later. But in its day, *The Greek Slave* was the subject of extraordinary praise and fame. Take, for example, the admiring 1850 sonnet "Hiram Powers' Greek Slave" by Elizabeth Barrett Browning, in which the slave's "Ideal beauty" is tied to tropes of Whiteness that even shadow cannot darken.[7]

> They say *Ideal beauty* cannot enter
> The house of anguish. On the threshold stands
> An alien Image with enshackled hands,

Called the Greek Slave! as if the artist meant her
(That passionless perfection which he lent her,
Shadowed not darkened where the sill expands)
To so confront man's crimes in different lands
With man's ideal sense. Pierce to the centre,
Art's fiery finger! and break up ere long
The serfdom of this world. *Appeal, fair stone,*
From God's pure heights of beauty against man's wrong!
Catch up in thy divine face, not alone
East griefs but west, and strike and shame the strong,
By *thunders of white silence*, overthrown.[8]

It is precisely those "thunders of white silence" that I want to summon—but not exactly the triumphant ones that Browning had in mind. Clearly, *The Greek Slave* had been training its Victorian public for years before the Great Exhibition opened, instructing White art lovers in how to view a female nude, inspiring them to transcend lust in order to empathize with a virtuous, evidently Christian girl exhibiting the stoic demeanor of a democratic people in duress.[9] Of Black views and viewers we hear nothing—it is *this* White silence that I want to address.

The Greek Slave got around. The second marble version was trundled down the Eastern seaboard of the United States beginning in 1847—tours managed for Powers by an artist friend, Miner Kellogg. They produced a pamphlet compiling the encomia flowing from the pens of poets and the press to accompany these appearances. The sculpture's zigzagging itinerary has recently been mapped by art historian Martina Droth in a fascinating interactive web publication. Marshaling drifts of documents about the tours (press clippings, poems, engravings, and other evidence), Droth insists that we reexamine the materiality of the sculpture itself, asserting that such nineteenth-century viewing conditions (preceded by subscriptions by art unions and ticketed admission to displays) were physical encounters that produced specific kinds of embodied receptions. She also rightly surfaces the occasional conjunction of the sculpture with humans on display—and for sale.[10] To this essential juxtaposition, I will return.

The *Slave* was so successful that it was quickly replicated in variants both midsize and tabletop, in plaster, clay, and metal, made in Germany's foundries in dark-bronze versions as well as poured into molds as white "Parian marble" (high-fired ceramics) in England and France. So popular was Powers's motif as a collectible—and so variable the quality of the knockoffs—that Powers tried to patent the very design itself (because there was no other protection, since visual art as yet had no copyright).[11]

Yet what I am here to report is that despite the work's fame, its countless replicas, and its tours through cities with active slave markets, there was near total silence within elite cultural discourse of the time regarding the obvious connection between this *Greek* slave and the *African* ones at the shameful heart of the U.S. economy. Moreover, my own education as an art historian never included the evidence for Black viewers of the sculpture, nor did graduate school readings investigate its specific uptake by varieties of abolitionists despite my mentors' firm commitments to a social history of art. I am making distinctions, then, between the work's cultural celebration by White abolitionists (as a condemnation of all slavery), such as Browning, and the *specific* attachment of Powers's work to the odious American institution by Black abolitionists of the time.

To be sure, the thunderous art historical White silence on this matter has recently been punctured by hints and revisions.[12] To my reading, these only confirm that we must take up agnotology for an expanded account, particularly inquiring after Black agnotology as a tool for counteracting White ignorance.[13] Specifically, I want to make agnotology both polyvocal and historically situated, positing a particular Black agnotology *from the period in question*, using this heuristic to understand the operations of historical actors. If White interpreters haloed Powers's sculpture through high-cultural allusions (a Whitening discourse), others with African forebears and abolitionist intent forced it into an agonistic contrast with specifically American forms of enslavement (a Black retort, which I am labeling "Black agnotology").

The art historical reticence begins with that pamphlet of "authorized" but anonymous responses to the *Slave*.[14] Although the publication offers a rich palimpsest of authored texts, Powers and Kellogg simply acknowledge the ineffable qualities of the sculpture that "must be seen to be appreciated." The compilers keep themselves distant from explicit links to abolitionism—this was key to their goal of achieving positive reception for the sculpture among elites in slave states. Ironically, while the eminent scholar Droth at Yale has punctiliously reestablished these viewing conditions, she shades into a White art historical operation by suggesting that the linkage of the *Slave*'s iconography to abolition debates is a "dematerialization" of the sculpture that merely turns it into an image or illustration.[15] Respectfully, I suggest that Black agnotology would see the sculpture's meticulous fashioning as the very *materialization* of White ignorance. From the choice of marble to the shaping of shackles, material *is* narrative here. Exclusions or elisions of the American slavery debate from the art historical literature on Powers and the reception of his most famous sculpture is *part of what we now have to understand*. If Droth finds the statue "instrumental to its own visibility," I want to see the marble as complicit in actual slavery's *invisibility*. The two come in together. The white sculptural body contributes to the Whitening of art historical discourse as a whole, an echoing monochrome silence around the abolitionist reception by free Black people of Powers's *Slave*.

AGNOTOLOGY FOR POWERS'S SLAVE

Let us return to that first carved marble version made by Powers in 1844 and borrowed from its British collector for the U.S. pavilion at the Great Exhibition in 1851 (it now resides in the aristocratic Raby Castle). We can consider exemplary both this first version of the sculpture and its display condition within a nationalist U.S. context at the first world's fair.

None of the Powers literature that I was taught in the late 1980s addressed the impact of this work on diasporic Africans in the communities where it was shown, either in Britain or the United States. This classic case of apparent archival absence—at least in the discipline of art history—allows one to imagine simply that no such people encountered Powers's sculpture (or, worse, that no such people existed). Perhaps the discipline wishes us to imagine there were no African diasporic responses to this once famous American sculpture that is indisputably about the "peculiar institution" of modern slavery.

Such an account began to seem unlikely as I returned to the scholarship on Powers around 2010. What I found at that time needs to be framed as a case of *segregation* in academic and curatorial art history, calling for agnotology as the study of how ignorance is arranged and produced. Semantics of our agnotological distinctions are important. "Ignorance" by itself is innocent. By dubbing art history's White silence as segregation, I am invoking *racialized* ignorance, no longer entirely innocent but part of a pattern determined by implicit bias. On this account, Rawlsian philosopher Charles W. Mills, in the first volume on agnotology edited by Robert N. Proctor and Londa Schiebinger, *Agnotology: The Making and Unmaking of Ignorance*, and again in Mills's later book *Black Rights / White Wrongs*, provides a gloss on how "white ignorance" works to protect itself from knowing the facts about the systems of racialized oppression that secure its own privilege: "*White ignorance* [is] an example of a particular kind of systemic group-based miscognition. . . . [White ignorance] plays itself out in the complex interaction of Eurocentric perception and categorization [and functions through] the derogation of non-white testimony."[16] As I took up Mills's practice in favor of what I now want to call Black agnotology, I had to identify myself as a White subject of the segregation that art history performs. I held the unconscious privilege that comfortably shielded my White ignorance from close scrutiny. The present chapter can thus be described within the scandalous vocabulary of eighteenth- and nineteenth-century slave narratives upended: these are the confessions of a White academic attempting to understand, via historical methods of Black agnotology, the segregation of her own education, opening out structures and operations hidden within the field of art history as a whole.

Agnotology often begins its sleuthing with an empirical leak that hints at the territory occluded by not knowing. Very late in my own research for a 2016 book that tracked art in global contexts, I happened to attend an art history talk by Lisa Volpe, a museum curator from Kansas who had just finished her dissertation on abolitionist photography. The reason I was interested in the talk was her promise to speak about photographs shown at the Great Exhibition. I had already known a little bit about how Powers's *Greek Slave* had been mocked by the satirical magazine *Punch* during the Great Exhibition both in an 1851 caricature by John Tenniel that depicted the dark-skinned *Virginian Slave* (*Intended as a Companion to Power's* [*sic*] *"Greek Slave"*) and via an earlier squib in that journal questioning, "Why have [the Americans] not sent us some choice specimens of slaves? We have the Greek Captive in dead stone—why not the Virginian slave in living ebony?"[17] What Volpe described in her talk, however, was something far more significant than this well-known framing by White British satirists in the popular press. What she taught me had been radically excluded from art history up to that point—at least, *my* art history.

Drawing on extensive discussions in the discipline of African American studies and historical abolitionist texts, Volpe recounted how, on Saturday, June 21, 1851, under the "pasteboard eagle" of the U.S. display at the London Great Exhibition and adjacent to the costumed mannikins of Native Americans, a small group assembled: African Americans and White British abolitionists making common cause with these self-emancipated escapees.[18] The group proceeded to stage an educational pantomime around Powers's statue. There is no visual record of the performance, which is instead deposited in retrospective accounts by formerly enslaved Black abolitionist performer William Wells Brown and his White British abolitionist ally William Farmer. According to Farmer's account published in the Boston abolitionist newspaper *The Liberator*, the assembled party chose a particularly busy time to stage their intervention in front of Powers's sculpture. It is tempting to imagine that a particular engraving in the *Illustrated London News*, published the day after Farmer's narrative appeared in *The Liberator*, provides a bird's-eye view of the Brown-Farmer group as they had perambulated through the Crystal Palace a month before.

Performance studies scholar Lisa Merrill, who reads this illustration as showing mixed-race publics at the fair,[19] is clearly doing what visitors to the exhibition did in person, scanning the crowd to parse the many differences and ethnicities on display. As Brown himself put it in his memoir of this time in London, the Great Exhibition was "one great theatre, with thousands of performers, each playing his own part."[20]

In Farmer's write-up for *The Liberator*, it was clear what the group in-

FIGURE 13.2. Detail of July 1851 engraving in the *Illustrated London News* showing what appear to be mixed-race visitors to the Crystal Palace.

Source: https://www.britishnewspaperarchive.co.uk/titles/illustrated-london-news.

tended: "Side by side with [U.S.] specimens of cotton, sugar and tobacco, ought to have been placed the human instruments of their production." Let me note how the never-enslaved Farmer worked to deploy the now-free but formerly enslaved Black abolitionists to "place" the appropriate White agnotological argument before the public (and arrogates for himself the speech about that act). In *The Liberator*'s pages, Farmer argued that since the British had displayed their national products along with the proud factory men who had made them, shouldn't the Americans have shown "some specimens, not merely of hams, locks, revolvers and firearms, but of the more peculiar staple produce of America—Slavery"?[21] He suggested with sarcasm that doubtless any displayed slaves' evident happiness—much touted by apologists for the peculiar institution—would have shown the "white [wage] slaves" oppressed by British factory production how much worse off they were than their chattel analogues across the sea.[22]

What can we reconstruct of the performance itself? We know that the *Greek Slave* display was chosen as the focal point of the group's intervention. Powers's marble statue was on a pedestal and inside a velvet-lined niche but one that allowed visitors to *rotate* the statue. Implicitly, this was an invitation to participatory spectatorship that art history has failed to address.[23] Farmer described that Brown placed the Tenniel engraving of the *Virginian Slave* from *Punch* directly in this niche, loudly proclaiming to startled passersby, "As an American fugitive slave, I place this Virginia Slave by the side of the Greek Slave, as its most fitting companion." Perhaps we can imagine it propped up against the slave's marble shin, a small flat black figure—almost a silhouette—placed well below the white marble crucifix and chains dangling above. Standing alongside Brown during this declaration was the rest of the group, linked arm in arm.[24] Among them was a well-dressed couple of color—none other than fugitives Ellen and William Craft, who remained silent (according to published accounts). Ellen was as close to a celebrity as the abolitionist community in Britain then possessed (Frederick Douglass's two-year sojourn had ended in 1847). Her notorious escape from Macon, Georgia, in December 1848 had involved her dressing up in men's clothing, covering her hairless chin with a bandage, and pretending to be the injured White owner of her darker-skinned husband, William Craft. The continuity with racialized *performance* is significant here—an activated agency that art history of the nineteenth century doesn't know how to handle.

Craft's performative code switching allowed her to alter class, gender, and racial identity. All were alluded to in a portrait engraving of her transvestism that cemented her sensational fame.[25] Based on an outfit she seems to have worn during her time in Scotland to evoke her escape costume, the well-known image (based on a daguerreotype) shows her wearing a top hat, a cravat, and a tasseled Scottish plaid sash that parallels the discarded "bandage" (from her escape) worn over a shapeless overcoat. Reproduced in many forms, this portrait appeared in the *Illustrated London News*, was offered for sale at the abolitionist gatherings at which she appeared, and formed the frontispiece to the Crafts' own book of their adventurous escape. The widespread circulation of this visual imagery also performs Black agnotology as a shadow narrative to the *Greek Slave* aesthetic of womanhood. Against the white marble maiden's surrender to nudity, the outrageous portrait of Ellen Craft in manly garb is a recollection of rebellion and performative agency: the misery of enslaved people will drive them to extraordinary lengths to leave the odious condition of bondage (with Craft muting herself literally and figuratively to justify why her husband, William, was the only one who could speak during the perilous journey to freedom).

Did Craft appear in this costume at the abolitionists' intervention at the fair? More likely, she would have been dressed in proper female attire (like those bonneted ladies perambulating in the *Illustrated London News* in fig. 13.2). Nevertheless, she was a key player in the pantomime. Her performances with

William Wells Brown were already well honed from travels around the United Kingdom in which she and her husband appeared as living exempla of the horrors of slavery. Since proper women didn't speak in public, William Craft narrated the details of their harrowing escape while Ellen stood by in silent modesty, eerily mirroring the muteness required during their escape. In these public pantomimes, Brown's role was to explain the features of an elaborate twenty-four-scene canvas panorama he had commissioned in 1850 to be set up behind the couple. The complex semiotics of Ellen's pale-skinned appearance set within this White abolitionist theater have been elegantly analyzed as "the Negro exhibit," referencing decades of scholarship by Houston Baker, Lerone Bennett, Saidiya V. Hartman, and now Lisa Mellin and Teresa Zackodnik (none of them, I note, art historians). While her skin was to suggest the complexities of her ambiguous racial status, the important feature I want to emphasize here is Ellen Craft's *silence*—as silent as a statue.

Sometimes silence is an accomplice in the work of ignorance production, which agnotology must address. However, in this case, Ellen Craft's ambiguously Black (or mixed-race) silence was set up to break the White silence and ignorance of *The Greek Slave*. It was an agnotological rebuke. Even or perhaps especially when mute, Craft's decorous body would be enlisted to enact the most vivid lesson in this moral theater, standing as a comparative instance of the *Virginian Slave*—or, in this case, Georgian slave. As Volpe puts it, "Standing silent before the audience and with downcast eyes, signaling her modesty . . . Ellen embodied the 'tragic octoroon.'"[26] (Such period terms as "mulatta" and "octoroon" are themselves performative speech acts of White racialization.) The silent Negro exhibit, per Baker and Bennett, stirred White abolitionist sentiment even as it participated in what, for contemporary critics, is an almost pornographic circuitry. The racialized body, according to Baker, offered "an erotic sign of servitude in the social, liberational discourses of white abolitionists."[27] Baker, summoning what I am dubbing "Black agnotology," questions the motivations of White abolitionists themselves. Craft's own skin testified to the miscegenation that Southern slaveholders claimed to fear—a product of the rape culture they had promulgated within and as the violence of enslavement. Adjacent to Powers's otherwise unmarked production of *Whiteness* around the issue of slavery, Craft's presence would suddenly have made this theater of race and sexual violence obvious in its national particulars. Although visually revealed, it was further *announced* by these unruly actors through performative acts of embodied juxtaposition of gradations of Blackness—whose apogee *Punch* had already called for—"living ebony"—the closest to that condition being Ellen's husband, William Craft. By agreeing to put herself on display adjacent to Powers's sculpture of chaste white nudity, Ellen Craft intensified the already-erotic theater of slavery named by Houston Baker even as she forced

Powers's artificially segregated discourse to begin revealing its Whitening racial politics.

I am drawing on the historical evidence of an active agnotology, obviously not named as such, on the part of free Black people in 1851. There is evidence of the Crafts going back to the fair multiple times to hang around and *talk* with visitors to the U.S. display, particularly those visiting Southerners whom they hoped to trap in "recourse to lying" about the slave conditions they had personally experienced.[28] These historical traces of active Black agnotology consistently unveiled the mechanisms of White ignorance that were circulating around *The Greek Slave* and the entire U.S. display. I resurrect what was not then called "agnotology" for the purposes of my own present-day opposition to the disciplinary segregation of this history from *art* history. What readings become possible when this eventful agnogenesis is brought to bear? Certainly, *The Greek Slave*'s "thunders of white silence," for me, begin to be accompanied by Black performance, abolitionist speech, and mixed-race muteness in very complicated ways. Can we begin to understand what kind of *affect* was produced by the Crafts' performance, animated arguing with Southern visitors, and/or mute presence? Would nineteenth-century international viewers of *The Greek Slave*, coming upon its juxtaposition with the performative or exhorting bodies of escaped African American exiles, experience the sculpture differently than we do today? There is, of course, no single receptive field for knowing *The Greek Slave* in the context of her far more numerous African sisters; waves of interpretation include the present illustrated text by which I aim to produce something akin to Saidiya Hartman's "critical fabulation" around these sparsely documented events.[29] Just as the production of doubt or ignorance comes from many acts, knowledge splits into multiplied registers—for those then still enslaved, for those who were not, for those who had escaped, for those inheriting the intergenerational trauma of enslavement even today, and for segregated art history in this unfolding agnotological frame.

DISPLAY AND THE RACIAL MÉLODRAME

Breaking the White ignorance around *The Greek Slave* required me to move intentionally outside art history to African American studies, literature, performance, and media theory. Art history is capable of tackling the volatile display conventions of *The Greek Slave* but prefers the stable moments documented in prints and photographs. Can we hybridize art history's excellent deployment of *Rezeptionsgeschichte* (reception history) with other disciplines' more capacious embrace of performative dynamics?[30] We must. Take this indisputable fact: the full-scale marble *Greek Slave* was, in every setting, elevated well above the visitors' sight line. Powers controlled every display and ensured the virtuosity of the early versions of the carving itself, with their delicate marble links of a chain miraculously liberated from a single slab of marble.[31] Following these conventions, the sculptured nude is positioned as "high" art—not of our world.[32] Browning's sonnet touches on this, emphasizing the liminality of the display's "threshold" and "sill," producing this "alien . . . with enshackled hands." Whites with elite education would have viewed the marble material itself as a high-cultural material, connected to classical Greece (a semiotic of pure white marble, even if ancient objects of marble were in fact mostly Roman copies of "original" Greek bronzes). Notably, Droth argues that the sculpture's virtuosity and modern subject would hold allusions to the ancient world in tension. Disputing the dismissive "neoclassical" label and digging back into the contemporaneous (White) accounts, she notes, "The skillfully wrought chains, fully freed from the marble block, even in the narrow space behind the left hand, disrupted the fantasy [of a distant past]. As one commentator put it, Powers offered the 'highest idealized conception of female loveliness' only to 'pull down our fancy' back to earth: 'The chain and manacles, when the eye does steal a glimpse of them, produce strange contrarieties of feeling and emotion.' "[33]

Hold onto that brief mention of "manacles" in this surprisingly early (1853) account; I will come back to them. Suffice it to say that Droth's revisionary rescue of this note of contemporaneous affect ("strange contrarieties") was not present in Powers's own pamphlet. For African Americans, experiencing the aura of elevation and sanctification would be even more *contrary* to their experiences at the ticketed art venues where *The Greek Slave* was shown. How different, then, would have been the *tableau vivant* presented by Ellen Craft on that June day at the Great Exhibition. Methodologically, theater history and narratology contribute to a conjuring of Black reception and performative Black agnotology that chastise art history's segregation of these dynamics. Craft's active calling out of ignorance also upended the "racial optics" that Saidiya Hartman rightly called out two decades ago in her pioneering book *Scenes of Subjection*.[34] Hartman's quarry—the White reader deploying empathy to project White melancholy and erotics into the Black subject of the printed slave narrative (such as Brown's or the Crafts')—was, in 1851, confronted by the living juxtaposition of the Crafts, the flat print of Tenniel's, and the oratory of Brown against the pretentions of the marble sculpture. Black agnotology definitely maps out a complex emotional cartography. The fetishistically white Seravezza marble that Powers used is now brought down graphically by the lowbrow newsprint of the *Punch* engraving. The scurrilous Virginian paper doll slave also contrasts to the solemn silent figure of Ellen Craft, the (former) Georgian slave. Even the male orators fade away from the emergent visual tableau as the racialized mulatta (that odious period term) produces a silent but vivid triangulation among the Georgian, the "Virginian," and the "Greek" slaves.

In my attempt to critically fabulate this fleshy, mediatic, and material conjunction, I want to highlight how Black abolitionist performers were well versed in the silent pantomime and its variants in the static tableau. Film scholar Linda Williams has given us an analysis of racialized affect in U.S. literature and theater.[35] She identifies such displays as early instances of American *racial melodrama*—the form itself emerging from the French postrevolutionary *mélodrame*, constituting a new form combining music (*melos*) and theater to handle the hot politics of the moment. In the melodramatic tableau as Williams analyzes it, a charged composition of classed or racialized bodies, arrested at the very peak of theatrical action to highlight the cathartic climax, both announces the moral of the story and freezes time to allow affect to bloom at its own pace. Per Williams, American racial melodrama takes up this structure over centuries to organize national feelings around race, violence, and victimhood in history and at present.

The ingredients of the racial melodrama involved Black abolitionists as well as Whites. British abolitionist John Bishop Estlin was one of the men who accompanied Brown and the Crafts to that June performance at the Great Exhibition.[36] Estlin had earlier complained to a fellow White abolitionist that the Crafts were "excessively" theatrical in their handbills and traveling show. Writing to a Miss Wigham in Bristol on May 3, 1851 (a month before the *Slave* demonstration), Estlin confided that he and others "have been endeavouring to improve the tone of Brown and Crafts' *Exhibition* altering their too *showman-like* handbills."[37] What I wouldn't give to see the visual evidence of that excessive melodrama and those showy handbills!

We have something like a handbill that is associated with Ellen Craft from her own letter to Estlin from 1853 (fig. 13.3). It was sent from Ockham, England, on March 14, 1853, in a floridly illustrated abolitionist envelope. There is nothing in Craft's letter that mentions the decorative tableau—she merely thanks Estlin for urging that her baby get vaccinated and wishes good health for him and Mrs. Estlin in an impeccable copperplate hand.

I speculate that the two images at the left and right of this handbill-like envelope would sort differently for a White abolitionist, such as Estlin, than they would for Black abolitionists, such as the Crafts. The image on the left is *not* "excessively melodramatic" for White abolitionists, focusing as it does on the righteous and charitable White manumission of the humble, supplicant Black person. This tracks perfectly with Josiah Wedgwood's decades-old "suppliant-slave" tokens of abolition circulating across the Atlantic and promoted by complex figures such as Benjamin Franklin (who began his career as a profitable slave owner but later tacked to abolitionism).[38] The iconography on the left also tracks perfectly, even in its left-to-right orientation, with the London edition of William Wells Brown's sensational *Panoramic Views* of his birth in slavery

FIGURE 13.3. Envelope printed with abolitionist tableau from the letter by Ellen Craft to John Bishop Estlin, 1853.

Source: Boston Public Library, Rare Books Department. Digital file courtesy of Digital Commonwealth, https://www.digitalcommonwealth.org/search/commonwealth:2z10z 7257.

and escape to freedom, published by White abolitionist presses just a few years before Brown joined the Crafts' performance at the Great Exhibition. Far different, however, is the dramatic scene going on in the tableau at the right of the envelope. Contrasting with the redemption of Black people by White lions of Britain is an action scene set in a Spanish-looking colony in which multiple forms of violence are being waged against Black enslaved bodies. Here, in unenlightened lands far, far away, whitish slavers wear sombreros as they shackle, whip, and brand African men. One African man lies senseless on the ground, presumably dead. A presumed African woman holding a baby looks helplessly on while a dynamically slanting slaver's ship rushes away from the setting sun. I suspect *this* is the kind of showman-like melodrama that Estlin was worried about in his letter. Stirring up White guilt is a delicate matter that must be managed (by Whites) in the theater of abolitionism. Estlin had a month to coach Brown and the Crafts before their trip to the Great Exhibition. Would they get it right?

The frozen stillness of Powers's neoclassical sculpture was a clue to the resolution of the melodrama's "right" configuration. Stasis was framed as an ideal of art by Johann Joachim Winckelmann in his 1764 history of the classical ideal in ancient art, which he held to be an art of "stillness and repose."[39] The

German critic's further emphasis on *"edle einfalt und stille größe"* (noble sim-
plicity and quiet grandeur) could justify the stasis of melodrama tableau as
linked to the "quiet" of classical (and neoclassical) sculpture. Both style and
subject of Powers's *Greek Slave* thus offered coaching in how a White European
audience preferred to deal with violent situations.[40] Such values guided how
Ellen Craft would appear at the fair: "The tableau was used theatrically as a
silent, bodily expression of what words could not fully say."[41]

But stasis also served Black agnotology, in my argument. Yes, the Crafts
might have needed to attend to Estlin's urging to tone it down. But by performing her forceful symmetry with the white marble maiden on that June day at the
Crystal Palace, the decorous Ellen Craft also broke the agnogenesis of Powers's
sculpture and its obvious misdirection about the American system of slavery.
The tableau as Ellen staged it would have been more complex than the White
abolitionist's savior narrative might wish. Craft's silenced body had already
liberated itself through transgressive transvestism. Nonetheless, that body, even
in gender-appropriate Victorian dress, would have manifested the unspeakable
trace of actual sexual violence in juxtaposition with Powers's implied future
ravishing. That is, *The Greek Slave* presents the *before* of virginity. This is
forced into comparisons with Tenniel's "Virginian" and Craft's "Georgian"
slaves. Tenniel's engraving caricatures the penultimate moment of (sexual)
bondage, while Craft's body provides the "after" of that defilement and the
miscegenated results.

If the racial categories of "Christian slave" defiled by "Moorish" captors is
reversed into the mixed-race body made by a Christian master defiling Craft's
(Moorish) mother, this is just one of the many loopings and inversions that the
melodrama's symmetrical staging performed.

THE SLAVE IN ART HISTORY / FOR BLACK ABOLITIONISM

The work of agnotology has led, in this chapter, to a quite specific query about
how ignorance is produced. First, the misdirection of *The Greek Slave* had
nineteenth-century viewers worrying about White slavery somewhere far away.
Second, knowledge of Black resistance to actual U.S. conditions becomes segregated from the sculpture's history (as does the sculptural narrative's obfuscation of these realities). That segregation begins by expunging certain forms of
reception from the history of this art, but such a segregation also echoes larger
procedures of White ignorance that art history enacts. On the one hand, segregation leaves Black performative agency out of art's history (bodies in time
are not a "medium" of art in the museological sense). Even as aesthetic theory
from Europe privileged stillness and formal beauty, it also chose to ignore the
politics and poses of the Crafts' action, no matter how still Ellen stood. The pat-

tern of this segregation of reception history is interpretable. Powers celebrated Browning's sonnet (White literary reception); no mention was made of Black performance.

There is a larger Whitening in art history that plays a role. The aesthetic of white marble that Powers fetishized had been dominant for centuries in Europe (and was brought to the New World in the form of white plaster copies of revered Greco-Roman sculptures). This was so dominant an aesthetic that the incontrovertible evidence of actual polychromy on ancient Greek sculpture and architectural reliefs was constantly being denied. After the Crystal Palace opened, British architect Owen Jones had to write an "apology" for the poly-chromy he put into the decorations, identifying the scholarly sources informing his "Greek" architectural color at the Great Exhibition! On this account, the fetishized whiteness of Powers's carefully selected Italian marble is analogous to the generalized Whitening that banished the Crafts' performance from art his-tory. Both are processes of an art historical "universalism" that eliminates both ancient Greek polychromy and an African Egypt, each forced into a thunder-ously silent White monochrome ideal.[42] Powers himself wrote Elizabeth Barrett Browning in opposition to color in sculpture, for it "detracts from the purity of the spirit revealed by the forms."[43] Art history would have no place for ac-tivated Black interpreters of Powers's *Slave*, just as it still has little room for a polychromed classical past.[44]

I elsewhere propose that the dark bronze multiples of *The Greek Slave* offer a repressed counternarrative of Blackness as marketed for the tabletops of White abolitionist elites.[45] Yale art historian Martina Droth continues to help open such questions, reminding us of other sculptors' period bronzes of *African* slaves; she works as a curator to install them in potent confrontation with *The Greek Slave*.[46] These new stagings are important, but as with this chapter itself, they are only the tentative beginnings of the necessary agnotological work. To open art history to the evidence for Black reception of *The Greek Slave* for future scholars, I want to spend the remainder of this essay on an investigation of one diminutive copy of the *Slave* that I first noticed in the background of a period photograph of the camp table in Frederick Douglass's study during his appointment as U.S. minister to Haiti from 1889 to 1891.

It is remarkable that Douglass thought enough of this sculpture to bring it to Haiti. This supposed "Parian marble" ceramic cast survives on display in the preserved family parlor of Douglass's Washington home on Cedar Hill (as it was at his death) in the traditionally Black Anacostia neighborhood of DC (fig. 13.5). By calling it "Parian ware,"[47] scholars are referring to the process for replicas first developed by curator of the Great Exhibition Henry Cole and later made proprietary by British firms such as Minton and W. T. Copeland, both of whom began to produce replicas of *The Greek Slave* (Minton in 1848;

FIGURE 13.4. Frederick Douglass at his desk in Haiti, sometime between October 1889 and July 1891.

Source: Photographer unknown, n.d., collection no. FRDO 3899, Frederick Douglass National Historic Site, Cedar Hill, Washington, DC. Courtesy of U.S. National Park Service.

FIGURE 13.5. The surviving statuette *The Greek Slave*, made by an unidentified firm ca. 1870s. The sculpture is made of high-fired ceramic, to which metal shackles (painted white) have been added, estimated 14–20 in. in height. From the collection of Frederick Douglass held at Cedar Hill, his final home in Washington, DC.

Source: The Greek Slave, n.d., collection no. FRDO 1678. Courtesy of Frederick Douglass National Historic Site, U.S. National Park Service.

Copeland in 1851). But to be honest, the label for this object cannot be definitive since no technical analysis has been done of Douglass's *Slave*.

The form of the small statue matches neither Minton nor Copeland versions. Sharpening the art historical puzzle about this particular piece (and how and when it entered Douglass's collection) are its unconventional manacles (the Douglass statuette has no fewer than five of these linked together). Art historian R. Tess Korobkin notes the shift in iconography from traditional oval chain links to these strange linear manacles. As she unearths in Powers's archives, these long links were used on later variants because they hewed more closely to what had actually been used on enslaved African Americans—a small truth Powers wanted following the conclusion of the U.S. Civil War: Korobkin writes of *three* manacles in the full-scale sculptures authorized by Powers (instead of the five we see in Douglass's replica): "Three long, straight, linked bars [are in] the last full-scale marble that Powers produced of *The Greek Slave* (1866, sixth version) and sold to Edwin W. Stoughton in 1869 (now in the Brooklyn Museum, New York). In a letter to Stoughton, Powers explained the substitution of 'regular manacles for the rather ornamental than real chain in former repetitions of the 'Greek Slave' as 'more to the purpose.'" Korobkin interprets the even more numerous manacles on the Douglass replica as significant: "The fact that Douglass owned this particular version indicates that he acquired the statuette in the 1870s, after abolition was an accomplished fact, and that he chose to own the version of *The Greek Slave* that was the most explicit in its reference to the slavery in America."[48]

How, then, does this detail relate to the Black agnotology I have been identifying in nineteenth-century Black abolitionist receptions of *The Greek Slave*? Korobkin's essay gives Douglass agency in selecting this later version of Powers's work as appropriate to a settled abolitionist accomplishment ("accomplished fact") during Reconstruction. There might be an agnotology of sorts in requiring more accurate U.S. manacles for this figurine from years of White abolitionism and its "universalizing" of slavery à la Browning.

But Douglass never had a smooth or settled relationship with White abolitionism, particularly not during Reconstruction, as examined by historian Robert S. Levine in his 2016 biography.[49] Indeed, Douglass made a public and radical rejection of the "suppliant-slave" tropes of White abolitionism that were still rampant in the 1870s. He did so in the highly public context of the monument to Abraham Lincoln paid for by formerly enslaved Black persons and unveiled as part of the official national celebrations of the U.S. centennial in 1876: Thomas Ball's Freedmen's Monument (now known as the Emancipation Memorial, located in Washington, DC).

Douglass agreed to give a speech on the occasion of the sculpture's dedication before a large audience, including Supreme Court judges and Reconstruction executives joining an exuberant mixed-race crowd. As befitting the occasion, Douglass praised the monument's very existence as the first public memorial to emancipation and the first public statue in the United States of a Black man. Ornamented with patriotic emblems, one depicting (slave-owning) George Washington, and featuring a Christlike Lincoln, Ball's memorial actually appalled Douglass. Privately, he complained about why President Ulysses S. Grant, as engineer of the actual Reconstruction of American governance, was not included. Publicly, Douglass used the event of the unveiling to frame Lincoln squarely as the ambivalent leader that he was: "the white man's President."[50]

Douglass would turn directly to Black agnotology a few days later. As he addressed the actual iconography of Ball's monument in the *National Republi-*

can, he chastised Americans' White ignorance and asked his readers to consider the deeply conservative politics manifested in Freedmen's Monument, which recycled the century-long trope of the Christlike White man as the only agent of Black liberation: "The negro here, though rising, is still on his knees and nude. What I want to see before I die is a monument representing the negro, not couchant on his knees like a four-footed animal, but erect on his feet like a man. There is room in Lincoln park for another monument, and I throw out this suggestion to the end that it may be taken up and acted upon."[51] How could the sentimental, luminously pale, female, naked *Greek Slave* escape from such a righteous Black agnotological judgment of White abolitionism?

A telling detail of the Powers replica in Douglass's collection may provide a clue, for it departs in significant ways from the original 1866 "manacled" marble. Despite the recent increase in art historical scholarship on this statuette, something has gone unnoticed: Douglass's replica of *The Greek Slave* has *no crucifix*. This stunning omission certainly distinguishes the statuette, numbered FRDO 1678 in the Douglass estate, from the authorized and unauthorized versions we know. Was this excision of Christianity intentional, or was it just part of the slow slippage of detail in a mesh of ever-more efficient and mechanized reproductions? Given that the cast had to have entered Douglass's collection at least by the 1880s (when he took it to Haiti) and that the rendition *in metal* of the "manacles" is clearly present in Powers's 1866 marble, I accept, with Korobkin, that the reproductive statuette is in dialogue with that full-sized carving and probably became available as a reproduction around 1870. Thus, it is even more significant that during Reconstruction, the "knowledge" of the slave's obeisance to a Christian god is omitted from the statuette. (The crucifix is still fully visible in the Powers 1866 marble, even in the original plaster.)

FRDO 1678, as I'm affectionately calling it, gives us an absolute blank where the *Slave*'s religion used to be. Can absence become productive fuel for agnotology? We can use Douglass's replica and its shocking absence to critique the "Christian slave" as itself a lie in the United States, where both Christianization and enslavement had to be *applied* to African bodies stolen for the slave trade or raised up in its detestable domestic trade in bodies *born* as enslaved Americans.

To be scrupulously historical, I have no idea whether Douglass even noticed this detail as the sculpture settled into its packing for the boat ride or when it was unpacked and installed behind his camp table in Haiti. If he had, would it have allowed him to mobilize her for reception by a Haitian Vodou practitioner as readily as a Catholic emissary from the Dominican Republic? Perhaps only subliminally, removal of the slave's religion—this sculptural absence—releases her for multiple purposes in Douglass's context. Analytically, the very subtraction of the insistent signaling of the crucifix reveals an unexamined thread of Magdalenian "forgiveness," which had always been built into Powers's discur-

FIGURE 13.6. Comparative details of the final marble version of *The Greek Slave* from 1866 by Powers (*left*) and the ca. 1870s replica statuette in the Douglass collection (*right*). *Left*: Hiram Powers, *The Greek Slave*, n.d., marble, 65.5 × 19.25 × 18.75 in., Brooklyn Museum. *Right*: ceramic replica, Cedar Hill, Washington, DC.

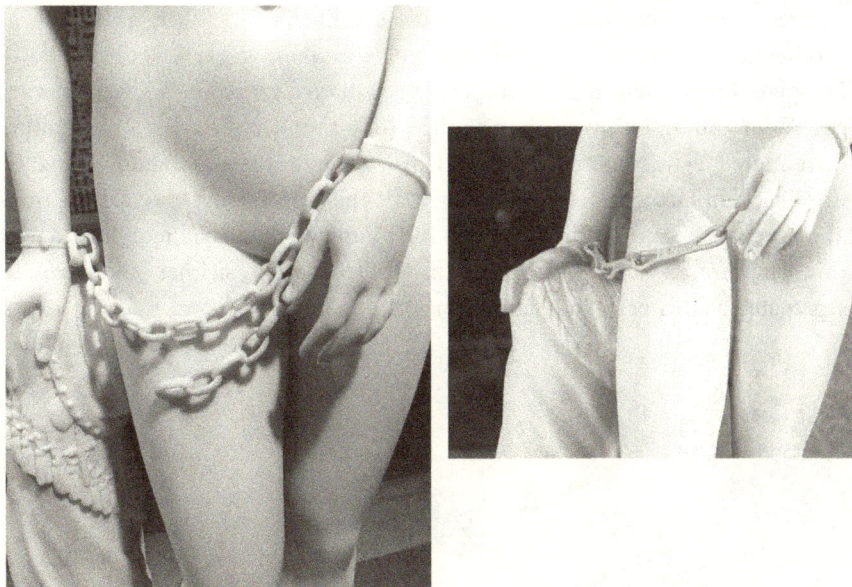

Sources: Brooklyn Museum, Gift of Charles; Bound, 55.14. Creative Commons-BY (Photo: Brooklyn Museum, 55.14_front_PS20.jpg), collection no. FRDO 1678. Courtesy of Frederick Douglass National Historic Site, U.S. National Park Service.

sive production of Christian chastity for the *Slave*. Where the Magdalen has sinned and must receive Christ's blessing to be cleansed and become "Christian," Powers's maiden is *preforgiven*. In the phrasing of Browning, the *Slave*'s "passionless perfection" was clear.[52] This freedom from passion would be analogous to the Catholic Church's theorization of a lust-free Immaculate Conception (sinful sex and its turbulent affects converted to divine and miraculous generation).[53] FRDO 1678 takes no position on such matters.

Douglass's *Slave* statuette is as ambiguous and unsettled as American reality. In scuffed ceramic, her arm at some point broken and visibly repaired, she stands for us as a material witness to Douglass's legacy—demanding that we examine more of the variable viewers who would have encountered her over time. We can speculate that the more accurate manacles together with the ambiguity of her religion might have allowed Douglass to more actively *recast* the *Slave* sculpture as the product of a system of oppression on a global scale rather than a Christianized instance Whitened for easy Orientalist consumption.

I have, admittedly, engaged in critical fabulation, imagining situated

nineteenth-century viewers who might practice what I am calling Black agnotology. This fabulation works against the segregation of knowledge that has
kept activated Black responses from the art history of *The Greek Slave* for so
long. The sculpture's insipid White lies about American slavery were first un-
covered by the mute mixed-race pantomime of Ellen Craft. Then, as its replicas
proliferated in delirious variation, the uncompromising Black abolitionism of
Douglass further rejected the tropes and palliative forms of White ignorance
and White abolitionism: a slave is the product of a political system, neither
instantiated by race nor saved by a Christian faith. The ambiguous *absence*
of religion in Douglass's *Slave* replica seals the case against easy assumptions
about her meaning in Haiti. If we can fully animate the varied mobilizations of
this sculpture for audiences of color in all their variety, neither ignorance nor
segregated art history can hold out for long.

FOURTEEN

"Civilian" Ignorance, American Militarism, and the Post-9/11 Wars

NADIA ABU EL-HAJ

"I FEAR THEY DO not know us. I fear they do not comprehend the full weight of the burden we carry or the price we pay when we return from battle. . . . We must help them understand, our fellow citizens who so desperately want to help us."[1] Spoken by Admiral Mike Mullen, former chair of the U.S. Joint Chiefs of Staff, these words frame the opening chapter of a handbook designed for "civilian" clinicians treating (ex-)military personnel. In a nation where less than 1 percent of its population enlists, we are often told that a cultural abyss isolates military personnel from the rest of the citizenry. In the words of an influential text on treating military trauma, "Members of the warrior profession subscribe, more or less, to shared values and guiding ideals that are uniquely theirs and indispensable to their way of life." Drawing on assumptions that ground the discipline of cultural psychology, whose advocates focus on the mental health of indigenous persons, immigrants, and refugees and call for respecting cultural diversity in clinical practice, in *Adaptive Disclosure*, "warriors" stand in as the diverse subject: soldiers have their own *identity*, formed by the values and virtues of "military culture" and the experiences of war.[2] Mental health practitioners cannot hope to treat soldiers returning from war without taking into account that essential difference.

In framing "warrior culture" as distinct, the authors of *Adaptive Disclosure* echo a pervasive public discourse that, in the United States during the forever wars (after 9/11), is referenced most frequently as the "civil-military divide." What's more, in speaking to mental health practitioners, these trauma professionals link that civil-military divide to a second prominent trope—the figure of the traumatized soldier/veteran. Contrary to a prevailing common sense, the post-9/11 wars play an important role in American public culture; they have not been ignored or simply erased. Indeed, if one pays attention to the figure of the *traumatized soldier* as it has emerged over the past twenty-odd years, the wars

209

are anything but forgotten. In news coverage, film and television series, long-form journalism, and the growing canon of memoirs written by American vets, the soldier suffering from trauma is a ubiquitous figure. The national conversation about those wars has been mediated to a great extent through representations of the psychic life of the American soldier/veteran—on the home front, one might argue, he *is* the war.[3]

Here, I want to explore the mythos of the traumatized soldier as juxtaposed to a second figure: that of the "civilian" citizen—one who never enlisted, one who does not or *cannot comprehend* the truths of war. Ignorance—or more accurately, the *charge* of ignorance—plays a foundational role in constituting this second subject and, as such, the civil-military divide itself. In what follows, I describe a public discourse about imputed civilian "ignorance"—the figure *who does not "know"*—and consider its ethical and political force. To be clear, my argument is not that the American public is in fact ignorant or that they do not or cannot know the violence of war. Instead, I show the ways in which the claim of civilian ignorance is wielded *as an accusation* by vets and (especially) by those who purport to speak on their behalf, with the effect of rendering all but invisible the death and suffering of foreign victims of America's wars.

What I hope to show is that ignorance as accusation operates within the frame of identity politics insofar as it links "ontological essentialism"—who I am—to "epistemological foundationalism"—what I (can) know.[4] The traumatized soldier/veteran in this sense functions as an *identity*, a distinct kind of citizen subject produced by an enduring and constitutive injury. Within this narrative frame, the experience of war becomes the exclusive truth of war: only the soldier knows war and, as such, alone has the authority to speak about or judge (that) war. The consequences of this grammar, which operates as a moral command, are profound insofar as it has become difficult to mount a sustained anti-imperial, antimilitarist critique of these latest American wars, at least among those who were never there. In the U.S. public domain, the post-9/11 wars circulate as an almost entirely "American tragedy," within which most of the actual (i.e., foreign) victims rarely appear, let alone matter.[5]

THE AMERICAN CIVILIAN

Standing in the White House briefing room on October 19, 2017, White House chief of staff John Kelly delivered a defense of his commander in chief, President Donald Trump. Responding to criticism of the president for his insensitive comments during a phone call with the wife of a soldier killed in Niger, Kelly's remarks were steeped in references to the purported civil-military divide, even if he did not name it as such: "Who are these young men and women? They are the best one percent this country produces. Most of you, as Americans, don't know them. Many of you don't know anyone who knows any one of them. But they . . . volunteer to protect our country when there's nothing in our country anymore that seems to suggest that selfless service to the nation is not only appropriate but required." At the end of his statement, Kelly announced that he would take questions only from "someone who knows," meaning a gold star parent or sibling or someone who knows one. (A gold star parent is one who has lost a child in military service.) Kelly then closed his briefing with the following words: "We don't look down upon those of you who haven't served. In fact, in a way we're a little bit sorry because you'll never have experienced the wonderful joy you get in your heart when you do the kinds of things our service men and women do—not for any other reason than they love this country. So just think of that."[6]

In the shadow of our now decades-long war on terror, the first war ever on a human emotion, public discourse in the United States has been saturated with references to a civil-military divide. And in addressing the apparent abyss between those who have served and those who have not, a lot of ink has been spilled explaining who soldiers are, what they have experienced, and how they are purportedly different from (better than?) the rest of the American populace, even if not in ways that suggest stereotypes. But what about the "civilian"? Who are these Americans who "don't know"?

In *The Image before the Weapon*, Helen M. Kinsella develops a genealogy of "the civilian" in both common parlance and the laws of war. While the meaning of "the combatant" has been quite stable since the twelfth century, she argues, not so for the term "civilian." Only in the nineteenth century, for example, did the word come to mean someone "who is not a member of the armed forces." But as Kinsella rightly insists, focusing on *who* a civilian is belies a more fundamental question as to *what* a civilian is and that cannot be answered merely by referencing one's status as a noncombatant. Kinsella traces the emergence of the figure of the civilian—in different times, in different contexts— out of threads that weave together specific discourses about gender, innocence, and civilization.[7] In what follows, I focus on one thread, that of innocence, or, more specifically, the ways in which the ostensible "innocence" of the civilian is translated into a kind of *ignorance*. With this translation, the presumed purity of the civilian emerges as a sign of ethical failure and unvirtuous citizenship.

If the civilian is one who is not a combatant, that means something quite specific in the context of a state that for well over a century now, has waged all its wars on foreign soil. In the American vernacular and especially in conversations about the post-9/11 forever wars, the term "civilian" is widely used not with reference to noncombatants in a war zone but rather to describe Americans who have never experienced war. The claim to experience matters greatly here insofar as the claim is that never having lived (or, rather, fought in) war, civilians *can never truly know war.* Among U.S. citizens are those who know war—soldiers and veterans and, in a different way, their families—and those who do not. The divide is stark. An epistemological and moral abyss is said to separate these two kinds of citizens, and all sorts of work about patriotism, responsibility, self-sacrifice, and civic virtue gets done in the space in between. American civilians are "innocent," albeit innocent with a twist. Their innocence is suspect, even pernicious. It is, perhaps, a guilty innocence. The innocence of American civilians operates as a kind of accusation: ignorant about what war is really like, civilians are naïve about the evils that exist in the world. Civilian innocence signals a kind of privilege that those sent overseas to fight on their behalf do not have.

COMBAT TRAUMA

At the center of this discourse on civilian innocence stands the figure of the traumatized soldier, who is situated within a broader trauma imaginary—within a set of assumptions about the nature of trauma that extend well beyond the world of combat. As Didier Fassin and Richard Rechtman trace in *The Empire of Trauma*, until the 1980s, the traumatized subject—the soldier returning from war, women in the aftermath of rape, or children in the aftermath of incest—was often an object of suspicion.[8] Was the suffering soldier unpatriotic, a malingerer, a drug addict? Did incest or rape really precipitate long-term psychological distress? Were female victims of sexual assault really all that innocent?

By the time of the 9/11 attacks on the Twin Towers and the Pentagon, however, a different clinical truth and cultural sensibility had taken hold. The prevailing common sense was now that people who are subjected to violence suffer trauma and therefore require—indeed they *deserve*—immediate attention and care. That was evident in one of the initial steps taken following the collapse of the Twin Towers in New York City. The city mobilized thousands of mental health personnel to attend to its citizens—residents and first responders alike. As Fassin and Rechtman write, "After the mourning, the trauma remains."[9] And so the psychological scars left by violence have to be attended to.

Attending to the psychological scars left by violence would take on a different meaning after the United States launched its post-9/11 wars. While it would be naïve to claim that the military now unflinchingly accepts posttraumatic stress disorder (PTSD) as a legitimate malady or that there is no longer any suspicion or stigma attached to its diagnosis, it is nonetheless true that the military is paying far more attention to the psychological wounds of combat than during any of the pre-9/11 wars.[10] Institutional attention to the mental health of soldiers (and vets) has reached a level never seen before—part of a more general shift in how trauma has come to be recognized in worlds beyond the military.[11] That attention is evident not just inside the military and the Department of Veterans Affairs (VA) but also in American public culture writ large and in how journalists and essayists and clinical psychologists, among others, talk about the trauma caused by war.

A GOOD WAR STORY

"Five months after carrying Sergeant Emory down the stairs in Kamaliyah," he could no longer shake off all the hurt and harm of war. On his third deployment in Iraq, Adam Schumann was done.

> His war had become unbearable. He was seeing over and over again his first kill disappearing into a mud puddle, looking at him as he sank. He was seeing a house that had just been obliterated by gunfire, a gate slowly opening, and a wide-eyed little girl about the age of his daughter peering out. He was seeing another gate, another child, and this time a dead-aim soldier firing. He was seeing another soldier, also firing, who afterward vomited as he described watching head spray after head spray through his magnifying scope. He was seeing himself watching the vomiting soldier while casually eating a chicken-and-salsa MRE [meal ready to eat].
> He was still tasting the MRE.
> He was still tasting Sergeant Emory's blood.[12]

Schumann, diagnosed with depression and PTSD and thinking about suicide, was sent home.

David Finkel, an American journalist embedded within the U.S. military during the "surge" in Iraq in 2007, tells Schumann's story in his book *The Good Soldiers*. This Pulitzer Prize–winning account of war reads like a diary of specific days and moments: the excitement and the patriotism, the killing, the gore, the anger, and the hatred, along with counterinsurgency tasks, like handing out soccer balls and the psychological pain of the (mostly young) soldiers assigned such tasks. In Finkel's retelling, Schumann is not exceptional. Injuries for which there is no visible evidence are common among the veterans of these wars. And, in a discourse that passes over the military's concern with "force protection" in favor of talk of a national moral obligation, those so-called invisible injuries are examined and explained to the public at large—so they can be understood and healed, so that proper attention to "the war" is paid on the home front. (Finkel's book has now been made into a film, *Thank You for Your Service*, starring Miles Teller as Adam Schumann with Amy Schumer in a supporting role.)

War stories told about combat trauma written by scholars and journalists "embedded" with American soldiers and those back home are largely told from the soldiers' points of view, reproducing the basic terms of what we have come to expect of a "good war story"—that is, one that keeps the soldier (our side) front and center.[13] Readers are typically admonished to recognize their obligation to face the *realities* of war—that is, the *soldier's* reality—and to confront rather than evade the question of what "we" have sent "them" to do. In the words of one retired general, "What does the nation owe its citizens who-become-soldiers?"[14]

What, though, might any of this have to do with the cultural production of ignorance? Civilian ignorance and civilian obligation, I will suggest, are two sides of the same coin: insofar as civilians are saddled with the assumption that they *do not* and *cannot* know the most fundamental truths of war, they are tasked with the obligation to listen but never judge. Civilians must defer to the soldier's experience and point of view. In order to explicate this dynamic, let me turn to an account of "moral injury"—a particular understanding of combat trauma borne not of victimhood but of real or perceived moral transgression in the heat of battle.[15]

DAMNED IF THEY KILL, DAMNED IF THEY DON'T?

David Wood, an American journalist writing in the *Huffington Post*, opens his 2014 article on moral injury with the following words: "How do we begin to accept that Nick Rudolph, a thoughtful, sandy-haired Californian, was sent to war as a 22-year-old Marine, and in a desperate gun battle outside Marjah, Afghanistan, found himself killing an Afghan boy? That when Nick came home, strangers thanked him for his service and politicians lauded him as a hero?" Wood continues, "*Can we imagine ourselves* back on that awful day in the summer of 2010, in the hot firefight that went on for nine hours? Nick . . . spots somebody darting around the corner of an adobe wall, firing assault rifles at him and his Marines. Nick raises his M-4 carbine. He sees the shooter is a child, maybe 13. With only a split second to decide, he squeezes the trigger and ends the boy's life." As for Nick, he struggles with what he's done: the boy he shot "was just a kid." And he struggles with a kind of rationale: "I'm trying not to get shot and I don't want any of my brothers getting hurt. . . . You know it's wrong. But . . . you don't have a choice." This, Wood writes, is the essence of war trauma as it appears among soldiers who have fought in these latest American wars.[16]

In telling Nick's story, Wood wants his readers to grasp the depths of the chasm that separates noncombatants from "those who have served." He quotes a navy psychologist: "Civilians are lucky, they still have a sense of naïveté about what the world is like." Average Americans, this psychologist contends, need to know that these military men and women "are seeing incredible evil, and coming home with that weighing on them." Wood also recounts one veteran's experience of civilians: "People say, 'Thanks for your service.' Do you know what I did over there?" He quotes a second vet: "Even though you're home, you don't feel at home. When you try to talk to someone, they don't understand, they start making their own assumptions about, like, 'Well, I don't think we should have been fighting that war in the first place.' Really?"[17]

In accounts by journalists and scholars, in movies, and on TV, much of what we read or see about the post-9/11 wars focuses on the figure of the traumatized soldier. In these depictions, talk of combat trauma as moral injury presents an opportunity for rebuilding community and repairing a divided nation by getting "civilians" to engage with soldiers returning from war. For one ex-marine who returned from Iraq and became a critic of the war, moral injury opens up a space for political engagement: "It takes the problem [of trauma] out of the hands of the mental health profession and the military and attempts to place it where it belongs—in society, in the community, and in the family. . . . It transforms 'patients' back into citizens, and 'diagnoses' into dialogue."[18] But such calls for dialogue and mutual engagement among citizens—that is, among all of "us"—are rare. Another frame is far more prevalent—one that foregrounds claims of the exceptionalism of "the warrior": "Do they know what we did over there?" "They don't understand." Someone who has not been to war cannot know or understand. Combat is held to give those who experience it a unique lens on the world, to which the noncombatant has no access.[19]

"YOU WHO NEVER WAS THERE"

According to the historian Annette Wieviorka, the 1961 trial of Adolph Eichmann in a Jerusalem courtroom marked the advent of the survivor-witness. By the end of that trial, the "survivor" had acquired a social identity at the heart of which stood "a new function: to be the bearer of history." As a moral and political figure, the "survivor as witness" was born of twentieth-century encounters with extreme violence—the Holocaust especially but also Hiroshima. The survivor, we are told, knows truths the rest of us cannot possibly fathom.[20] Toward the end of the twentieth century, this truth-bearing figure converged with the growing recognition that encounters with violence generate trauma: the survivor as witness becomes not just someone who knows but also someone who needs to speak their truth and to be listened to and believed—not just so that truth can be spoken but also so that healing can begin. In post-9/11 America, we find this grammar of trauma, of witnessing, and of truth telling and healing in public discourse about soldiers, combat, and war.

Of course, the "soldier as witness" did not suddenly emerge after 9/11. As recounted by the historian and philosopher Yuval Noah Harari, our quintessentially modern Western war story describes the experience of war as an experience of "learning the truth about oneself and about the world." But Harari shows that war was not always represented as revelatory. Rather, a "master narrative" of war as a transformational revelation emerged during the long Romantic period, which he dates from 1740 to 1865. In contrast to the "tame experiences of peace" and to life experienced by the civilian, "war experiences reveal the truth precisely by blowing apart all prior cultural constructions." The epistemological authority of war, Harari continues, rests on a particular kind of witnessing: eye witnessing, which is taken to be factual, can be "easily conveyed to other people." But being at war, he writes, also involves "flesh-witnessing," which emerges as a potent new source of authority. A flesh witness can never really transmit what they know to other people; they have felt it, but that feeling cannot be fully conveyed to others. And if war is impossible to understand for those who were never there, it is also impossible for them to judge: only he or she who has "flesh witnessed" war's horrors can "speak about *and judge* what she witnessed."[21]

Even if not unprecedented, the (American) soldier as witness—as truth teller—has become especially powerful in a contemporary social imaginary. The U.S. invasion and occupation of Iraq and Afghanistan are the first sustained combat-intensive wars launched in an era in which the reality of posttraumatic stress disorder is a broadly accepted common sense in the United States and beyond. What's more, in an era of wars fought by an all-volunteer force, the soldier has become (in Catherine Lutz's words) a "super-citizen,"[22] who has the power to kill and who can suffer from being ordered to kill. His trauma and testimonials to that trauma carry a particularly weighty moral force.

This supercitizen, the soldier-citizen who stepped up and volunteered, is a survivor as witness who inhabits different political and ethical fields than do survivors of Hiroshima, the Holocaust, or sexual assault. No mere victim, the soldier is, to borrow Roy Scranton's term, a "trauma hero." The young soldier is imagined to be excited, patriotic, and in search of adventure; he returns from war in possession of insights that others cannot possibly fathom—and suffers psychological pain as a result. Upon return, this trauma hero "struggles to turn the inassimilable reality of the traumatic event into narrative but finds himself blocked at every turn: the memories slip from his grasp, no one wants to hear about the horrors he's seen." And crucially for our purposes, the story goes, "it is impossible for people who were not there to understand."[23] Enter "the civilian"—one who was not there—with ignorance, innocence, privilege, and obligations toward "those who have served."

I am an anthropologist and have done fieldwork among practitioners treating soldier trauma. During a workshop I participated in on PTSD and moral injury led by a VA psychologist in 2015, one veteran's father told the rest of us, "If a combat veteran is gracious enough to share his or her story, be very careful to respond in a nonjudgmental way. You weren't there." At nineteen years old, he told us, his son deployed to Afghanistan at a time when it was "the Wild West." His son lost thirty of his platoon's one hundred members. One day, his unit was taking mortar fire; he was ordered to "put steel into that village until they quit shooting at us." They flattened the village and then went in. "There's not a living thing in it: no Taliban, no man, woman, child, goat—nothing." What should he say to his son? this father wondered. You're a terrible human being? You killed a bunch of civilians? How could you do that? The workshop leader responded to this father's obvious pain: "Yeah, that's a good question." Judgment was "absolutely not the way to go because if people grace you with telling you bad things that happened to them, it's very easy to think, 'Oh, I would've behaved differently,' or 'How could you have done that?' And that is incredibly disrespectful and absolutely not therapeutic for the person." The psychologist also cautioned against easy answers, including "You were just following orders" or "You did what you had to do." Another participant added that, while he hadn't been in that situation, the only thing he would want to hear is "I cannot *imagine* what that would be like. Because it's the fucking truth."[24]

For his part, David Wood advises his (presumably civilian) readers how to support our returning warriors. "Let's set aside the question of war itself," he begins. He recommends a program developed by a clinical psychologist, whose idea was "to match veterans with volunteer civilian listeners" to facilitate "uninterrupted, intentional listening"—that is, "listening with validation," without judgment, offering sympathy and gratitude, as in "Yeah, that was fucked up. But also . . . I honor your service."[25] In a similar vein, the philosopher Nancy Sherman writes that "we have a sacred obligation to those who serve, whether or not we agree with the cause of those wars and whether or not those who serve agree with them." We must engage one-on-one with fellow citizens who have returned from war, and we must *listen without judgment.*[26]

Joan W. Scott writes critically about a movement among social and cultural historians in the 1990s to rest their claims to "legitimacy on the authority of experience, the direct experience of others." Given history's referential understanding of evidence, she argues, "what could be truer . . . than a subject's own account of what he or she has lived through?"[27] This privileging of experience as epistemological ground has not been limited to the discipline of history, of course, nor has it been limited to the academy. Identity politics has also assumed and valorized essentialism tethered to the authority of experience. The "way we argue now,"[28] to quote Amanda Anderson, is often key to claims about "who we are."

It is from within these moral imaginaries that calls for the American public to just listen to those "who have served" ring true, that they resonate and make sense. Such talk also draws on a longstanding sensibility about extreme violence that has been central to trauma theory (and culture) for decades now: that its very extremity renders it incomprehensible and incommunicable among those who were never there.[29] And yet the violence of combat also occupies a distinct and more ambivalent space in the trauma imaginary: combat is horrifying and yet also invigorating or even sublime. The traumatized soldier suffers, but he is no (simple) victim; he is a hero, a supercitizen, to return to Lutz's words. Civilian citizens, for their part, are called upon to listen to all these experiences of violence and to do so without judgment. At the same time, the prevailing conviction among many psychiatrists and psychologists, anthropologists and philosophers, and soldiers and veterans is that civilians will never really be able to hear what someone who has been in combat has to say if hearing implies the possibility of coming to understand. The difference between what the soldier knows and what most American civilians know is "often at the center of their encounters with each other," per the observation of anthropologist Zoë Wool. For Wool, "It is only through a particularly *privileged position of civilian safety* that we are able to make claims about the appropriateness or veracity of soldiers' knowledge about the world."[30] To borrow Walter Benn Michaels's apt turn of phrase, "You who never was there" can never know.[31] Civilian citizens are called upon to attend to the suffering of returning troops, to listen to their moral and psychological pain, all the while refraining from judgment—Did he enjoy the violence? Was he involved in war crimes? Was the war just?—because they cannot know. An entire politics about war, knowing, truth, ignorance, and citizenship lives in and through the distinction between the soldier who knows and the civilian who is decidedly and forever ignorant.

STAGING THE SCENE OF SUFFERING AND HEALING

Founded in 2009, the Theater of War project was born of director Brian Doerries's belief that ancient Greek tragedies were produced for the purposes of catharsis: the purification of "dangerous emotions" borne of ongoing wars. The project stages scenes of two of Sophocles's plays because, Doerries believes, they resonate with postcombat experiences of our own millennium—including the unprecedented survival rates for wounded soldiers, who often face years of suffering and pain and high rates of suicide. Drawing on the talents of a long list of established actors who volunteer their time, each reading is followed by a panel assigned to respond to the play—a combination of veterans, spouses of veterans, and mental healthcare practitioners who serve the role of the ancient Greek chorus, intermediating between the play and the audience. Following the panel's intervention, Doerries poses questions to the audience and engages in an extended give and take.[32]

At one performance at Columbia University, the actors presented two scenes from *Ajax*: one following upon Ajax's slaughter of the animals and a second dramatizing his suicide born of shame and melancholy for violence he had perpetrated. The audience was transfixed. Doerries had introduced the scenes before the actors began, and when they finished, he summarized the rest of the play. He then asked, "How do we honor and show respect for people like Ajax who sacrifice everything, including sometimes their mental health, without honoring the violence that sometimes takes over their lives?" Doerries then posed a more specific question to the audience: "Given everything you know about the play, written 2,500 years ago by a general, performed for seventeen thousand soldiers . . . what do you think Sophocles was up to when he staged this play? What was he trying to say?" A few students raised their hands, with one offering that "Sophocles is seeing the experiences of soldiers. He thought they needed to be seen on a larger stage and to get it out there and put it down for generations."[33]

Many of those who intervened—young and old, students and community members—identified themselves as veterans before they spoke. Doerries asked, "If Ajax were someone you knew, loved, cared about, swore you would never leave behind . . . and you knew that person was on a sand dune with thoughts and demons struggling in the way Ajax was at end of play and you had a chance to be with that person, what might you say or what would you do?" A veteran's hand shot up: "I would listen. In the play everyone was trying to be direct and give advice . . . sometimes what a person needs is somebody just to listen." Someone else jumped in: "I think we don't give enough recognition to our veterans. There are some veterans out there who don't get as much love as they deserve." Yes, Doerries responded, "the problem Sophocles was dealing with is compounded today since only about 1 percent serve in the military. That creates an opportunity for people to feel not respected or cared for." Another self-identified veteran and student then spoke up: "We call on soldiers to engage in violence but then belittle the violence." Visibly struggling to contain his anger, he continued, "Maybe we are all antiwar here," but rather than pathologizing the soldier, "we should be saying, you are reacting to this situation in a way that . . . [anyone might while being put] in these extreme, impossible circumstances."[34]

In an audience that included scholars of the classics, not everyone was so ready to get on board with Doerries's therapeutic reading of *Ajax*. One professor intervened: Sophocles was living in "a society and culture defined by nationalism. . . . I register this play as a critique of that and of the fame and glory [attached to war] while neglecting those who suffer." Another older member of

the audience chimed in: "A lot of what we've heard is heartfelt but . . . I found
myself looking for clarity, and what came back to me was a memorandum by
a Colonel Robert Heimer, in 1971, at the height of Vietnam conflict, where he
said the U.S. Army was no longer able to fight because it was either in rebellion, on drugs, or other elements that kept it from being a fighting force." The
United States withdrew "because the army rebelled." Doerries intervened. He
insisted that "it's not the job of the veterans to stop the war," and continued,
"There is nothing more powerful for talking about war and condemning war
than hearing veterans talk about war and share their story." A self-identified
student and marine veteran quickly contradicted Doerries, pointing out that
while there was certainly stigmatization around the war in Vietnam, the world
had changed. We are no longer in Vietnam; these are "totally different circumstances." As tension in the room rose, over and over again, Doerries reminded
the audience to *listen* to the veterans since *they* are the ones who really know
war. What's more, he assumed that it would be from their war stories that an
authentic antiwar politics would emerge—studiously ignoring comments made
by most veterans who spoke that were anything but antiwar. Despite its claims
to the contrary, this Theater of War performance did not stage a dialogue or
discussion between "civilians" and "soldiers." Veterans and soldiers spoke, and
the rest of us were enjoined to listen. This was to be *their* cathartic moment.[35]

Many decades ago, Hannah Arendt faced an onslaught of criticism for her
coverage of the Eichmann trial. Under the terrifying conditions produced by
the Nazi regime, her critics suggested, can anyone be sure that they would have
acted differently? (This was primarily with reference to the Jewish councils
tasked with horrific complicity in mass murder.)[36] A similar logic structures
the demands the American public is subjected to vis-à-vis the post-9/11 wars:
refrain from judging the actions of U.S. soldiers abroad and even the war itself.
There are those who "serve" and those who do not, those who may speak and
those who must listen (at least when face-to-face with a soldier or veteran),
those who know and those who *cannot* know and therefore must remain silent
and, certainly, must not judge.

THE POWER AND PRIVILEGE OF IMPERIAL INATTENTION

Allow me to return to that 2015 VA workshop on PTSD, where one of the
participants cited as "the fucking truth" that no one not in that young soldier's
position could "imagine what it would be like." Insisting on this inability to
imagine is also, of course, a call to pay attention to those who have suffered
after having returned from war. That imperative has a less virtuous flip side,
however. It generates *inattention* to the suffering of those on the receiving end
of American guns, the Afghan villagers whose lives were lost and lifeworlds de-

stroyed in the action that the former soldier recounts. In other words, entangled with the operation of ignorance as accusation is the production of inattention and indifference. Ignorance in this sense emerges not so much from the absence of the war on the home front, but from the home front's saturation with a distinct war imaginary—that of "the good war story," the story told (solely) from the (American) soldier's point of view.

With all the talk of killing and moral transgression that appears within this combat-trauma imaginary, the violated Other does appear. To quote the aforementioned marine once again, "Without the Iraqi people, the troops can have no moral injuries to speak of." Keep in mind that we are talking about a form of combat trauma that emerges primarily *from acts of harming others*. The call to attend to the trauma of the American soldier references the harms done to others but then asks us, as American citizens, to set aside that reality—along with the whole question of the grounds for the war itself. This drama of injury and suffering revolves entirely around the American national self. We are told stories of Iraqi doctors shot at checkpoints, nameless children killed, people humiliated, heads that "spray," but we are admonished to listen only for the purpose of comprehending the moral pain suffered by American soldiers for having carried out or witnessed such brutalities. These war stories call on the public to identify with—to listen to, to help—those whom Allan Young termed "self-traumatized perpetrators."[37] ("Shooting and crying" is the term used in Israeli society.)

I want to be clear: the facts of the Other's suffering are "out there." Sometimes they are even discussed. This is not a problem of state secrecy or an occlusion of knowledge or a void in need of being filled in—say, by yet another Wikileaks-type download of classified information. The U.S. press has covered civilian casualties in Iraq and Afghanistan and in other countries subjected to America drone warfare and special ops, at least to some extent. We know of depleted uranium weapons and burn pits; and while the public conversation focuses on the harms American soldiers suffer from their effects, it should not be hard to imagine the extent of environmental harms the war has left behind, which Iraqis and Afghans will suffer for decades to come. The problem, the stumbling block, resides elsewhere. Inattention—indifference—to the suffering of those on the receiving end of American military violence is produced and sustained by an endless call for attention to and care for those who enacted the harms. In turning these forever wars into an American tragedy, there is little space for a *political* critique of American power as it operates across the globe, let alone for even "humanitarian" concern for those who have and continue to suffer its effects on the ever-expanding battlefields of the ongoing "War on Terror."

As I have argued here, the making of this American tragedy depends, in part, on the production of ignorance—not on the making of actual ignorance, on secrets and lies, but on the power and ubiquity of the figure of "the civilian" who is accused of ignorance, a morally dubious citizen who is juxtaposed to the virtuous soldier citizen who stepped up and "served" ("the best one percent this country produces," in Kelly's words). To return to Wool's admonition, it is, purportedly, only from the "privileged position of civilian safety that we are able to make claims about the appropriateness or veracity of soldiers' knowledge about the world." But what if we refuse her admonition? What if we recognize the ways in which this stereotype of the civilian—the one who did not sacrifice, the one who is ignorant of the horrors of war—perpetuates American nationalist and imperial myopia and its attendant militarism? After all, only a nation with the power to fight its wars on someone else's soil can so commonly associate the term "civilian" with someone who is innocent of the violence and brutality of war. The constant reiteration of civilian ignorance and the attendant call to defer to the soldiers' "knowledge of the world" are together but one more weapon in the arsenal of imperial warfare and privilege. A "good war story" is not the truth of war. As Viet Thanh Nguyen writes, "A true war story should also tell of the civilian, the refugee, the enemy, and, most importantly, the war machine that encompasses them all."[38] The demand to defer to the "knowledge of the world" held by American soldiers, then, is best understood as a pervasive and powerful exercise in agnotology.

Contributors

NADIA ABU EL-HAJ is the Ann Whitney Olin Professor in the Departments of Anthropology at Barnard College and Columbia University. The recipient of numerous awards, including from the MacArthur and Guggenheim Foundations, Abu El-Haj has published three books: *Facts on the Ground: Archaeological Practice and Territorial Self-Fashioning in Israeli Society* (2001), which won the Albert Hourani Annual Book Award from the Middle East Studies Association; *The Genealogical Science: The Search for Jewish Origins and the Politics of Epistemology* (2012); and *Combat Trauma: Imaginaries of War and Citizenship in Post-9/11 America* (2022).

DANIEL AKSELRAD is a doctoral candidate in the Department of Communication at Stanford University and a Stanford Institute for Research in the Social Sciences Dissertation Fellow. His research focuses on organizational ideology, infrastructures of atrocity, and legal-rhetorical culture. He has used this lens to examine distributed decision-making in fighter jet cockpits, Nazi bureaucratic obfuscation, and the internal communications of the global cigarette industry. Akselrad's most recent article, with Robert N. Proctor, is "Why Did Philip Morris Stop Making Cigarettes at Auschwitz? An Essay on the Geometry and Kinetics of Atrocity" (2024).

ERIK M. CONWAY is the historian of the Jet Propulsion Laboratory at the California Institute of Technology. He works at the intersection of science and technology in the later twentieth century, specializing in aerospace. He is coauthor with Naomi Oreskes of *The Big Myth: How American Business Taught Us to Loathe Government and Love the Free Market* (2023) and *Merchants of Doubt: How a Handful of Scientists Obscured the Truth on Issues from Tobacco Smoke to Global Warming* (2010).

JOHN J. DONOHUE, the Carlsmith Professor at Stanford Law School, has written extensively on the relationship between guns and crime. Donohue, who holds a JD (Harvard) and a PhD in economics (Yale), is an elected member of the American

Academy of Arts and Sciences. He has edited the *American Law and Economics Review*, has served as the president of the American Law and Economics Association and the Society for Empirical Legal Studies, and is a former member of the Committee on Law and Justice of the National Academy of Sciences.

HANY FARID is professor at the University of California, Berkeley, with a joint appointment in Electrical Engineering and Computer Sciences and the School of Information. His research focuses on digital forensics, forensic science, misinformation, image analysis, and human perception. He is the recipient of an Alfred P. Sloan Fellowship and a John Simon Guggenheim Fellowship and is a fellow of the National Academy of Inventors. He is the author of *Photo Forensics* (2016) and *Fake Photos* (2019).

BENJAMIN FRANTA is the Titular Associate Professor of Climate Litigation at the University of Oxford and the founding head of the Climate Litigation Lab at the Oxford Sustainable Law Programme. His research and writing have been featured in the television series *Black Gold*, cited in the U.S. *Congressional Record*, and published in *Nature Climate Change*, *The Guardian*, and other platforms around the world. He has a PhD in applied physics from Harvard University, a PhD in the history of science from Stanford University, and a JD from Stanford Law School.

PETER GALISON is the Joseph Pellegrino University Professor at Harvard University. He directs the Black Hole Initiative, a leading center for interdisciplinary research on black holes, and is the Science Team's lead for the Black Hole Explorer. Galison's work explores the interaction between the principal subcultures of physics and the embedding of physics in the world, alongside the visibilization of scientific discovery. His books include *How Experiments End* (1987); *Einstein's Clocks, Poincaré's Maps* (2004); and, with Lorraine Daston, *Objectivity* (2010). Galison's latest feature film is *Black Holes: The Edge of All We Know.*

JENNIFER JACQUET is an interdisciplinary scientist interested in the natural science of environmental problems and the social questions of how to address them. She is a professor of environmental science and policy at the Rosenstiel School of Marine, Atmospheric, and Earth Science at the University of Miami and affiliated faculty with the Abess Center for Ecosystem Science and Policy. From 2012 to 2022, she worked in the Department of Environmental Studies at New York University. She has written two books: *Is Shame Necessary?* (2015) and *The Playbook* (2022).

CAROLINE A. JONES teaches in the History, Theory, and Criticism program at MIT and serves as associate dean in the School of Architecture and Planning. Her research focuses on art and its technological modes of production, as well as its interfaces with science. Jones's essays appear in a range of publications from *Artforum* to *Critical Inquiry*; her solo and coedited books examine technology and the senses, art and neuroscience, and the art-science nexus. Most recent is the 2022–2023 exhibition *Symbionts: Contemporary Artists and the Biosphere* and its accompanying publication (2022).

ROBERT H. LUSTIG is Emeritus Professor of Pediatric Endocrinology and member of the Institute for Health Policy Studies at the University of California, San Francisco. Dr. Lustig is a neuroendocrinologist, with expertise in obesity, diabetes, metabolism, and nutrition. Dr. Lustig received an SB from MIT in 1976, MD from Cornell University Medical College in 1980, and a masters of studies in law at UC College of the Law, San Francisco, in 2013. He is the author of the popular books *Fat Chance* (2012), *The Hacking of the American Mind* (2017), and *Metabolical* (2021).

NAOMI ORESKES is the Henry Charles Lea Professor of the History of Science and affiliated professor of Earth and planetary sciences at Harvard University. She is an elected fellow of the American Philosophical Society and the American Academy of Arts and Sciences and recipient of a 2018 Guggenheim Fellowship, the 2019 British Academy Medal, and the Nonino Foundation (Italy) Master of Our Time prize. Her most recent book, with Erik M. Conway, is *The Big Myth: How American Business Taught Us to Loathe Government and Love the Free Market* (2023).

ROBERT N. PROCTOR is professor of history at Stanford, where he is also professor by courtesy of pulmonary medicine. He writes on human origins, Nazi medicine, rock hound aesthetics, cigarette design, and the history of ignorance. His books include *Racial Hygiene* (1990), *Cancer Wars* (1995), *The Nazi War on Cancer* (2000), and *Golden Holocaust* (2011). He was the first Senior Scholar in Residence at the U.S. Holocaust Museum and the first historian to testify against the cigarette industry in court, earning him the title of Hero of Science from the Max Planck Institute in Berlin.

ROSEMARY SAYIGH is a British-born journalist and scholar of Middle Eastern history. She taught oral history at Birzeit University, Palestine, and at the American University of Beirut, Lebanon, and is author of *Palestinians: From Peasants to Revolutionaries* (2007) and *Too Many Enemies: The Palestinian Experience in Lebanon* (2015). In 2024, she was presented the Lifetime Achievement Award for her work recording the displacement narratives of Palestinian women from the Palestine Book Awards.

LONDA SCHIEBINGER is the John L. Hinds Professor of History of Science at Stanford University and founding director of Gendered Innovations in Science, Health and Medicine, Engineering, and Environment. She is an elected member of the American Academy of Arts and Sciences and the recipient of numerous prizes and awards, including the Alexander von Humboldt Research Prize. Schiebinger is most recently author of *The Secret Cures of Slaves: People, Plants, and Medicine in the Eighteenth-Century Atlantic World* (2017), as well as articles in *Nature* and *Science*.

NANNA BONDE THYLSTRUP is associate professor at the Department of Arts and Cultural Studies, University of Copenhagen. Thylstrup is most recently coeditor of *Uncertain Archives: Critical Keywords for Big Data* (2021) and *(W)archives: Archival Imaginaries, War, and Contemporary Art* (2021) and author of *The Politics of Mass Digitization* (2018). She is the recipient of the prestigious European Research Council Starting grant for a project devoted to researching the politics of data loss in digital societies.

Notes

Chapter 1

1. "Agnotology" is one of a half a dozen words Iain Boal has helped me coin—the most important others being "eavescasting," "the ectopic fallacy," "long tobacco," "catarheumatics," and "anadogmatics."

2. "Smoking and Health Proposal," 1969, https://www.industrydocuments.ucsf.edu/tobacco/docs/#id=psdw0147.

3. Evidentiary bar raising was part of this: cigarette makers paid to have scholars develop statistical tests, which could then be used to say, "You haven't used the most recent or sophisticated tests." Donald Rubin at Harvard helped the industry with this, as did Alvan Feinstein at Yale.

4. See the essays in Janet Kourany and Martin Carrier, eds., *Science and the Production of Ignorance* (Cambridge, MA: MIT Press, 2020).

5. Charles Golden, "Cigaret Cancer Link Is Bunk," *National Enquirer*, March 3, 1968, https://www.industrydocuments.ucsf.edu/tobacco/docs/#id=xxyf0145.

6. On Ancel Keys's role in blaming fat for heart attacks, see Nina Teicholz, *The Big Fat Surprise: Why Butter, Meat and Cheese Belong in a Healthy Diet* (New York: Scribe, 2014).

7. "March Birth, Lung Cancer Linked," *Reports on Tobacco and Health Research* 6, no. 3 (Nov.–Dec. 1963): 1, 4, https://www.industrydocuments.ucsf.edu/tobacco/docs/#id=fyym0146.

8. Cigarette makers typically did not *deny* that cigarettes cause cancer; the more common claim was that it had not been *proved*—there was honest doubt. This was to avoid legal liability: by stressing doubt rather than denying harm, cigarette makers could claim there was no breach of "express warranty." Climate denial is likewise not so much *anti*science (or pseudoscience) as *ortho*science: evidence-based distraction combined with pseudo-solutionism.

9. At the Twenty-Eighth Conference of the Parties to the UN Framework Convention on Climate Change in 2023, a record 2,456 lobbyists helped fifty oil producers craft the language and priorities of the pact that is supposed to guide us through the climate catastrophe. See Nina Lakhani, "Record Number of Fossil Fuel Lobby-

ists Get Access to Cop28 Climate Talks," *The Guardian*, Dec. 5, 2023, https://www
.theguardian.com/environment/2023/dec/05/record-number-of-fossil-fuel-lobbyists
-get-access-to-cop28-climate-talks.

10. Tobacco's two closest allies on the Supreme Court were Abe Fortas and
Lewis Powell. Philip Morris celebrated Powell's 1971 appointment to the court at a
party, where he was given (and donned) a judicial robe emblazoned with the Marlboro brand and the logo of the company. See "The Memo That Changed America,"
Aug. 27, 2024, in *The Master Plan*, produced by David Sirota, podcast, 51:09, The
Lever, https://www.levernews.com/master-plan-episode-3-the-memo-that-changed
-america/.

11. See, e.g., U.S. Department for Health and Human Services, *How Tobacco
Smoke Causes Disease: The Biology and Behavioral Basis for Smoking-Attributable
Disease; A Report of the Surgeon General* (Washington, DC: U.S. Government
Publishing Office, 2010), 6, https://www.ncbi.nlm.nih.gov/books/NBK53017/. It
was first in 2014 that a U.S. surgeon general's report concluded that the industry
had "deliberately misled the public on the risks of smoking cigarettes."

12. Cigarette makers have weaponized complexity for similar purposes, claiming that since cancer is a "complex" disease we cannot trace it to any single culpable
agent.

13. An interesting example of rhetorical reframing is cardiologist Olivier Milleron's book whose title translates as "Why smoking is right wing." See Milleron,
Pourquoi fumer, c'est de droite (Paris: Éditions Textuel, 2022).

14. Zenn Kaufman, "Vitamin M," Dec. 1948, https://www.industrydocuments
.ucsf.edu/tobacco/docs/#id=fxwp0125.

15. David R. Hardy to DeBaun Bryant, letter, July 21, 1970, https://www.indus
trydocuments.ucsf.edu/tobacco/docs/#id=jydd0024.

16. Thomas D. Bakker to J. W. Burgard, "Final Report: Copy Test of Two Kool
Prints with Editorial Copy on the Smoking-Health Relationship," memorandum,
Oct. 7, 1969, https://www.industrydocuments.ucsf.edu/tobacco/docs/#id=yyd
p0099.

17. Anne H. Duffin to William Kloepfer, "Audience Testing of 'Smoking &
Health: The Need to Know,'" memorandum, June 29, 1973, https://www.industry
documents.ucsf.edu/tobacco/docs/#id=thdy0146.

18. Bruno Latour, "The Affects of Capitalism" (lecture, Royal Danish Academy of Sciences, Copenhagen, Denmark, Feb. 26, 2014), posted Feb. 28, 2014, by
Videnskabernes Selskab, YouTube, 1:00:46, at 16:00, https://www.youtube.com/
watch?v=8i-ZKfShovs. For a transcript, see John Keane, "Rethinking Capitalism,"
The Conversation, April 20, 2014, https://theconversation.com/rethinking-capital
ism-20995. See also Christophe Bonneuil and Jean-Baptiste Fressoz, *The Shock of
the Anthropocene* (London: Verso Books, 2017), 198–221.

19. See, e.g., Peter Burke, *Ignorance: A Global History* (New Haven, CT: Yale
University Press, 2023); Shannon Sullivan and Nancy Tuana, *Race and Epistemolo-
gies of Ignorance* (New York: State University of New York Press, 2007); Rick Peels,
Ignorance: A Philosophical Study (New York: Oxford University Press, 2023).

20. Matthias Gross and Linsey McGoey, eds., *Routledge International Hand-*

book of Ignorance Studies (New York: Routledge, 2015); and for their updated volume, which contains essays from scholars in philosophy, psychology, sociology, anthropology, and political theory, see Gross and McGoey, eds., *Routledge International Handbook of Ignorance Studies*, 2nd ed. (New York: Routledge, 2023).

21. There is some debate about the degree to which plants exhibit true mimicry. See Simon Henderson, "The Incredible Mimicry of *Boquila trifoliolata*," Deception by Design, Jan. 8, 2023, https://deceptionbydesign.com/the-incredible-plant-mimic -boquila-trifoliolata/; H. Martin Schaefer and Graeme D. Ruxton, "Deception in Plants: Mimicry or Perceptual Exploitation?," *Trends in Ecology and Evolution* 24, no. 12 (2009): 676–85; Ernesto Gianoli and Fernando Carrasco-Urra, "Leaf Mimicry in a Climbing Plant Protects against Herbivory," *Current Biology* 24, no. 9 (2014): 984–87. For a critique, see Christie Wilcox, "Can Plants See? In the Wake of a Controversial Study, the Answer's Still Unclear," *The Scientist*, Nov. 30, 2022, https://www.the-scientist.com/news-opinion/can-plants-see-in-the-wake-of -a-controversial-study-the-answer-is-still-unclear-70796.

22. Joseph Bondy-Denomy discovered a bacteriophage that evades immune nucleases by pretending to be a eukaryote; some bacteria are also known to survive in the human gut by camouflaging themselves with a coat of familiar sugars (to avoid the immune system). See (reporting on the work of Laurie Comstock et al.) Mason Inman, "Bacteria's Sweet Deception," *Science*, March 18, 2005; Ursula W. Goodenough, "Deception by Pathogens," *American Scientist* 79, no. 4 (1991): 344–55, https://www.jstor.org/stable/29774426.

23. Agnotology is sometimes misunderstood to be *only* about the deliberate spread of doubt or confusion by corporate malefactors. This misimpression is perpetrated by Wikipedia, which states (as of 2024) that agnotology is "the study of deliberate, culturally induced ignorance or doubt." It is broader than this!

24. Ian Leslie, "The Sugar Conspiracy," *The Guardian*, April 7, 2016, https: //www.theguardian.com/society/2016/apr/07/the-sugar-conspiracy-robert-lustig -john-yudkin.

25. Londa Schiebinger explores the nontransfer of knowledge in Schiebinger, *Plants and Empire: Colonial Bioprospecting in the Atlantic World* (Cambridge, MA: Harvard University Press, 2004).

26. On forgetting as a survival strategy, see Evelyn Conlon, *Not the Same Sky* (Adelaide: Wakefield, 2013). This novel asks, "Why do we yearn to remember what we did not have to endure?" Jorge Borges explores the pathology of infinite recall in his "Funes the Memorious."

27. The work of Michael Smithson is key here. See Smithson, *Ignorance and Uncertainty: Emerging Paradigms* (New York: Springer, 1989); Smithson, "The Future of Ignorance Studies," in *Routledge International Handbook of Ignorance Studies*, ed. Matthias Gross and Linsey McGoey (2015; New York: Routledge, 2023). E. E. Evans-Pritchard also talks about "structural amnesia." See Evans-Pritchard, *The Nuer* (Oxford: Clarendon, 1940).

28. Ventriloquy is key to the industry's use of "third parties" to carry its messages, but proxemics are also used to pretty up cigarette suffering with their talk of "smoking and health" and the like. Phonaesthemics are at work: Philip Morris

became Altria, for example, to hide behind the pastoral fog of altruism and trees.

29. One could list thousands of obfuscating euphemisms—think of the duplicity underlying expressions like "filtered cigarette," "healthcare provider," "hopelessly addicted," "Truth Social," "tort reform," "trade association," "air freshener," "comfort food," "partial-zero-emissions vehicles" (PZEVs), etc.

30. The classic source is George Orwell's 1946 "Politics and the English Language," which talks about the use of language to defend "the indefensible." Recall the motto of his totalitarian society: "War is peace, freedom is slavery, ignorance is strength." See also Dwight Bolinger, *Language: The Loaded Weapon; The Use and Abuse of Language Today* (London: Longman, 1980).

31. Carrie Halperin and Natalia V. Osipova, "'Cutting the Grass' of Hamas's Militancy," *New York Times*, July 8, 2014. Some euphemisms are state mandated: in Russia, you're supposed to call the invasion of Ukraine a "special operation," and in China, you have to be careful talking about Tibet, Tiananmen, or Taiwan.

32. Gerald Markowitz and David Rosner, *Deceit and Denial: The Deadly Politics of Industrial Pollution* (Berkeley: University of California Press, 2002), 43–44.

33. Rebecca Solnit, "Big Oil Coined 'Carbon Footprints' to Blame Us for Their Greed," *The Guardian*, Aug. 23, 2021.

34. There are several variants of this metaphor, dating mostly from the early 1970s, though the idea is also captured in the Renaissance maxim that it is better to prevent a malady than to cure it—which we find already in the writings of the Dutch humanist Desiderius Erasmus (1466–1536). And for "the causes of causes," see Robert N. Proctor, *Cancer Wars: How Politics Shapes What We Know and Don't Know about Cancer* (New York: Basic Books, 1995), 270–71.

35. Sara D. Guardino and Richard A. Daynard, "Tobacco Industry Lawyers as 'Disease Vectors,'" *Tobacco Control* 16, no. 4 (2007): 224–28, https://www.ncbi.nlm.nih.gov/pmc/articles/PMC2598535/.

36. A comparable mistake (hyperopia?) one could call "anadogmatic": the poverty (or paralysis) of inappropriate *upstream* thinking. Here, the (stupid) idea is that you can't fix anything until you've fixed everything. So tobacco isn't the problem; the real problem is, say, capitalism.

37. On the agnotology of atrocity, see Daniel Akselrad and Robert N. Proctor, "Why Did Philip Morris Stop Making Cigarettes at Auschwitz? An Essay on the Geometry and Kinetics of Atrocity," *Public Culture* 36, no. 1 (2024): 47–74.

38. An example of "attribution agnotology" would be cigarette makers insisting that a warning on packs of cigarettes be attributed to the surgeon general, rendering it an opinion instead of a fact—and in no way an admission. The journalistic counterpart would be the statement "Environmentalists say that . . ." (rather than simply stating the fact).

39. Seth Mnookin, *The Panic Virus: A True Story of Medicine, Science, and Fear* (New York: Simon and Schuster, 2011).

40. Mark Fainaru-Wada and Steve Fainaru, *League of Denial* (New York: Crown, 2014). See also the 2013 PBS documentary by the same name: Michael Kirk, dir., "League of Denial: The NFL's Concussion Crisis," *Frontline*, season 2013, episode 2, aired Oct. 8, 2013, on PBS, posted Feb. 1, 2019, by FRONTLINE PBS, YouTube, 1:53:55, https://www.youtube.com/watch?v=SedClkAnclk.

41. "Cigarette Citadels: The Map Project," Stanford University, n.d., accessed Jan. 10, 2025, https://cigarettecitadels.stanford.edu/.

42. Proctor, *Cancer Wars*, 256.

43. Michael E. Mann, *The Hockey Stick and the Climate Wars* (New York: Columbia University Press, 2012); Chris Mooney, "The Hockey Stick: The Most Controversial Chart in Science, Explained," *The Atlantic*, May 10, 2013, https://www.theatlantic.com/technology/archive/2013/05/the-hockey-stick-the-most-con troversial-chart-in-science-explained/275753/.

44. Denis Wood, *The Power of Maps* (New York: Guilford, 1992). For his fol-low-up work from 2010, see Wood, *Rethinking the Power of Maps* (New York: Guilford, 2010).

45. Nadia Abu El-Haj, *Facts on the Ground: Archaeological Practice and Terri-torial Self-Fashioning in Israeli Society* (Chicago: University of Chicago Press, 2002).

46. Nikita Vladimirov, "CNN Chyron Challenges Trump on 'Sarcasm' about Obama and ISIS," *The Hill*, Aug. 12, 2016, https://thehill.com/blogs/ballot-box/ presidential-races/291245-cnn-chyron-challenges-trump-on-sarcasm-about-obama -and/.

47. Stephen J. Pyne, "Vestal Fires and Virgin Lands," in *World Fire: The Culture of Fire on Earth* (New York: Henry Holt, 1995), 238–55.

48. Maris Fessenden, "Why So Few Scientists Are Studying the Causes of Gun Violence," *Smithsonian Magazine*, July 13, 2015, https://www.smithsonianmag .com/smart-news/cdc-still-cant-study-causes-gun-violence-180955884/.

49. David Michaels, *Doubt Is Their Product: How Industry's Assault on Sci-ence Threatens Your Health* (New York: Oxford University Press, 2008). See also Michaels, *The Triumph of Doubt: Dark Money and the Science of Deception* (New York: Oxford University Press, 2020).

50. On agnotology in pedagogy, see A. J. Angulo, ed., *Miseducation: A History of Ignorance Making in America and Abroad* (Baltimore: Johns Hopkins Univer-sity Press, 2016).

51. See, esp., Gross and McGoey, *Handbook of Ignorance Studies*.

52. See, e.g., Stephan Lewandowsky and John Cook, *The Conspiracy Theory Handbook* (self-pub., Skeptical Science, 2020), http://sks.to/conspiracy.

53. Wikipedia has a site listing hundreds of conspiracy theories. See https://en .wikipedia.org/wiki/List_of_conspiracy_theories. And on conspiracy theory more generally, see https://en.wikipedia.org/wiki/Conspiracy_theory.

54. Agnotologists have been accused of conspiratorial thinking. For a rebuttal, see Naomi Oreskes, "Four Obstacles to the Embrace of Agnotology in the History of Science," in *Debating Contemporary Approaches to the History of Science*, ed. Lukas M. Verburgt (London: Bloomsbury, 2024), chap. 13.

55. The cigarette conspiracy began at the Plaza Hotel on December 14, 1953. For the finding of RICO racketeering, see "Amended Final Opinion," U.S. District Court for the District of Columbia, USA vs. Philip Morris USA, http://www.public healthlawcenter.org/sites/default/files/resources/doj-final-opinion.pdf.

56. Guy Rosen, "How We're Tackling Misinformation across Our Apps," *Meta*, Mar. 22, 2021, https://about.fb.com/news/2021/03/how-were-tackling-misinforma tion-across-our-apps/.

57. We also need more work on "attribution agnotology"—that is, how citing an authority can raise doubts about a fact. Attributing a warning on a pack of cigarettes to the U.S. surgeon general, e.g., transforms it from a fact to an opinion. Something similar happens when journalists write that "environmentalists say" that carbon is contributing to climate change.

58. Roper Organization, *A Study of Public Attitudes toward Cigarette Smoking and the Tobacco Industry in 1978*, vol. 1 (n.p.: Roper Organization, May 1978), https://www.industrydocuments.ucsf.edu/tobacco/docs/#id=fgbf0086.

59. In parts of East Africa, some "witch doctors" still today kill people with albinism to use their body parts in magical rituals. See https://en.wikipedia.org/wiki/Persecution_of_people_with_albinism.

60. Nico Grant, "Google Chatbot's A.I. Images Put People of Color in Nazi-Era Uniforms," *New York Times*, Feb. 22, 2024.

61. A credit to my brother, John Proctor, for this provocative aphorism.

62. Gary S. Cross and Robert N. Proctor, *Packaged Pleasures: How Technology and Marketing Revolutionized Desire* (Chicago: University of Chicago Press, 2014).

63. For the contorted origins of this quote, see "Quote Origin: A Lie Can Travel Halfway around the World While the Truth Is Putting On Its Shoes," Quote Investigator, July 13, 2014, https://quoteinvestigator.com/2014/07/13/truth/.

64. For a diagnosis, see Jonathan Haidt, *The Anxious Generation: How the Great Rewiring of Childhood Is Causing an Epidemic of Mental Illness* (New York: Penguin, 2024).

Chapter 2

1. Copernicus Climate Change Service, "Copernicus: Summer 2024—Hottest on Record Globally and for Europe," Sept. 6, 2024, https://climate.copernicus.eu/copernicus-summer-2024-hottest-record-globally-and-europe.

2. Benjamin Franta, "Early Oil Industry Knowledge of CO_2 and Global Warming," *Nature Climate Change* 8 (2018): 1024–25.

3. Benjamin Franta, "A Future Foreseen and Transition Delayed: The American Petroleum Industry and Global Warming, 1959–1986," in *Energy Transitions*, ed. Stephen G. Gross and Andrew Needham (Pittsburgh: University of Pittsburgh Press, 2023), 55–58.

4. Franta, "Early Oil Industry Knowledge," 13.

5. Franta, "Future Foreseen," 112.

6. Christophe Bonneuil, Pierre-Louis Choquet, and Benjamin Franta, "Early Warnings and Emerging Accountability: Total's Responses to Global Warming, 1971–2021," *Global Environmental Change* 71 (2021): 102386.

7. Bonneuil, Choquet, and Franta, "Early Warnings," 19.

8. Franta, "Future Foreseen," 99.

9. Warren E. Morrison, "Application of Contingency Forecasting for the Determination of Prospective Long Run Energy Conditions in the United States," in *Proceedings of the American Institute of Mining, Metallurgical and Petroleum Engineers* (Laramie: Geology Library, University of Wyoming, 1970), 127.

10. Franta, "Future Foreseen," 26.

11. Jule G. Charney et al., *Carbon Dioxide and Climate: A Scientific Assessment* (Washington, DC: National Academy of Sciences, 1979), 1–2.

12. W. Häfele and W. Sassin, "Energy Strategies," *Energy* 1, no. 2 (1976): 147–63, https://doi.org/10.1016/0360-5442(76)90014-1.

13. John A. Laurmann, "Fossil Fuel Utilization Policy Assessment and CO_2 Induced Climate Change," in *Carbon Dioxide, Climate and Society*, ed. Jill Williams, International Institute for Applied Systems Analysis (hereafter cited as IIASA) Proceedings Series: Environment 1 (Laxenburg: IIASA, 1978), 253, 259.

14. John A. Laurmann, "Market Penetration Characteristics for Energy Production and Atmospheric Carbon Dioxide Growth," *Science* 205, no. 4409 (1979): 896.

15. Laurmann, "Market Penetration Characteristics," 897–98.

16. International Energy Agency, "World Energy Demand Growth Rate, 2011–2019," last modified April 28, 2020, https://www.iea.org/data-and-statistics/charts/world-energy-demand-growth-rate-2011-2019.

17. F. Niehaus and J. Williams, "Studies of Different Energy Strategies in Terms of Their Effects on the Atmospheric CO_2 Concentration," *Journal of Geophysical Research* 84, no. C6 (June 1979): 3123–29.

18. Jeanne Anderer, Alan McDonald, and Nebojsa Nakicenovic, *Energy in a Finite World: Paths to a Sustainable Future* (Cambridge, MA: Ballinger, 1981).

19. A. M. Perry et al., "Energy Supply and Demand Implications of CO_2," *Energy* 7, no. 12 (1982): 991.

20. Perry et al., "Energy Supply and Demand," 992, 1000.

21. Using a climate sensitivity of three degrees Celsius of warming per doubling of carbon dioxide.

22. J. F. Black to F. G. Turpin, letter and report (describing the presentation "The Greenhouse Effect," July 1977), Exxon internal documents, June 6, 1978, ID no. xqwl0228, Climate Investigations Center Collection, University of California, San Francisco (hereafter UCSF), https://www.industrydocuments.ucsf.edu/fossilfuel/docs/#id=xqwl0228.

23. Black to Turpin, vu-graph 3.

24. W. L. Ferrall to R. L. Hirsch, "Re: Controlling Atmospheric CO_2," Exxon Research and Engineering internal memorandum, Oct. 16, 1979, ID no. mqwl0228, p. 1., Climate Investigations Center Collection, UCSF, https://www.industrydocuments.ucsf.edu/fossilfuel/docs/#id=mqwl0228; Steve Knisely, *Controlling the CO_2 Concentration in the Atmosphere*, Exxon internal report, Oct. 16, 1979, ID no. mqwl0228, Climate Investigations Center Collection, UCSF, https://www.industrydocuments.ucsf.edu/fossilfuel/docs/#id=mqwl0228.

25. Knisely, *Controlling the CO_2 Concentration*, 3, appendix A.

26. Knisely, *Controlling the CO_2 Concentration*, 1, 5.

27. National Aeronautics and Space Administration, "The Relentless Rise of Carbon Dioxide," Aug. 29, 2013, https://climate.nasa.gov/climate_resources/24/graphic-the-relentless-rise-of-carbon-dioxide/.

28. Knisely, *Controlling the CO_2 Concentration*, 7.

29. Royal Dutch Shell, *Scenarios: 1989–2010; Challenge and Response* ([London]: Shell Confidential Group Planning, Oct. 1989), 36, https://s3.documentcloud.org/

documents/23735737/1989-oct-confidential-shell-group-planning-scenarios-1989
-2010-challenge-and-response-disc-climate-refugees-and-shift-to-non-fossil-fuels
.pdf.

30. R. E. Barnum, *Scoping Study on CO_2* ([Spring, TX]: Exxon, Contract Research Office, Jan. 1981), in G. H. Long to P. J. Lucchesi et al., letter and report, Exxon internal document, Feb. 5, 1981, ID no. yxfl0228, p. 13, Climate Investigations Center Collection, UCSF, https://www.industrydocuments.ucsf.edu/fossilfuel/docs/#id=yxfl0228.

31. Isabell Schrickell, "Control versus Complexity: Approaches to the Carbon Dioxide Problem at IIASA," *Berichte zur Wissenschaftsgeschichte* 40, no. 2 (2017): 140–59.

32. Benjamin Franta, "Big Carbon's Strategic Response to Global Warming, 1950–2020" (PhD diss., Stanford University, 2022), 129.

33. Franta, "Big Carbon's Strategic Response," 130–33.

34. Franta, "Big Carbon's Strategic Response," 131.

35. Naomi Oreskes and Erik M. Conway, *Merchants of Doubt: How a Handful of Scientists Obscured the Truth on Issues from Tobacco Smoke to Global Warming* (New York: Bloomsbury, 2010), 182.

36. Bonneuil, Choquet, and Franta, "Early Warnings."

37. Royal Dutch Shell, *Scenarios*, 36.

38. R. P. W. M. Jacobs et al., *The Greenhouse Effect* (The Hague: Shell Internationale Petroleum Maatschappij B.V., 1988), ID no. khfl0228, p. 29, Climate Investigations Center Collection, UCSF, https://www.industrydocuments.ucsf.edu/fossilfuel/docs/#id=khfl0228.

39. United Nations, *United Nations Framework Convention on Climate Change* ([New York]: UN, 1992), 6, 15, http://unfccc.int/files/essential_background/background_publications_htmlpdf/application/pdf/conveng.pdf.

40. Franta, "Big Carbon's Strategic Response," chaps. 8, 9, 11.

41. Robert J. Brulle, "Advocating Inaction: A Historical Analysis of the Global Climate Coalition," *Environmental Politics* 32, no. 2 (2022): 185–206.

42. Benjamin Franta, "Weaponizing Economics: Big Oil, Economic Consultants, and Climate Policy Delay," *Environmental Politics* 31, no. 4 (2021): 555–75.

43. See Sofia Hiltner et al., "Fossil Fuel Industry Influence in Higher Education: A Review and a Research Agenda," *Wiley Interdisciplinary Reviews: Climate Change* 15, no. 6 (Nov.–Dec. 2024): e904.

44. Joe Walker to Global Climate Science Team, "Draft Global Climate Science Communications Plan," email and report, April 3, 1998, ID no. jtwl0228, Climate Investigations Center Collection, UCSF, https://www.industrydocuments.ucsf.edu/fossilfuel/docs/#id=jtwl0228.

45. Some materials in this section are drawn from an article in preparation coauthored with Geoffrey Supran. See Franta, "Big Carbon's Strategic Response," 213.

46. High Meadows Environmental Institute, "About CMI," Carbon Mitigation Initiative, Princeton University, accessed July 5, 2022, https://cmi.princeton.edu/about/.

47. See S. Pacala and R. Socolow, "Stabilization Wedges: Solving the Climate Problem for the Next 50 Years with Current Technologies," *Science* 305, no. 5686 (2004): 968–72.

48. Precourt Institute for Energy, "Global Climate and Energy Project at Stanford University," Stanford University, accessed July 5, 2022, https://gcep.stanford.edu/.

49. Franta, "Big Carbon's Strategic Response," 214.

50. Robert N. Proctor, *Golden Holocaust: Origins of the Cigarette Catastrophe and the Case for Abolition* (Berkeley: University of California Press, 2011).

51. Sydney Marketing Society, "BP Rebranding in 2000: Marketing Campaign Fails #2," University of Sydney, Sept. 2, 2021, https://www.smsusyd.com/post/bp-rebranding-in-2000-marketing-campaign-fails-2.

52. Mark Kaufman, "The Carbon Footprint Sham," *Mashable*, 2021, https://mashable.com/feature/carbon-footprint-pr-campaign-sham.

53. A 2008 analysis by MIT students found that even homeless Americans have carbon footprints twice the global average. See David Chandler, "Leaving Our Mark," *MIT News*, April 16, 2008, http://news.mit.edu/2008/footprint-tt0416.

54. Franta, "Big Carbon's Strategic Response," 236, 238, 240.

55. Franta, "Big Carbon's Strategic Response," 251.

56. Franta, "Big Carbon's Strategic Response," 255, 256.

57. Franta, "Big Carbon's Strategic Response," 274–78.

58. Alternative Fuels Data Center, "Biodiesel Blends," U.S. Department of Energy, accessed Sept. 3, 2024, https://afdc.energy.gov/fuels/biodiesel_blends.html; Anjli Raval and Leslie Hook, "Oil and Gas Advertising Spree Signals Industry's Dilemma," *Financial Times*, Mar. 6, 2019, https://www.ft.com/content/5ab7edb2-3366-11e9-bd3a-8b2a211d90d5.

59. Ben Elgin and Kevin Crowley, "Exxon Retreats from Major Climate Effort to Make Biofuels from Algae," Bloomberg, Feb. 10, 2023, https://www.bloomberg.com/news/articles/2023-02-10/exxon-retreats-from-major-climate-effort-to-make-biofuels-from-algae?in_source=embedded-checkout-banner.

60. Fred Panzer to Horace R. Kornegay, "The Roper Proposal," memorandum, May 1, 1972, UCSF, http://legacy.library.ucsf.edu/tid/whz50e00.

Chapter 3

1. For background data, see Gilda Sedgh, Susheela Singh, and Rubina Hussain, "Intended and Unintended Pregnancies Worldwide in 2012 and Recent Trends," *Studies in Family Planning* 45, no. 3 (2014): 301–14, esp. 310.

2. The climate impact is as great in wealthy as in poor parts of the world because, while unwanted births are more common in poorer parts of the world, wealthier people produce far more emissions. In the United States, e.g., a remarkable 5 percent of all women of reproductive age have an unintended pregnancy *every year*. See Guttmacher Institute, "Unintended Pregnancy in the United States," Jan. 2019, https://www.guttmacher.org/sites/default/files/factsheet/fb-unintended-pregnancy-us.pdf; Jonathan M. Bearak et al., "Country-Specific Estimates of Unintended Pregnancy and Abortion Incidence: A Global Comparative Analysis of Levels in 2015–2019," *BMJ Global Health* 7 (2022): e007151.

3. U.S. Centers for Disease Control and Prevention, "Sexual Health Education," Nov. 29, 2024, https://www.cdc.gov/healthyyouth/whatworks/what-works-sexual-health-education.htm.

4. John Guillebaud, "Voluntary Family Planning to Minimise and Mitigate Climate Change," *BMJ* 353 (2016): 1–5.

5. According to the United Nations, one in every five girls is married prior to the age of eighteen; more than one in ten is married before age fifteen. See UN Human Rights Office of the High Commissioner, "Child and Forced Marriage, Including in Humanitarian Settings," United Nations, accessed Jan 27, 2025, https://www.ohchr.org/en/women/child-and-forced-marriage-including-humanitarian-set tings#:~:text=Forced%20marriage%20is%20a%20marriage,full%2C%20free%20and%20informed%20consent.

6. U.S. Centers for Disease Control and Prevention, "Understanding Pregnancy Resulting from Rape in the United States," 2020, https://www.cdc.gov/violenceprevention/sexualviolence/understanding-RRP-inUS.html (site discontinued).

7. "Distribution of Carbon Dioxide Emissions Worldwide in 2023, by Select Country," Statistica, 2021, https://www.statista.com/statistics/271748/the-largest -emitters-of-co2-in-the-world/.

8. U.S. Environmental Protection Agency, "Greenhouse Gas Emissions from a Typical Passenger Vehicle," last modified Aug. 23, 2024, https://www.epa.gov/greenvehicles/greenhouse-gas-emissions-typical-passenger-vehicle.

9. John Bongaarts and Brian C. O'Neill, "Global Warming Policy: Is Population Left Out in the Cold?" *Science* 361, no. 6403 (2018): 650–52.

10. Kirk R. Smith et al., "Human Health: Impacts, Adaptation, and Co-benefits," in *Climate Change 2014: Impacts, Adaptation, and Vulnerability*, pt. A, *Global and Sectoral Aspects: Contribution of Working Group II to the Fifth Assessment Report of the Intergovernmental Panel on Climate Change*, ed. C. B. Field et al. (Cambridge: Cambridge University Press, 2014), 740, https://archive.ipcc.ch/pdf/assessment-report /ar5/wg2/WGIIAR5-Chap11_FINAL.pdf. See also Brian C. O'Neill et al., "Demographic Change and Carbon Dioxide Emissions," *The Lancet* 380 (2012): 157–64.

11. Intergovernmental Panel on Climate Change, *Climate Change 2014: Synthesis Report; Summary for Policymakers*, Fifth Assessment Report ([Geneva]: IPCC, 2014), https://www.ipcc.ch/site/assets/uploads/2018/02/AR5_SYR_FINAL_SPM .pdf.

12. Intergovernmental Panel on Climate Change, *Climate Change 2022: Impacts, Adaptation and Vulnerability*, Sixth Assessment Report ([Geneva]: IPCC, 2022), 2702, https://www.ipcc.ch/report/sixth-assessment-report-working-group-ii/.

13. Intergovernmental Panel on Climate Change, *Climate Change 2022: Mitigation of Climate Change*, Sixth Assessment Report ([Geneva]: IPCC, 2022), 548, https://www.ipcc.ch/report/ar6/wg3/downloads/report/IPCC_AR6_WGIII_Full Report.pdf.

14. For UN Climate Change press releases, see UN Climate Change, "UN Climate Change News," accessed Nov. 6, 2022, https://unfccc.int/news?field_page_type_of _news_target_id=1842&field_page_main_text_body_value=&sort_bef_combine= created+DESC.

15. UN Department of Economic and Social Affairs, *Family Planning and the 2030 Agenda for Sustainable Development: Data Booklet* ([New York]: United Nations, 2019), 1–2, https://www.un.org/en/development/desa/population/publica tions/pdf/family/familyPlanning_DataBooklet_2019.pdf.

16. Johan Rockström et al., "A Safe Operating Space for Humanity," *Nature* 461, no. 7263 (2009): 472–75.

17. Kate Raworth, *Doughnut Economics: Seven Ways to Think like a 21st-Century Economist* (White River Junction, VT: Chelsea Green, 2017), 49.

18. Mariana Mazzucato, *Mission Economy: A Moonshot Guide to Changing Capitalism* (London: Penguin, 2021).

19. European Commission, "EU Missions in Horizon Europe," accessed Jan. 29, 2025, https://research-and-innovation.ec.europa.eu/funding/funding-opportunities/funding-programmes-and-open-calls/horizon-europe/eu-missions-horizon-europe_en.

20. Giorgos Kallis, Christian Kerschner, and Joan Martinez-Alier, "The Economics of Degrowth," *Ecological Economics* 84 (2012): 172–80.

21. Christophe Bonneuil and Jean-Baptiste Fressoz, *The Shock of the Anthropocene: The Earth, History and Us* (New York: Verso Books, 2016).

22. Paul N. Edwards, *A Vast Machine: Computer Models, Climate Data, and the Politics of Global Warming* (Cambridge, MA: MIT Press, 2010).

23. On the "causes of causes," see Robert N. Proctor, *Cancer Wars: How Politics Shapes What We Know and Don't Know about Cancer* (New York: Basic Books, 1995), 270.

24. Robert N. Proctor, *Racial Hygiene: Medicine under the Nazis* (Cambridge, MA: Harvard University Press, 1988).

25. Alexandra M. Stern, *Eugenic Nation: Faults and Frontiers of Better Breeding in Modern America* (Berkeley: University of California Press, 2016).

26. Helen Rodriguez-Trias, "The Women's Health Movement: Women Take Power," in *Reforming Medicine: Lessons of the Last Quarter Century*, ed. Victor W. Sidel and Ruth Sidel (New York: Pantheon Books, 1984), 107–26. See also Iris Ofelia López, *Matters of Choice: Puerto Rican Women's Struggle for Reproductive Freedom* (New Brunswick: Rutgers University Press, 2008).

27. Betsy Hartmann, *Reproductive Rights and Wrongs: The Global Politics of Population Control*, 3rd ed. (Chicago: Haymarket Books, 2016).

28. Lucía Berro Pizzarossa, "Here to Stay: The Evolution of Sexual and Reproductive Health and Rights in International Human Rights Law," *Laws* 7, no. 3 (2018): 29.

29. Zara Ahmed, "The Unprecedented Expansion of the Global Gag Rule: Trampling Rights, Health, and Free Speech," Guttmacher Institute, April 28, 2020, https://www.guttmacher.org/gpr/2020/04/unprecedented-expansion-global-gag-rule-trampling-rights-health-and-free-speech.

30. Ann M. Starrs, "The Trump Global Gag Rule: An Attack on U.S. Family Planning and Global Health Aid," *The Lancet* 389 (2017): 485–86.

31. Thomas McKeown, *The Modern Rise of Population* (London: Edward Arnold, 1976).

32. Frank W. Notestein, "Population: The Long View," in *Food for the World*, ed. Theodore W. Schultz (Chicago: University of Chicago Press, 1945), 36–57.

33. Saul E. Halfon, *The Cairo Consensus: Demographic Surveys, Women's Empowerment, and Regime Change in Population Policy* (Lanham, MD: Lexington Books, 2007).

34. Sonia Corrêa, "From Reproductive Health to Sexual Rights Achievements and Future Challenges," *Reproductive Health Matters* 5, no. 10 (Nov. 1997): 107–16.

35. UN Department of Economic and Social Affairs, *Programme of Action of the International Conference on Population and Development, Cairo, 5–13 September 1994* ([New York]: United Nations, 1995). And for an update, see UN Population Fund, *Programme of Action Adopted at the International Conference on Population and Development, Cairo, 5–13 September 1994* ([New York]: United Nations, 2004).

36. John Bongaarts et al., *Family Planning Programs for the 21st Century: Rationale and Design* (New York: Population Council, 2012); Steven W. Sinding, "What Has Happened to Family Planning since Cairo and What Are the Prospects for the Future?," *Contraception* 78, no. 4 (2008): S3–S6.

37. UN Department of Economic and Social Affairs, *World Population Policies: 2021; Policies Related to Fertility* (New York: United Nations, 2021), https://www.un.org/development/desa/pd/sites/www.un.org.development.desa.pd/files/undesa_pd_2021_wpp-fertility_policies.pdf.

38. UNFPA (United Nations Population Fund), *Seeing the Unseen: The Case for Action in the Neglected Crisis of Unintended Pregnancy* ([New York]: UNFPA Division for Communications and Strategic Partnerships, 2022), chap. 5, https://eeca.unfpa.org/sites/default/files/pub-pdf/en_swp22_report_0_2.pdf.

39. Nandita Bajaj and Kirsten Stade, "Challenging Pronatalism Is Key to Advancing Reproductive Rights and a Sustainable Population," *Journal of Population and Sustainability* 7, no. 1 (2023): 39–70.

40. Joseph Raulin, *De la conservation des enfans* (Paris: 1768), vol. 1, under "Épître au roi."

41. Amanda S. Barusch, "The Aging Tsunami: Time for a New Metaphor," *Journal of Gerontological Social Work* 56, no. 3 (2013): 181–84; UN Department of Economic and Social Affairs, Population Division, *World Population Ageing 2020: Highlights; Living Arrangements of Older Persons* (New York: United Nations, 2020), https://www.un.org/development/desa/pd/sites/www.un.org.development.desa.pd/files/undesa_pd-2020_world_population_ageing_highlights.pdf.

42. UN Department of Economic and Social Affairs, *World Population Policies 2021*; Monica Scigliano, "Welcome to Gilead: Pro-natalism and the Threat to Reproductive Rights," *Population Matters*, Nov. 30, 2021, https://populationmatters.org/news/2021/11/welcome-to-gilead-how-population-fears-drive-womens-rights-abuses/.

43. Jeff Pao, "China's Demographic Timebomb Starts Ticking Down," *Asia Times*, Jan. 17, 2023, https://asiatimes.com/2023/01/chinas-demographic-timebomb-starts-ticking-down/.

44. "Hungary to Provide Free Fertility Treatment to Boost Population," BBC, Jan. 10, 2020, https://www.bbc.com/news/world-europe-51061499.

45. Dina Kraft, "Where Families Are Prized, Help Is Free," *New York Times*, July 17, 2011, https://www.nytimes.com/2011/07/18/world/middleeast/18israel.html.

46. Elon Musk (@elonmusk), "Population collapse due to low birth rates is a much bigger risk to civilization than global warming," Twitter (now X), Aug. 25, 2022, 9:27 p.m., https://x.com/elonmusk/status/1563020169160851456?lang=en.

47. "About Us," Pronatalist.org, accessed April 14, 2023, https://pronatalist.org/aboutus/.

48. Naomi Oreskes and Erik M. Conway, *The Big Myth: How American Business Taught Us to Loathe Government and Love the Free Market* (New York: Bloomsbury, 2023).

49. Tim Hains, "Thanos, like Malthus, Is Wrong about Population Control," Foundation for Economic Education, Aug. 18, 2018, https://fee.org/articles/thanos-like-malthus-is-wrong-about-population-control/.

50. Marian L. Tupy and Gale L. Pooley, *Superabundance: The Story of Population Growth, Innovation, and Human Flourishing on an Infinitely Bountiful Planet* (Washington, DC: Cato, 2022), front matter. And for a critique, see Naomi Oreskes, "Eight Billion People in the World Is a Crisis, Not an Achievement," *Scientific American*, Mar. 1, 2023, https://www.scientificamerican.com/article/eight-billion-people-in-the-world-is-a-crisis-not-an-achievement/.

51. Population Research Institute, accessed April 14, 2023, https://www.pop.org.

52. "Bernie Sanders in Climate Change 'Population Control' Uproar," BBC, Sept. 5, 2019, https://www.bbc.com/news/world-us-canada-49601678.

53. Vegard Skirbekk, *Decline and Prosper! Changing Global Birth Rates and the Advantages of Fewer Children* (Cham: Springer Nature, 2022).

54. Frank Götmark, Philip Cafaro, and Jane O'Sullivan, "Aging Human Populations: Good for Us, Good for the Earth," *Trends in Ecology and Evolution* 33, no. 11 (2018): 851–62.

55. "Countries with the Highest Military Spending Worldwide in 2023," Statista, April 2024, https://www.statista.com/statistics/262742/countries-with-the-highest-military-spending/.

56. Bongaarts and O'Neill, "Global Warming Policy," supp.

57. According to the World Bank, the average resident of a high-income nation contributes ten tons of carbon per year, while the average inhabitant of a low-income nation contributes only 0.2 tons. By one calculation, each new baby born in the United Kingdom will generate (on average) thirty-five times more greenhouse gas emissions than a baby born in Bangladesh. See Guillebaud, "Voluntary Family Planning," 1–3.

58. UN Department of Economic and Social Affairs, *Programme of Action*, 11.

59. Eileen Crist, Camilo Mora, and Robert Engelman, "The Interaction of Human Population, Food Production, and Biodiversity Protection," *Science* 356, no. 6335 (2017): 260–64; Eileen Crist et al., "Scientists' Warning on Population," *Science of the Total Environment* 845, no. 157166 (2022): 1–5.

60. Paul R. Ehrlich and Anne H. Ehrlich, "The Population Bomb Revisited," *Electronic Journal of Sustainable Development* 1, no. 3 (2009): 63–71.

61. Paul R. Ehrlich and John P. Holdren, "Population and Panaceas: A Technological Perspective," *BioScience* 19, no. 12 (Dec. 1969): 1070–71, https://academic.oup.com/bioscience/article/19/12/1065/497574.

62. Bongaarts and O'Neill, "Global Warming Policy," supp.

63. Center for Reproductive Rights, "The World's Abortion Laws," accessed June 2, 2023, https://reproductiverights.org/maps/worlds-abortion-laws/.

64. Center for Reproductive Rights, "Categories of Abortion Laws from Most to Least Restrictive," accessed June 2, 2023, https://reproductiverights.org/wp-con tent/uploads/2022/09/WALM_20220927_V1.pdf.

65. Robert Engelman, *More: Population, Nature, and What Women Want* (Washington, DC: Island, 2008).

66. Yoichi Kaya and Keiichi Yokobori, eds., *Environment, Energy, and Economy: Strategies for Sustainability* (Tokyo: United Nations University Press, 1997).

Chapter 4

1. Ann Rudinow Saetnan, Ingrid Schneider, and Nicola Green, eds., *The Politics and Policies of Big Data: Big Data, Big Brother?* (New York: Routledge, 2018).

2. The term "AI" is terribly overused. Although most of what I discuss here falls under the category of statistical machine learning (a subfield of AI), I will use the more popular and succinct term "AI."

3. Four petabytes is four million gigabytes, or approximately twenty-eight million hours of high-definition video.

4. Marc Faddoul, Guillaume Chaslot, and Hany Farid, "A Longitudinal Analysis of YouTube's Promotion of Conspiracy Videos," preprint, submitted Mar. 6, 2020, https://arxiv.org/abs/2003.03318.

5. Robert N. Proctor and Londa Schiebinger, eds., *Agnotology: The Making and Unmaking of Ignorance* (Stanford, CA: Stanford University Press, 2008).

6. Jeff Horwitz and Deepa Seetharaman, "Facebook Executives Shut Down Efforts to Make the Site Less Divisive," *Wall Street Journal*, May 26, 2020, https://www.wsj.com/articles/facebook-knows-it-encourages-division-top-executives -nixed-solutions-11590507499.

7. QAnon is a Far Right, far-reaching conspiracy that began with the claim that Satan-worshipping cannibalistic pedophiles plotted against Donald Trump and has since morphed into a movement promoting increasingly bizarre and fantastical prognostications.

8. Faddoul, Chaslot, and Farid, "Conspiracy Videos."

9. Renee DiResta, "Online Conspiracy Groups Are a Lot like Cults," *Wired*, Nov. 13, 2018, https://www.wired.com/story/online-conspiracy-groups-qanon-cults/.

10. Hany Farid, "Creating, Using, Misusing, and Detecting Deep Fakes," *Journal of Online Trust and Safety* 1, no. 4 (2022), https://doi.org/10.54501/jots.v1i4.56.

11. Sophie J. Nightingale and Hany Farid, "AI-Synthesized Faces Are Indistinguishable from Real Faces and More Trustworthy," *Proceedings of the National Academy of Sciences* 119, no. 8 (2022): e2120481119, https://doi.org/10.1073/pnas .2120481119.

12. BuzzFeedVideo, "You Won't Believe What Obama Says in This Video!," posted April 17, 2018, YouTube, https://www.youtube.com/watch?v=cQ54GDm1eL0.

13. Metaphysic.ai, "Deeptomcruise," TikTok, https://www.tiktok.com/@deep tomcruise.

14. Vaness Etienne, "Val Kilmer Gets His Voice Back after Throat Cancer Battle Using AI Technology: Hear the Results," *People*, August 19, 2021, https://people .com/movies/val-kilmer-gets-his-voice-back-after-throat-cancer-battle-using-ai -technology-hear-the-results/.

15. Zero Malaria Britain, "David Beckham Speaks Nine Languages to Launch Malaria Must Die Voice Petition," posted April 8, 2019, by Zero Malaria Britain, YouTube, https://www.youtube.com/watch?v=QiiSAvKJIHo.

16. Morgan Neville, dir., *Roadrunner: A Film about Anthony Bourdain* (CNN Films, HBO Max, Tremolo Productions, and Zero Point Zero, 2021).

17. Helen Rosner, "The Ethics of a Deepfake Anthony Bourdain Voice," *New Yorker*, July 17, 2021, https://www.newyorker.com/culture/annals-of-gastronomy/the-ethics-of-a-deepfake-anthony-bourdain-voice.

18. Bobby Chesney and Danielle Citron, "Deep Fakes: A Looming Challenge for Privacy, Democracy, and National Security," *California Law Review* 107 (Dec. 2019): 1753.

19. Shannon Bond, "People Are Trying to Claim Real Videos Are Deepfakes," NPR, May 8, 2023, https://www.npr.org/2023/05/08/1174132413/people-are-trying-to-claim-real-videos-are-deepfakes-the-courts-are-not-amused.

20. Ilia Shumailov et al., "The Curse of Recursion: Training on Generated Data Makes Models Forget," preprint, submitted May 27, 2023, https://arxiv.org/abs/2305.17493.

21. Nicholas Carlini et al., "Poisoning Web-Scale Training Datasets Is Practical," preprint, submitted Feb. 20, 2023, https://arxiv.org/abs/2302.10149.

22. I am an advisor to the CAI.

23. Jeff Kosseff, *The Twenty-Six Words That Created the Internet* (Cornell: Cornell University Press, 2019).

24. The libel suit against Prodigy was brought by Stratton Oakmont, a New York brokerage firm whose executives were found guilty and imprisoned for defrauding customers in 1999. The firm's escapades were immortalized in the blockbuster film *The Wolf of Wall Street* (2013).

25. Hany Farid and Brandie Nonnecke, "The Case for Regulating Platform Design," *Wired*, Mar. 13, 2023, https://www.wired.com/story/make-platforms-safer-regulate-design-section-230-gonzalez-google/.

Chapter 5

This work is supported by the European Research Council grant DALOSS, no. 101078386. Its completion is owed to the contributions of many. I am deeply grateful to Londa Schiebinger and Robert N. Proctor for inviting me into this project and placing me in the company of such inspiring thinkers, many of whom have profoundly influenced my work. I also extend sincere thanks to the archivists, librarians, and data managers from the Bodleian Library, the Danish Royal Library, and the Danish National Archive for generously sharing their insights on data loss. Special appreciation goes to Thomas Gammeltoft-Hansen, Louis Ravn, Katie MacKinnon, Esmée Colbourne, and Frederik Schade for their invaluable feedback. I am also thankful to the anonymous reviewers for their excellent suggestions on how to improve the text. Finally, my heartfelt thanks to Laura Clough, whose diligent copyediting supported me in completing this chapter during a time of mourning. I dedicate this chapter to my father, Asger.

1. See, e.g., Jean-François Blanchette and Deborah G. Johnson, "Data Retention and the Panoptic Society: The Social Benefits of Forgetfulness," *Information*

Society 18, no. 1 (2002): 33–45; Kate Eichhorn, *The End of Forgetting: Growing Up with Social Media* (Cambridge, MA: Harvard University Press, 2019); Viktor Mayer-Schönberger, *Delete: The Virtue of Forgetting in the Digital Age* (Princeton, NJ: Princeton University Press, 2009), 201; Michael A. Peters and Tina Besley, "Digital Archives in the Cloud: Collective Memory, Institutional Histories and the Politics of Information," *Educational Philosophy and Theory* 51, no. 10 (2019): 1020–29.

2. Thomas H. Lenhard, "Data Destruction," in *Data Security: Technical and Organizational Protection Measures against Data Loss and Computer Crime* (Wiesbaden: Springer, 2022), 55–60; David S. H. Rosenthal, "Format Obsolescence: Assessing the Threat and the Defenses," *Library Hi Tech* 28, no. 2 (2010): 195–210; Muira McCammon and Jessa Lingel, "Situating Dead-and-Dying Platforms: Technological Failure, Infrastructural Precarity, and Digital Decline," *Internet Histories* 6, no. 1–2 (2022): 1–13.

3. On outages, e.g., see Lawrence Abrams, "Amazon AWS Outage Shows Data in the Cloud Is Not Always Safe," Bleeping Computer, Sept. 5, 2019, https://www.bleepingcomputer.com/news/technology/amazon-aws-outage-shows-data-in-the-cloud-is-not-always-safe/.

4. Marcus Becker, "The European Commission Deletes Mass Amounts of Emails and Doesn't Archive Chats," Spiegel International, Nov. 12, 2021, https://www.spiegel.de/international/europe/a-new-controversy-erupts-around-ursula-von-der-leyen-s-text-messages-a-6510951f-e8dc-4468-a0af-2ecd60e77ed9; Haroon Siddique, "Cabinet Policy Obliges Ministers to Delete Instant Messages," *The Guardian*, Oct. 12, 2021, https://www.theguardian.com/politics/2021/oct/12/cabinet-policy-ministers-delete-whatsapp-messages; Lise Rønn Tofte and Søren Larsen, "Mette Frederiksen om at genskabe slettede sms'er: Det har desværre ikke været muligt," DR Nyheder, June 21, 2023; Nicholas Confessore, "Cambridge Analytica and Facebook: The Scandal and the Fallout So Far," *New York Times*, April 4, 2018, https://www.nytimes.com/2018/04/04/us/politics/cambridge-analytica-scandal-fallout.html.

5. The cultural history of knowledge is ripe with examples; see Proctor and Schiebinger, eds., *Agnotology: The Making and Unmaking of Ignorance* (Stanford, CA: Stanford University Press, 2008). See also Peter Burke, *What Is the History of Knowledge?* (Cambridge: Cambridge University Press, 2016); Burke, *Ignorance: A Global History* (New Haven, CT: Yale University Press, 2023). Also see, writing from the trenches of archives and libraries, Richard Ovenden, *Burning the Books* (Cambridge, MA: Harvard University Press, 2022). Each in their own ways offers excellent forays not only into the history and historiography of knowledge but also into how knowledge is hidden, discarded, and destroyed. Within the framework of the history of science, read also Londa Schiebinger's incisive analysis of the role of gender and enslavement in the status and suppression of scientific knowledge in her *Plants and Empire: Colonial Bioprospecting in the Atlantic World* (Cambridge, MA: Harvard University Press, 2004).

6. Arlette Farge, *The Allure of the Archives* (New Haven, CT: Yale University Press, 2014), 87.

7. Michelle Caswell, "Dusting for Fingerprints: Introducing Feminist Standpoint

Appraisal," *Journal of Critical Library and Information Studies* 3, no. 2 (2021): 1–36.

8. Ann Stoler, "Imperial Debris: Reflections on Ruins and Ruination," *Cultural Anthropology* 23, no. 2 (May 2008): 192.

9. Lindsey Dillon et al., "Environmental Data Justice and the Trump Administration: Reflections from Environmental Data and Governance Initiative," *Environmental Justice* 10, no. 6 (2017): 186–92. A recent study reveals that the median age of a given research-data repository is a little more than a decade. Daniel Strecker et al., "Disappearing Repositories: Taking an Infrastructure Perspective on the Long-Term Availability of Research Data," *Quantitative Science Studies* 4, no. 4 (2023): 839–56.

10. Tim Smith, "An Exabyte of Disk Storage at CERN: CERN Disk Storage Capacity Passes the Threshold of One Million Terabytes of Disk Space," CERN (blog), Sept. 29, 2023, https://home.web.cern.ch/news/news/computing/exabyte-disk-storage -cern.

11. Robert N. Proctor, "Agnotology: A Missing Term to Describe the Cultural Production of Ignorance (and Its Study)," in *Agnotology: The Making and Unmaking of Ignorance*, ed. Proctor and Londa Schiebinger (Stanford, CA: Stanford University Press, 2008), 10.

12. Proctor, "Agnotology," 14.

13. Ethan Siegel, "Has the Large Hadron Collider Accidentally Thrown Away the Evidence for New Physics?," *Forbes*, Sept. 13, 2018, https://www.forbes.com /sites/startswithabang/2018/09/13/has-the-large-hadron-collider-accidentally -thrown-away-the-evidence-for-new-physics.

14. Brewster Kahle, "Universal Access to All Knowledge," interview by Nanna Bonde Thylstrup, Long Now Foundation, Nov. 30, 2011, https://archive.org/details /brewsterkahlelongnowfoundation.

15. Nanna Bonde Thylstrup et al., "Politics of Data Reuse in Machine Learning Systems: Theorizing Reuse Entanglements," *Big Data and Society* 9, no. 2 (2022), https://doi.org/10.1177/20539517221139785.

16. Some of these services show just how much mediation is also a question of logistics, not least of which is the recently discontinued AWS Snowmobile—a road-worthy truck that was designed to transport exabyte-scale datasets.

17. Devika Narayan, "Platform Capitalism and Cloud Infrastructure: Theorizing a Hyper-scalable Computing Regime," *Environment and Planning A: Economy and Space* 54, no. 5 (2022): 911–29.

18. Dom Phillips, "Brazil's Museum Fire: 'Incalculable' Loss as 200-Year-Old Rio Institution Gutted," *The Guardian*, Sept. 3, 2018, quoted in Ofri Cnaani, "Museum and Its Afterness," Museum Why, Aug. 3, 2022, https://museumwhy.com/museums -and-its-afterness-by-ofri-cnaani/.

19. Anna Pingen, "Commission Proposes New Regulations to Improve Cybersecurity and Information Security of EU Administration," Eucrim, April 26, 2022, https://eucrim.eu/news/commission-proposes-new-regulations-to-improve -cybersecurity-and-information-security-of-eu-administration/.

20. Rachel Douglas-Jones, Antonia Walford, and Nick Seaver, introduction to "Towards an Anthropology of Data," ed. Douglas-Jones, Walford, and Seaver, special issue, *Journal of the Royal Anthropological Institute* 27, no. S1 (2021): 9–25.

21. Paul Baran, "On Distributed Communications Networks," *IEEE Transactions on Communications Systems* 12, no. 1 (1964): 1–9.

22. T. H. Hu, *A Prehistory of the Cloud* (Cambridge, MA: MIT Press, 2015); Alexander R. Galloway, "Global Networks and the Effects on Culture," *Annals of the American Academy of Political and Social Science* 597, no. 1 (2005): 19–31.

23. A. R. Taylor, "Future-Proof: Bunkered Data Centres and the Selling of Ultra-secure Cloud Storage," *Journal of the Royal Anthropological Institute* 27, no. S1 (2021): 76–94.

24. Chris Willman, "Musicians Freak Out as They Belatedly Learn Myspace Lost 50 Million Songs," *Variety*, Mar. 18, 2019, https://variety.com/2019/music/news/myspace-music-data-loss-50-million-songs-1203165649/.

25. Josh Taylor, "Google Cloud Accidentally Deletes UniSuper's Online Account Due to 'Unprecedented Misconfiguration,'" *The Guardian*, May 9, 2024, https://www.theguardian.com/australia-news/article/2024/may/09/unisuper-google-cloud-issue-account-access.

26. Nanna Bonde Thylstrup et al., "Infrapolitics, Archival Infrastructures and Digital Reparative Practices," in *Feminist Digital Humanities: Interventions in Praxis*, ed. Susan Schreibman and Lisa Rhody (Champaign: University of Illinois Press, forthcoming).

27. British Library, *Learning Lessons from the Cyber-Attack: British Library Cyber Incident Review* (London: British Library, Mar. 8, 2024), 1–18, https://www.bl.uk/home/british-library-cyber-incident-review-8-march-2024.pdf/.

28. Pallab Ghosh, "Google's Vint Cerf Warns of Digital Dark Age," BBC, Feb. 13, 2005, https://www.bbc.com/news/science-environment-31450389.

29. Jonas Palm, *The Digital Black Hole* (Stockholm: Riksarkivet, 2005), https://cdn-kb.avanet.nl/wp-content/uploads/2019/05/31173546/Palm-Black-Hole.pdf.

30. Jonathan Zittrain, Kendra Albert, and Lawrence Lessig, "Perma: Scoping and Addressing the Problem of Link and Reference Rot in Legal Citations," *Legal Information Management* 14, no. 2 (2014): 88–99.

31. Athena Chapekis et al., "When Online Content Disappears," Pew Research Center, May 17, 2024, https://www.pewresearch.org/data-labs/2024/05/17/when-online-content-disappears/.

32. Jeremy Bulow, "An Economic Theory of Planned Obsolescence," *Quarterly Journal of Economics* 101, no. 4 (Nov. 1986): 729–49.

33. Andy Klein, "Backblaze Drive Stats for Q1 2023," Backblaze, May 4, 2023, https://www.backblaze.com/blog/backblaze-drive-stats-for-q1-2023/#:~:text=As%20of%20the%20end%20of,Edition%3A%202022%20Drive%20Stats%20review.

34. Erik Poppe et al., "Is It a Bug or a Feature? The Concept of Software Obsolescence," paper presented at the Fourth Plate Virtual Conference, Limerick, Ireland, May 2021.

35. Amelia Acker and Adam Kreisberg, "Social Media Data Archives in an API-Driven World," *Archival Science* 20 (2020): 105–23.

36. Paul N. Edwards, "Platforms Are Infrastructures on Fire," in *Your Computer Is on Fire*, ed. Thomas S. Mullaney et al. (Cambridge, MA: MIT Press, 2021), 314–36.

37. Amelia Acker, "Accessing Software: Emulation in Information Institutions," *Information and Culture* 59, no. 1 (2024): 1–19.

38. Rowena Mason, "End Government by WhatsApp, Urges Former GCHQ Head," *The Guardian*, Dec. 26, 2023, https://www.theguardian.com/politics/2023/dec/26/end-government-by-whatsapp-urges-former-gchq-head.

39. In principle, the law is actually technology neutral, meaning that filing practices only take the informational content and not the format into account, but the case nevertheless led to the Ministry of Justice issuing new "guidelines for the storage of deleted emails by government authorities," which states that government authorities should ensure that deleted emails are stored for five years for employees, ten years for individuals in managerial positions, and twenty-five years for ministers, special advisors, and permanent secretaries. "Retningslinjer for statslige myndigheders opbevaring af slettede e-mails," Justitsministeriet, Sept. 28, 2021, https://www.justitsministeriet.dk/pressemeddelelse/nye-retningslinjer-for-statslige-myndigheders-opbevaring-af-slettede-e-mails/.

40. Peter Teffer, "European Commission Officials Admitted That Internal Record-Keeping Rules Were Vague," Follow the Money, Mar. 9, 2022, https://www.ftm.eu/articles/von-der-leyen-european-commission-internal-communication?share=feD67Gc8Y9ufrRxUlMgruOiSd%2FKvlBGg3FdrRdi8HyImFsmwW7u%2BNo8GapAQWLE%3D.

41. Take the uncomfortable scene in the now classic TV show *Succession* in which nephew Greg becomes complicit in the unlawful activities of his uncle's firm thanks to a late-night shredding session.

42. Kathleen F. Brickey, "Andersen's Fall from Grace," *Washington University Law Quarterly* 81, no. 4 (2003): 917.

43. Ashley Parker et al., " 'He Never Stopped Ripping Things Up': Inside Trump's Relentless Document Destruction Habits," *Washington Post*, Feb. 5, 2022, https://www.washingtonpost.com/politics/2022/02/05/trump-ripping-documents/.

44. Sarah Blacker, "Analogue Privacy: The Paper Shredder as a Technology for Knowledge Destruction," in *Boxes: A Field Guide*, ed. Susanne Bauer, Martina Schlünder, and Maria Rentetzi (Manchester: Mattering Press, 2020), 365.

45. Blacker, "Analogue Privacy."

46. Robert N. Proctor, *Golden Holocaust: Origins of the Cigarette Catastrophe and the Case for Abolition* (Berkeley: University of California Press, 2011).

47. Shred-it, "Secure Hard Drive Destruction Services for Your Business," accessed Aug. 7, 2024, https://www.shredit.com/en-us/secure-shredding-services/hard-drive-destruction.

48. Ministry of Justice, "Secure Disposal of IT—Public and Private Cloud," accessed Aug. 7, 2024, https://security-guidance.service.justice.gov.uk/secure-disposal-of-it-public-and-private-cloud/.

49. Ameera Ahmad and Ed Vulliamy, "In Gaza, the Schools Are Dying Too," *The Guardian*, Jan. 10, 2009, https://www.theguardian.com/world/2009/jan/10/gaza-schools.

50. Scholars against the War on Palestine, 2024, https://scholarsagainstwar.org. Taking stock, according to the assessment of UN experts, as of April 2024, over 80

percent of schools have been damaged or destroyed by the Israeli assault on Gaza, with thousands of students and hundreds of teachers killed, and many thousands more injured. Moreover, every university in Gaza has been partially or wholly destroyed, whether by bombing or demolition.

51. Achille Mbembe, "The Power of the Archive and Its Limits," in *Refiguring the Archive*, ed. Carolyn Hamilton et al. (Cape Town: David Philip, 2002), 19–26.

52. Susan Leigh Star, "The Ethnography of Infrastructure," *American Behavioral Scientist* 43, no. 3 (1999): 377–91; Heather Horst, Jolynna Sinanan, and Larissa Hjorth, "Storing and Sharing: Everyday Relationships with Digital Material," *New Media and Society* 23, no. 4 (2021): 657–71.

53. On their use as training data, see Louise Amoore, *Cloud Ethics: Algorithms and the Attributes of Ourselves and Others* (Durham, NC: Duke University Press, 2020).

Chapter 6

1. Peter Galison, "Removing Knowledge," *Critical Inquiry* 31, no. 1 (2004): 229–43; Galison, "Blacked-Out Spaces: Freud, Censorship, and the Re-territorialization of the Mind," *British Journal for the History of Science* 45, no. 2 (2012): 235–66; Galison and Robb Moss, dirs., *Secrecy* (Redacted Pictures, 2008), http://www.secrecymovie.net/.

2. Peter Galison, "Secrecy in Three Acts," *Social Research* 77, no. 2 (2010): 941–74.

3. Peter Galison and Caroline A. Jones, *Invisibilities: Seeing and Unseeing the Anthropocene* (Princeton, NJ: Zone Books, forthcoming).

4. Trespassing to Collect Data, State of Wyoming Legislature, Senate, SF0012, 63rd Legislature, 2015 General Session, introduced in Senate Jan. 14, 2015, §W.S. 6-3-414(c), accessed July 20, 2023, https://wyoleg.gov/Legislation/2015/SF0012 (hereafter cited as §W.S. 6-3-414); Trespassing to Collect Data-Civil Cause of Action, State of Wyoming Legislature, Senate, SF0080, 63rd Legislature, 2015 General Session, introduced in Senate Jan. 16, 2015, §W.S. 40-26-101(c), accessed July 20, 2023, https://wyoleg.gov/Legislation/2015/SF0080 (hereafter cited as §W.S. 40-26-101). The former law imposed criminal punishment, while the latter levied civil liability.

5. Congressional Research Service, *U.S. Department of the Interior: An Overview* ([Washington, DC]: CRS, last modified June 23, 2021), https://sgp.fas.org/crs/misc/R45480.pdf.

6. Jonathan Ratner, in discussion with the author, April 27, 2023.

7. Ratner, discussion.

8. It is an essential and common practice for the EPA and other governmental agencies to solicit and receive submissions of data from volunteers in the public, which they use to carry out their work. They provide instructions on how these submissions are best carried out. See Western Watersheds Project v. Michael, No. 16-8083 (10th Cir. 2017) at 11, accessed July 20, 2023, https://law.justia.com/cases/federal/appellate-courts/ca10/16-8083/16-8083-2017-09-07.html. See also U.S. Environmental Protection Agency, Office of Water, Volunteer Stream Monitoring: A

Methods Manual ([Washington, DC]: U.S. EPA, Nov. 1997), https://www.epa.gov/sites/default/files/2015-06/documents/stream.pdf.

9. Ratner, discussion.

10. For the standard analytical protocol for *E. coli* as promulgated by the EPA, see U.S. Environmental Protection Agency, *Standard Analytical Protocol for* Escherichia coli *O157:H7 in Water* ([Washington, DC]: U.S. EPA, Sept. 2010), https://cfpub.epa.gov/si/si_public_record_report.cfm?Lab=NHSRC&subject=Homeland%20Security%20Research&dirEntryId=224124.

11. Ratner, discussion. Collecting data from 2007 to 2014, Ratner and the WWP came into further conflict with activist ranchers. In 2006, the Bureau of Land Management (BLM) proposed revising their regulations to limit the BLM's enforcement powers and to increase ranchers' ownership rights to own and use improvements to public lands. The WWP pushed back, challenging the amendments both substantively and procedurally, contending that necessary consultations had not taken place as required, e.g., by the Endangered Species Act. The ranchers defended the loosening of BLM's power to regulate, but the district court found in favor of WWP and enjoined the BLM from making the changes. Back and forth the struggle went.

12. On the agreement, continuing disputes, and the funds, see Emilene Ostlind, "Surprises Flow from Ruby Pipeline," *High Country News*, Aug. 23, 2010, https://www.hcn.org/issues/42.15/surprises-flow-from-ruby-pipeline.

13. The bankruptcy was reported in the *Wall Street Journal*. See Jonathan Randles and Jodi Xu Klein, "Kinder Morgan's Ruby Pipeline Files for Bankruptcy," *Wall Street Journal*, April 1, 2022, https://www.wsj.com/articles/kinder-morgans-ruby-pipeline-files-for-bankruptcy-11648772226.

14. Erik Ryberg, "Protecting an Ancient Heritage From Grazing," *Watersheds Messenger* 20, no. 1 (Spring 2013): 5, https://westernwatersheds.org/wp-content/uploads/2013/04/WWPNews4-2013LR.pdf.

15. Larry Hicks, "Senate Afternoon Session," 5th General Session, Jan. 19, 2015, Archived Floor Debate, State of Wyoming 63rd Legislature, MP3 audio, 2:54:02, https://www.wyoleg.gov/Legislation/audioArchive/2015 (hereafter cited as Hicks, "Senate Afternoon Session").

16. Hicks, "Senate Afternoon Session."

17. Hicks, "Senate Afternoon Session."

18. Regarding the criminal version, the Senate passed it on March 2, 2015, and the House passed it the next day, March 3, 2015. The law went into effect March 5, 2015. See §W.S. 6-3-414. Regarding the civil version, the House passed it on March 4, 2015, and the Senate passed it on March 5, 2015. The law went into effect July 1, 2015. See §W.S. 40-26-101.

19. §W.S. 6-3-414 (d)(i).

20. For the percentage of U.S. ownership, see Scott [username], "Public and Private Land Percentages by US States," Summitpost, accessed July 16, 2023, https://www.summitpost.org/public-and-private-land-percentages-by-us-states/186111.

21. §W.S. 6-3-414 (a).

22. On a case of prosecution (and exculpation) of corner-hopping hunters, see Angus M. Theurmer Jr., "Jury Finds Four Corner-Crossing Hunters Not Guilty

of Trespass," WyoFile, April 29, 2022, https://wyofile.com/jury-finds-four-corner
-crossing-hunters-not-guilty-of-trespass/. See also *Iron Bar Holdings v Cape*, No.
23-8043 (10th Cir. 2025).

23. §W.S. 6-3-414(d)(iv).

24. *Western Watersheds Project*, No. 16–8083, 10.

25. §W.S. 6-3-414(f).

26. The state's motion to dismiss was denied in part on December 28, 2015. For
the original complaint filing date, see Western Watersheds Project v. Michael, No.
15–00169, "Complaint for Declaratory and Injunctive Relief," PACER Doc #1 (D.
Wyo. 2015) at 1, accessed July 18, 2023, https://storage.courtlistener.com/recap/gov
.uscourts.wyd.34510/gov.uscourts.wyd.34510.1.0.pdf. One aspect of the complain-
ants' case was their insistence that the supremacy clause (Article VI, clause 2) of the
Constitution was violated; the court did not accept this claim about preemption.

27. Chris Rothfuss, "Senate Morning Session," 19th Budget Session, Mar. 3,
2016, Archived Floor Debate, State of Wyoming 63rd Legislature, MP3 audio,
1:53:03, https://www.wyoleg.gov/Legislation/audioArchive/2016.

28. Charles Scott, "Senate Morning Session," 19th Budget Session, Mar. 3, 2016,
Archived Floor Debate, State of Wyoming 63rd Legislature, MP3 audio, 1:53:03,
https://www.wyoleg.gov/Legislation/audioArchive/2016.

29. University of Wyoming, *Trespassing to Unlawfully Collect Resource Data:
Frequently Asked Questions*, last modified May 14, 2015, 2–3, http://web.archive
.org/web/20230607105308/https://www.uwyo.edu/research/permission-to-collect
-resource-data/trespassfaqs.pdf. Another interesting question posed in the FAQ is,
"Does the new law only apply to private land, or does it also apply to public land?
Some concerns have been expressed regarding the terms 'open land' as distinct from
'private open land' in Wyoming statute 6-3-414 (a)(i). The legal advice provided to
UW [University of Wyoming] personnel is unchanged. Under the law, UW researchers must have statutory, contractual or other legal authority to enter or access public
land outside municipalities and subdivisions, as well as private land." Ibid., 2.

30. *Western Watersheds Project*, No. 16–8083, 9–10.

31. *Western Watersheds Project*, No. 16–8083, 11.

32. *Western Watersheds Project*, No. 16–8083, 11. "Other environmental statutes and regulations likewise require public input in crafting policy. See, e.g., 40
C.F.R. § 1500.1(b) (under the National Environmental Policy Act, federal agencies
must consider '[a]ccurate scientific analysis' in making environmental decisions,
subject to 'public scrutiny'); 43 USC § 1712(a), (f) (Federal Land Policy and Management Act requires land use plans be developed 'with public involvement')." Ibid.

33. Western Watersheds Project v. Michael, No. 15–00169, "ORDER by the Hon-
orable Scott W Skavdahl Granting 95 Motion for Summary Judgment; Denying 98
Motion for Summary Judgment," PACER Doc #113 (D. Wyo. 2018) at 11, accessed
July 17, 2023, https://www.publicjustice.net/wp-content/uploads/2018/10/WWP
-Decision.pdf (hereafter cited as Skavdahl, "Motion for Summary Judgement").

34. Skavdahl, "Motion for Summary Judgement," 2018, 24–25.

35. Adjacent Land Resource Data Trespass—Repeal, State of Wyoming Legislative Service Office, "Summary," SF0031, 67th Legislature, 2023 General Session,

introduced in Senate Jan. 10, 2023, §W.S. 6-3-414, accessed July 20, 2023, https://wyoleg.gov/2023/Summaries/SF0031.pdf.

36. Trespass by Small Unmanned Aircraft, State of Wyoming Legislature, Senate, SF0034, 67th Legislature, 2023 General Session, introduced in Senate Jan. 10, 2023, §W.S. 6-3-308 (a), accessed July 25, 2023, https://wyoleg.gov/Legislation/2023/SF0034; emphasis added.

37. Brett Moline, "Senate Judiciary Meeting, January 13, 2023," posted Jan. 13, 2023, by Wyoming Legislature, YouTube, 1:01:06, at 25:21–26:50, https://www.youtube.com/watch?v=_p5qmWC-GNA (hereafter cited as "Judiciary Meeting, January 13").

38. Wendy Schuler and Brett Moline, "Judiciary Meeting, January 13," at 34:31–36:52.

39. Jim Magagna and Travis Deti, "Judiciary Meeting, January 13," at 43:20–43:47, 53:55–54:13.

40. Bill Landen, "Senate Judiciary Meeting, January 16, 2023," posted Jan. 16, 2023, by Wyoming Legislature, YouTube, 44:57, at 12:24–13:30, https://www.youtube.com/watch?v=xhnblpGqdL4&t=637s.

41. U.S. Department of Defense, Office of Prepublication and Security Review, *Controlled Unclassified Information Markings* ([Washington, DC]: U.S. DoD, Oct. 23, 2020), 18, https://www.dcsa.mil/Portals/91/Documents/CTP/CUI/DOD-CUI_Marking_Handbook-DOD_(2020).pdf.

42. Unknown author to R. A. Pittman, memorandum, Aug. 21, 1969, https://www.industrydocuments.ucsf.edu/docs/xqkd0134, quoted in Robert N. Proctor, *Cancer Wars: How Politics Shapes What We Know and Don't Know about Cancer* (New York: Basic Books, 1995), 110. See also Stanton A. Glantz et al., eds., *The Cigarette Papers* (Berkeley: University of California Press, 1996), 190.

43. Naomi Oreskes and Erik M. Conway, *Merchants of Doubt: How a Handful of Scientists Obscured the Truth on Issues from Tobacco Smoke to Global Warming* (New York: Bloomsbury, 2010).

44. Jonathan Zittrain et al., "The Shifting Landscape of Global Internet Censorship," Berkman Klein Center for Internet and Society, June 2017, http://nrs.harvard.edu/urn-3:HUL.InstRepos:33084425.

Chapter 7

This chapter is adapted from *The Big Myth* © 2023 by Naomi Oreskes and Erik M. Conway with permission from Bloomsbury Publishing Inc. We are grateful to our research assistant, Gustave Lester.

2. Charles Mills, "White Ignorance," in *Agnotology: The Making and Unmaking of Ignorance*, ed. Robert N. Proctor and Londa Schiebinger (Stanford, CA: Stanford University Press, 2008), 230–49. See also Jennifer C. Mueller, "Imagine an Ignorance That Fights Back: Honoring Charles Mills, Our Inheritance and Charge," *Sociology of Race and Ethnicity* 8, no. 4 (2022): 443–50, https://doi.org/10.1177/23326492221119889; Andrew Seal, "Charles W. Mills's Challenge for International Historians," Rethinking Intellectual History, Aug. 3, 2022, https://metahistory.substack.com/p/charles-w-millss-challenge-for-intellectual.

3. For an expanded discussion of this issue, see Naomi Oreskes and Erik M. Conway, *The Big Myth: How American Business Taught Us to Loathe Government and Love the Free Market* (New York: Bloomsbury, 2023).

4. Martin Wolf, *The Crisis of Democratic Capitalism* (London: Penguin, 2023).

5. It is widely recognized that labor freedom has been an exception to the capitalist exhortations of the values of economic freedom, particularly when considering labor mobility between nation-states. See, e.g., Ha-Joon Chang, *23 Things They Don't Tell You about Capitalism* (London: Bloomsbury, 2012). However, there is a large and complex story of *internal* labor restrictions alongside the rise of capitalism. This is a major point in Karl Polanyi's famous indictment of capitalism, for which see Polanyi, *The Great Transformation* (1944; repr., Boston, MA: Beacon, 2001). Polanyi argues that English workers were immiserated not only because of inadequate wages but also because they were often prohibited from moving in search of work or better working conditions. Polanyi particularly stresses the 1662 Act of Settlement and the 1691 and 1697 amendments to it, which required workers to remain in their parishes under most conditions. There were at least half a dozen statutes, including various antivagrancy statutes, that limited worker mobility from the sixteenth to the eighteenth centuries in England. For a discussion of whether these statutes were enforced and how workers found ways around them, see Hilary Cooper and S. R. S. Szreter, *After the Virus: Lessons from the Past for a Better Future* (Cambridge: Cambridge University Press, 2021). While the details may be arguable, the basic point is not: free movement of labor was—and indeed remains—a vexed issue.

6. "La majestueuse égalité des lois, qui interdit au riche comme au pauvre de coucher sous les ponts, de mendier dans les rues et de voler du pain" (In its majestic equality, the law forbids rich and poor alike to sleep under bridges, beg in the streets, and steal loaves of bread). Anatole France, *The Red Lily*, trans. Winifred Stephens (London: John Lane, 1922), 95.

7. Oreskes and Conway, *Big Myth*, 1–15.

8. "George J. Stigler: Facts," The Nobel Prize (website), May 1, 2021, https://www.nobelprize.org/prizes/economic-sciences/1982/stigler/facts/.

9. A 1966 facsimile edition was published in two volumes, each over five hundred pages. See Adam Smith, *An Inquiry into the Nature and Causes of the Wealth of Nations*, 2 vols. (New York: Augustus M. Kelley, 1966).

10. Smith, *Wealth of Nations*, 250; emphasis added. This passage is particularly significant because it is one of the few places where Smith uses the word "restraint" in a positive context; most of his discussion of restraints—e.g., in the context of mercantilism in book 4—are critical. See particularly book 4, chapter 3, where he discusses the "unreasonableness" of mercantilist restraints on trade. I am indebted to Ron Hoffman, former member of the Council of Economic Advisors under President Richard Nixon, for calling my attention to this passage. I later found it discussed as well in Tahany Naggar, "Adam Smith's Laissez Faire," *American Economist* 21, no. 2 (1977): 35–38, esp. 36; and Jacob Viner, "Adam Smith and Laissez Faire," *Journal of Political Economy* 35, no. 2 (1927): 198–232, esp. 224–25.

11. Smith, *Wealth of Nations*, 224. This passage is also quoted by Amartya Sen

to argue that Smith emphasized the necessity of virtues other than self-interest for the market economy to function, as well as for the proper functioning of society as a whole. See Sen, "Uses and Abuses of Adam Smith," *History of Political Economy* 43, no. 2 (2011): 257–71, esp. 266.

12. Adam Smith, "On Money Considered as a Particular Branch of the General Stock of the Society, or of the Expense of Maintaining the National Capital," in *An Inquiry into the Nature and Causes of the Wealth of Nations*, bk. 2, *On the Nature, Accumulation, and Employment of Stock*, Marxists Internet Archive, accessed Feb. 19, 2025, https://www.marxists.org/reference/archive/smith-adam/works/wealth-of-nations/book02/ch02.htm#:~:text=A%20certain%20quantity%20of%20very,him%20in%20their%20proper%20proportions.

13. Smith, *Wealth of Nations*, 229, 228–30.

14. Smith, *Wealth of Nations*, 233–34.

15. Smith, *Wealth of Nations*, 247, 248–49.

16. Smith, *Wealth of Nations*, 250.

17. Friedrich von Hayek, *The Road to Serfdom: Text and Documents; the Definitive Edition*, ed. Bruce Caldwell (Chicago: University of Chicago Press, 2014). For a discussion of its significance, see, e.g., James Pethokoukis, "Taking a Step toward a Better Understanding of Friedrich Hayek's 'Road to Serfdom,'" AEIdeas, Jan. 25, 2023, https://www.aei.org/economics/taking-a-step-toward-a-better-understanding-of-friedrich-hayeks-road-to-serfdom/.

18. Hayek, *Road to Serfdom*, 20.

19. Adam Smith, *Wealth of Nations*, ed. Charles J. Bullock (New York: Cosimo Classics, 2007), 70.

20. Smith, *Wealth of Nations*, 61.

21. Smith, *Wealth of Nations*, 62.

22. Smith, *Wealth of Nations*, 63.

23. On variations in wages, see Smith, *Wealth of Nations*, 86–101.

24. Smith, *Wealth of Nations*, 70. This is discussed in Emma Rothschild, "Adam Smith and Conservative Economics," *Economic Historical Review* 45, no. 1 (1992): 74–96. See also Jules Steinberg, *"To Be Themselves Tolerably Well Fed, Cloathed, and Lodged": Liberalism, the Humanization of Labor, and Adam Smith's Protests against the Injustice of Working Class Poverty and Misery* (self-pub., BookLocker, 2019).

25. Smith, *Wealth of Nations*, 71.

26. Smith, *Wealth of Nations*, 70. For a discussion, see also Rothschild, "Adam Smith"; Steinberg, *"To Be Themselves."* This passage reminds us of the important point, stressed as well by Sen, that even though capitalism was in its infancy in 1776, many of its failures were already evident. Sen states, "Even to Adam Smith . . . the huge limitations of relying entirely on the market economy and only on the profit motive were absolutely clear." Sen, "Uses and Abuses," 260. Sen sees in this the explanation for how Marx could admire Smith yet revile John Stuart Mill.

27. Smith, *Wealth of Nations*, 73.

28. Smith, *Wealth of Nations*, 73–74.

29. Smith, *Wealth of Nations*, ed. Bullock, 151. See also the discussion in Sen, "Uses and Abuses," 262; emphasis added. Sen stresses that Smith believed that self-

interest was a powerful and useful motivator of economic activity, but by no means did he think it was the only valid—or even only existing—motivation for human action, as many latter-day libertarians weirdly insist.

30. Rothschild, "Adam Smith," 84.

31. Sen, "Uses and Abuses," 262.

32. Smith, *Wealth of Nations*, 625.

33. Smith, *Wealth of Nations*, 634–35.

34. E.g., Michael Huemer, "Is Taxation Theft?," Libertarianism.org, Mar. 16, 2017, https://www.libertarianism.org/columns/is-taxation-theft.

35. Smith, *Wealth of Nations*, 634, 625. We are not Marx scholars, but the language is so close that it seems reasonable to ponder whether Marx got his famous phrase ("From each according to . . .") from Smith.

36. Stigler's selective vision is evident in not least his attitude toward race. In 1962, he wrote an astonishing piece blaming the plight of African Americans on themselves, including this passage: "The Negro boy is excluded from many occupations by the varied barriers the prejudice can raise. . . . [B]ut he is excluded from more occupations by his own inferiority as a worker. . . . Consider the Negro as a neighbor. He is frequently repelled and avoided by the white man . . . because the Negro family is, on average, a loose, morally lax, group, and brings with its presence a rapid rise in crime and vandalism." The contrasting experiences of Jewish and Black people in America, Stigler holds, proved that the problems of the latter were mostly of their own making. George J. Stigler, "The Problem of the Negro," Grasping Reality on TypePad, accessed Aug. 19, 2024, https://www.bradford-delong.com/2019/05/weekend-reading-george-stigler-in-1962-on-the-problem-of-the-negro.html. The potential refutations of this position are numerous, but one excellent example is the driving out of Black homeowners and businesspeople and seizing of their property in Manhattan Beach, California. See Jacey Fortin, "This Black Family Ran a Thriving Beach Resort 100 Years Ago: They Want Their Land Back," *New York Times*, Mar. 11, 2021, https://www.nytimes.com/2021/03/11/us/bruce-family-manhattan-beach.html.

37. Viner, "Adam Smith," 227, 231 (also quoted in Naggar, "Adam Smith's Laissez Faire," 37), 218. For a discussion of Viner and his historical importance, see P. D. Groenewegen, "Jacob Viner and the History of Economic Thought," *Contributions to Political Economy* 13, no. 1 (1994): 69–86.

38. Viner, "Adam Smith," 218; italics in the original. Viner also notes that at times, Smith seems to suggest that government is part of the natural order—and not a distortion of it. However, his later venerators mostly rejected this view, framing "the free market" as part of the natural order of things and government as exogenous to it. Thus, government actions are viewed as "interventions" rather than part of the natural order of things.

39. Smith's hostility toward Oxford dons may have been linked to his hostility toward established religion since the universities and colleges were all ecclesiastical institutions at this time. On Smith's views of established religion, see Rothschild, "Adam Smith," 91. Or it may simply be that he was ill and unhappy while at Oxford. See John Rae, *Life of Adam Smith* (New York: Macmillan, 1895).

40. Adam Smith, *Selections from "The Wealth of Nations,"* ed. George J. Stigler (Wheeling, IL: Harlan Davidson, 1957), 108–9.

41. Smith, *Selections*, 113, 110.

42. There is, of course, a regulatory solution to this, which Smith hinted at: the state could mandate tranferability of credits and/or the transportability of scholarships.

43. This is the only use I found when reading the book. Searching Google Books reveals several additional uses of the verb "regulate" but no other use of the noun "regulation." *"Selections from 'The Wealth of Nations'* by Adam Smith," Google Books, accessed April 7, 2020, https://www.google.com/books/edition/Selections_from_The_Wealth_of_Nations/hPNjBAAAQBAJ?hl=en&gbpv=1&bsq=regulation.

44. Nathan Rosenberg, "George Stigler, Adam Smith's Best Friend," *Journal of Political Economy* 101, no. 5 (Oct. 1993): 833–48; Sherwin Rosen, "George J. Stigler and the Industrial Organization of Economic Thought," *Journal of Political Economy* 101, no. 5 (Oct. 1993): 809–17. The alleged dichotomy between the Adam Smith of *The Wealth of Nations* and the Adam Smith of *The Theory of Moral Sentiments* was known in nineteenth-century Germany as "Das Adam Smith Problem." The idea was that these books were contradictory—that there were, in effect, two Adam Smiths. In 2015, David Wilson and William Dixon suggested the problem was "to understand how Smith's postulation of self-interest as the organising principle of economic activity fits in with his wider moral-ethical concerns." They concluded that the solution is to recognize "the different levels of social reality to which Smith refers in his discourse." Wilson and Dixon, "Das Adam Smith Problem: A Critical Realist Perspective," *Journal of Critical Realism* 5, no. 2 (2006): 252. We see a simpler solution: to recognize that when one reads the whole of *The Wealth of Nations*, one sees that Smith is not the dogmatic polemicist that he has been taken to be and in fact allows for important exceptions to reliance on self-interest. Interestingly, a recent blog post from the Competitive Enterprise Institute agrees, although for different reasons than we offer here. See Ryan Young, "Das Adam Smith Problem? Nein!," Competitive Enterprise Institute, June 20, 2023, https://cei.org/blog/das-adam-smith-problem-nein/. See also Keith Tribe, " 'Das Adam Smith Problem' and the Origins of Modern Smith Scholarship," *History of European Ideas* 34, no. 4 (Dec. 2008): 514–25.

45. Rothschild, "Adam Smith."

46. See, for a starting point, Emma Rothschild, *"The Theory of Moral Sentiments* and the Inner Life," in *Essays on the Philosophy of Adam Smith*, ed. Vivienne Brown and Samuel Fleischacker (Oxfordshire: Routledge, 2014), 25–36.

47. Sen emphasizes that for Smith, the market was an important societal institution but by no means the only one. Nonmarket institutions particularly play a role in providing public goods, in addition to regulating the market when it needs to be regulated. See Sen, "Uses and Abuses," 258–59.

48. Sen, "Uses and Abuses," 258.

49. Craig Freedman, "Was George Stigler Adam Smith's Best Friend? Studying the History of Economic Thought," *Journal of the History of Economic Thought* 29, no. 2 (2007): 223.

50. Sen, "Uses and Abuses," 258, 257–71 passim.

51. Jeff Madrick, "How Alan Greenspan Helped Wreck the Economy," *Rolling Stone*, June 16, 2011, https://www.rollingstone.com/politics/politics-news/how -alan-greenspan-helped-wreck-the-economy-231162/.

52. Gregory Teddy Eow, "Fighting a New Deal: Intellectual Origins of the Reagan Revolution,1932–1952" (PhD diss., Rice University, 2007), 83.

53. Sen, "Uses and Abuses," 258.

54. See, e.g., the Amazon page for Stigler's version: *"Selections from 'The Wealth of Nations'* by Adam Smith," Amazon, accessed Aug. 19, 2024, https://tiny url.com/mv86dsnr.

55. Sen, "Uses and Abuses," 259. Sen observes Smith was impressed by the power and dynamism of markets, and he gave a compelling account of "how that dynamism" worked, but that is a far cry from viewing markets as comprehensive, and there is nothing in Smith to indicate that he believed in the self-sufficiency of the market economy.

56. Viner, "Adam Smith," 227. See also the discussion in Groenewegen, "Jacob Viner."

Chapter 8

1. John R. Lott and Carlisle E. Moody, "Brought into the Open: How the US Compares to Other Countries in the Rate of Public Mass Shooters," *Econ Journal Watch* 17, no. 1 (2020): 38.

2. John R. Lott, "Comparing Murder Rates and Gun Ownership across Countries," Crime Prevention Research Center, Mar. 31, 2014, https://crimeresearch.org /2014/03/comparing-murder-rates-across-countries/.

3. Darrell Huff, *How to Lie with Statistics*, ill. Irving Geis (New York: W. W. Norton, 1954).

4. "CPRC: Comparing Murder Rates and Gun Ownership across Countries," Firearms Owners against Crime, Oct. 9, 2017, https://foac-pac.org/Cprc:-Com paring-Murder-Rates-And-Gun-Ownership-Across-Countries/News-Item/7188; Robert Farago, "Gun Control Advocates' Logic in a Nutshell," The Truth about Guns, Sept. 11, 2015, https://www.thetruthaboutguns.com/gun-control-advocates -logic-in-a-nutshell/.

5. Michael S. Rosenwald, "The NRA Once Believed in Gun Control and Had a Leader Who Pushed for It," *Washington Post*, Feb. 22, 2018, https://www.wash ingtonpost.com/news/retropolis/wp/2017/10/05/the-forgotten-nra-leader-who -despised-the-promiscuous-toting-of-guns/.

6. Warren Burger, interview by Charlayne Hunter-Gault, *MacNeil/Lehrer NewsHour*, PBS, Dec. 16, 1991.

7. The first Trump administration aggressively pursued this strategy. A recent study found that "more Trump judges were NRA members (9.3%) than non-Trump Republican (1.6%) and Democratic judges (0.7%) were, differences significant at the 1% levels." The authors note, "In light of the polarizing nature of gun rights and the NRA's association with extreme views on gun ownership, jurists who seek a reputation for impartiality would normally want to avoid membership in the NRA.

That changed during the Trump administration, whose judicial appointments were represented in the NRA at a rate four times greater than that of the adult population." Stephen J. Choi, Mitu Gulati, and Eric A. Posner, "Trump's Lower-Court Judges and Religion: An Initial Appraisal," *Virginia Public Law and Legal Theory Research Paper* 49 (2023): 10–12.

8. Maggie Koerth and Amelia Thomson-DeVeaux, "Many Americans Are Convinced Crime Is Rising in the U.S.: They're Wrong," FiveThirtyEight, Aug. 3, 2020, https://fivethirtyeight.com/features/many-americans-are-convinced-crime-is-rising -in-the-u-s-theyre-wrong/.

9. Lincoln Quillian and Devah Pager, "Estimating Risk: Stereotype Amplification and the Perceived Risk of Criminal Victimization," *Social Psychology Quarterly* 73, no. 1 (2010): 79–104.

10. Nina Totenberg, "Justice Scalia, the Great Dissenter, Opens Up," *Morning Edition*, NPR, April 28, 2008, https://www.npr.org/2008/04/28/89986017/justice -scalia-the-great-dissenter-opens-up.

11. Richard A. Posner, "The Incoherence of Antonin Scalia," *New Republic*, Aug. 24, 2012.

12. John Paul Stevens, "The Supreme Court's Worst Decision of My Tenure," *The Atlantic*, May 14, 2019, https://www.theatlantic.com/ideas/archive/2019/05/john-paul -stevens-court-failed-gun-control/587272/. For two more of the many strong criticisms of Scalia's decision and its "faux originalism," both by conservative Reagan-appointed federal court of appeals judges, see Richard A. Posner, "In Defense of Looseness," *New Republic*, Aug. 27, 2008; and J. Harvie Wilkinson III, "Of Guns, Abortions, and the Unraveling Rule of Law," *Virginia Law Review* 95, no. 2 (2009): 253–323.

13. District of Columbia v. Heller, 554 U.S. 570 (2008); emphasis added.

14. Anne Swearer and Scott French, "Brett Kavanaugh's Defense of Second Amendment Is Hardly 'Extremist,'" Heritage Foundation, July 18, 2018, https: //www.heritage.org/courts/commentary/brett-kavanaughs-defense-second -amendment-hardly-extremist.

15. Jeffrey M. Jones, "Majority in U.S. Continues to Favor Stricter Gun Laws," Gallup, Oct. 31, 2023, https://news.gallup.com/poll/513623/majority-continues -favor-stricter-gun-laws.aspx.

16. Victoria Balara, "Fox News Poll: Voters Favor Gun Limits over Arming Citizens to Reduce Gun Violence," Fox News, April 27, 2023, https://www.foxnews .com/official-polls/fox-news-poll-voters-favor-gun-limits-arming-citizens-reduce -gun-violence. The McCourtney Institute for Democracy's May 2023 Mood of the Nation poll found that "86 percent of American adults support U.S. Congress mandating background checks for all firearm sales and transfers." Craig Helmstetter, Eric Plutzer, and Rithwik Kalale, "Poll: A Majority of Americans Support Universal Background Checks, Gun Licensing and an Assault Weapons Ban," MPR News, July 25, 2023, https://www.mprnews.org/story/2023/07/25/poll-majority--support -universal-background-checks-gun-licensing-assault-weapons-ban.

17. Polling Institute, "Gun Owners Divided on Gun Policy: Parkland Students Having an Impact," Monmouth University, Mar. 8, 2018, https://www.monmouth .edu/polling-institute/reports/MonmouthPoll_US_030818/.

18. Indeed, the threat goes beyond overriding the will of the people. When Russian president Vladimir Putin sought to exercise influence over the American government, his agent used the gun lobby to infiltrate Republican circles of power. See Associated Press, "Timeline of Suspected Russian Plot to Infiltrate NRA, GOP," Sept. 10, 2018, https://apnews.com/article/d83211350d2743d9a22b9c72ee048b4f.

19. John R. Lott and David B. Mustard, "Crime, Deterrence, and Right-to-Carry Concealed Handguns," *Journal of Legal Studies* 26, no. 1 (1997): 1–68.

20. Frank Zimring and Gordon Hawkins, "Concealed Handguns: The Counterfeit Deterrent," *Responsive Community* 7, no. 2 (Spring 1997): 46–60.

21. John R. Lott, "Simple Tests for the Extent of Vote Fraud with Absentee and Provisional Ballots in the 2020 US Presidential Election" (working paper, Crime Prevention Research Center, 2020); Lott, "A Suggestion and Some Evidence" (presentation, Presidential Advisory Commission on Election Integrity, Washington, DC, Sept. 12, 2017), Trump White House Archives, https://trumpwhitehouse.archives .gov/articles/presidential-advisory-commission-election-integrity-resources-2/.

22. On January 22, 2021, John R. Lott was deposed in *Miller v. Becerra*, calling it "highly likely" that Donald Trump won the 2020 presidential election. Miller v. Cal. Attorney Gen. Xavier Becerra, 488 F. Supp. 3d 949 (S.D. Cal. 2020) at esp. 80–81.

23. National Research Council, *Firearms and Violence: A Critical Review* (Washington, DC: National Academies, 2005), https://www.ojp.gov/ncjrs/virtual -library/abstracts/firearms-and-violence-critical-review.

24. James Q. Wilson, "Appendix A: Dissent," in *Firearms and Violence: A Critical Review*, ed. Charles F. Wellford, John V. Pepper, and Carol V. Petrie (Washington, DC: National Academies Press, 2005), 269, https://www.doi.org/10.17226 /10881.

25. James Q. Wilson, "Appendix B: Committee Response to Wilson's Dissent," in Wellford, Pepper, and Petrie, *Firearms and Violence*, 275.

26. John J. Donohue, "Applying What We Know and Building an Evidence Base: Reducing Gun Violence," in *The Oxford Handbook of Evidence-Based Crime and Justice Policy*, ed. Brandon C. Welsh, Steven N. Zane, and Daniel P. Mears (New York: Oxford University Press, 2024), table 28.1.

27. John J. Donohue, Abhay Aneja, and Kyle D. Weber, "Right-to-Carry Laws and Violent Crime: A Comprehensive Assessment Using Panel Data and a State-Level Synthetic Control Analysis," *Journal of Empirical Legal Studies* 16, no. 2 (2019): 198–247.

28. The figure supports the parallel-trends assumption, critical to identifying the causal impact of the intervention in question (here, the adoption of an RTC law).

29. John J. Donohue et al., "More Guns, More Unintended Consequences: The Effects of Right-to-Carry on Criminal Behavior and Policing in US Cities" (NBER Working Paper Series 30190, National Bureau of Economic Research, Cambridge, MA, June 2022), https://www.nber.org/papers/w30190.

30. Mitchell L. Doucette et al., "Officer-Involved Shootings and Concealed Carry Weapons Permitting Laws: Analysis of Gun Violence Archive Data, 2014–2020," *Journal of Urban Health* 99, no. 3 (2022): 373–84.

31. William English, "The Right to Carry Has Not Increased Crime: Improving an Old Debate through Better Data on Permit Growth over Time" (working paper,

McDonough School of Business Research Paper Series 3887151, Georgetown University, Washington, DC, July 2021), https://dx.doi.org/10.2139/ssrn.3887151.

32. New York State Rifle & Pistol Assn., Inc. v. Bruen, 597 U.S. 1 (2022) (No. 20–843), transcript of oral argument, 35, https://www.supremecourt.gov/DocketPDF/20/20-843/184380/20210720135957258_20-843%20Amicus%20Brief%20of%20William%20English%20et%20al..PDF.

33. *Bruen*, transcript, 119.

34. RAND, *Effects of Concealed-Carry Laws on Violent Crime*," last modified July 16, 2024, https://www.rand.org/research/gun-policy/analysis/concealed-carry/violent-crime.html.

35. *Bruen*, transcript, 3.

36. Congress passed the Dickey Amendment in 1996, which stopped federal funding of gun-violence research for more than twenty years. Notably, the Republican congressman from Arkansas who wrote the amendment stated twenty years later, "[I] now believe strongly that federal funding for research into gun-violence prevention should be dramatically increased." See Jay Dickey and Mark Rosenberg, "How to Protect Gun Rights While Reducing the Toll of Gun Violence," *Washington Post*, Dec. 25, 2015, https://www.washingtonpost.com/opinions/time-for-collaboration-on-gun-research/2015/12/25/f989cd1a-a819-11e5-bff5-905b92f5f94b_story.html. Of course, now that the *Bruen* decision has altogether dismissed empirical evidence in favor of pure constitutionality, the value of such research may only be to document, rather than prevent, harm.

37. Brennan Gardner Rivas, "When Texas Was the National Leader in Gun Control," *Washington Post*, Sept. 12, 2019, https://www.washingtonpost.com/outlook/2019/09/12/when-texas-was-national-leader-gun-control/.

38. Andrew Anglemyer, Tara Horvath, and George Rutherford, "The Accessibility of Firearms and Risk for Suicide and Homicide Victimization among Household Members: A Systematic Review and Meta-analysis," *Annals of Internal Medicine* 160, no. 2 (2014): 101–10.

39. David M. Studdert et al., "Homicide Deaths among Adult Cohabitants of Handgun Owners in California, 2004 to 2016: A Cohort Study," *Annals of Internal Medicine* 175, no. 6 (2022): 804–11.

40. "We are going to defeat virtually every gun control on the books—assault weapons bans, large capacity magazine bans, ammunition registration, rosters of approved handguns for sale, limitations on how many guns you can buy in a month," Sam Paredes, the executive director of the Gun Owners of California, said. "The courts have held that these laws don't have an analogous law to 1791 when the Second Amendment was written, so they are by definition unconstitutional." See Shawn Hubler, "In the Gun Law Fights of 2023, a Need for Experts on the Weapons of 1791," *New York Times*, last modified Mar. 16, 2023, https://www.nytimes.com/2023/03/14/us/gun-law-1791-supreme-court.html#:~:text=supreme%2Dcourt.html-,In%20the%20Gun%20Law%20Fights%20of%202023%2C%20a%20Need%20for,sending%20demand%20soaring%20for%20historians.

41. Duncan v. Becerra, 265 F. Supp. 3d 1106, 1139 (S.D. Cal. 2017) at 1135.

42. David Hemenway, "The Myth of Millions of Annual Self-Defense Gun Uses:

A Case Study of Survey Overestimates of Rare Events," *Chance* 10, no. 3 (1997): 6–10.

43. Miller v. Bonta, 542 F. Supp. 3d 1009 (S.D. Cal. 2021) at 1061, 1041.

44. John J. Donohue, "The Effect of Permissive Gun Laws on Crime," in "Preventing Gun Violence in America: What Works and What Is Possible," ed. Cassandra Crifasi, Jennifer Necci Dineen, and Kerri M. Raissian, special issue, *Annals of the American Academy of Political and Social Science* 704, no. 1 (2022): 92–117.

45. The federal statute in question was 18 U.S.C. § 922(g)(8).

46. The opinion striking down the federal law reiterated *Bruen*'s assertion that courts were not allowed to weigh the benefits of the law against its burdens. The Fifth Circuit claimed it was only relevant that "our ancestors would never have accepted" the law on domestic violence restraining orders. See United States v. Rahimi, 59 F.4th 163 (5th Cir. 2023) at 12.

47. Paul Waldman, "Jason Aldean Cashes in on the Right-Wing Fantasy of Violent Retribution," *Washington Post*, July 20, 2023, https://www.washingtonpost.com/people/paul-waldman/?itid=ai_top_waldmanp.

48. James Bickerton, "Fox News Guest's Advice after Texas Mass Shooting Goes Viral," *Newsweek*, last modified May 8, 2023, https://www.newsweek.com/fox-news-guests-advice-after-texas-mass-shooting-goes-viral-1798868.

49. Rep. Marjorie Taylor Greene (@RepMTG), "The federal government must partner with states for mental hospitals and drug rehab centers for the good of our society," X, May 6, 2023, https://x.com/RepMTG/status/1655004445229367296?ref_src=twsrc%5Etfw%7Ctwcamp%5Etweetembed%7Ctwterm%5E1655004445229367296%7Ctwgr%5Eb777e1d7e80ce97ca2d27faeae0d3ce8d86c02bf%7Ctwcon%5Es1_&ref_url=https%3A%2F%2Fwww.newsweek.com%2Fmarjorie-taylor-greene-says-us-must-study-factors-behind-mass-shootings-1798863.

50. Tal Axelrod, "After Mall Shooting, Abbott Says 'Mental Health Crisis' Must Be Solved; Biden Calls for Weapons Ban," *ABC News*, May 7, 2023, https://abcnews.go.com/Politics/after-mall-shooting-abbott-mental-health-crisis-solved/story?id=99152931.

51. As Anthony A. Braga and Philip J. Cook find, "Intrinsic power . . . had a direct effect on the likelihood that a victim of a criminal shooting died. For Boston, in the period studied here, simply replacing larger-caliber guns with small-caliber guns with no change in location or number of wounds would have reduced the gun homicide rate by 39.5%. It is plausible that larger reductions would be associated with replacing all types of guns with knives or clubs." Braga and Cook, "The Association of Firearm Caliber with Likelihood of Death from Gunshot Injury in Criminal Assaults," *JAMA Network Open* 1, no. 3 (2018): e180833.

Chapter 9

1. Global Burden of Disease 2021 Diabetes Collaborators, "Global, Regional, and National Burden of Diabetes from 1990 to 2021, with Projections of Prevalence to 2050: A Systematic Analysis for the Global Burden of Disease Study 2021," *Lancet* 402, no. 10397 (2023): 203–34.

2. Benjamin Koh et al., "Patterns in Cancer Incidence among People Younger than 50 Years in the US, 2010 to 2019," *JAMA Network Open* 6, no. 8 (2023):

e2328171; Global Burden of Disease 2019 Dementia Forecasting Collaborators, "Estimation of the Global Prevalence of Dementia in 2019 and Forecasted Prevalence in 2050: An Analysis for the Global Burden of Disease Study 2019," *Lancet Public Health* 7, no. 2 (2022): e105–25.

3. Halsted R. Holman, "The Relation of the Chronic Disease Epidemic to the Health Care Crisis," *ACR Open Rheumatology* 2, no. 3 (2020): 167–73.

4. Nikolaj Nottelmann, "Ignorance," in *Cambridge Dictionary of Philosophy*, 3rd ed., ed. Robert Audi (Cambridge: Cambridge University Press, 2015).

5. Tori Rodriguez, "Emotional Ignorance Harms Health," *Scientific American*, July 1, 2014, https://www.scientificamerican.com/article/emotional-ignorance -harms-health/.

6. Gerald Dworkin, "Voluntary Health Risks and Public Policy," *Hastings Center Report* 11, no. 5 (1981): 26–31; Dan Wikler, "Who Should Be Blamed for Being Sick?," *Health Education Quarterly* 14, no. 1 (1987): 11–25.

7. Anthony R. Cashmore, "The Lucretian Swerve: The Biological Basis of Human Behavior and the Criminal Justice System," *Proceedings of the National Academy of Sciences USA* 107, no. 10 (2010): 4499–504.

8. Ellen P. Williams et al., "Overweight and Obesity: Prevalence, Consequences, and Causes of a Growing Public Health Problem," *Current Obesity Reports* 4, no. 3 (2015): 363–70.

9. Centers for Disease Control, "Prevalence of Childhood Obesity in the United States," 2022, https://www.cdc.gov/obesity/php/data-research/childhood-obesity -facts.html#:~:text=Expand%20All-,Age,%25%20among%20adolescents%2012% E2%80%9319 (site discontinued); Miriam Garrido-Miguel et al., "Prevalence and Trends of Overweight and Obesity in European Children from 1999 to 2016: A Systematic Review and Meta-analysis," *JAMA Pediatrics* 173, no. 10 (2019): e192430.

10. Scott Howell and Richard Kones, " 'Calories In, Calories Out' and Macronutrient Intake: The Hope, Hype, and Science of Calories," *American Journal of Physiology: Endocrinology and Metabolism* 313, no. 5 (2017): e608–12.

11. Lawrence Grobel, "Playboy Interview: Jesse Ventura," *Playboy*, Nov. 1999, 55–66, 184–85.

12. James Sensenbrenner, "Personal Responsibility in Food Consumption Act," *Congressional Record Volume* 150, no. 30 (2004): h946–81, https://www.govinfo .gov/content/pkg/CREC-2004-03-10/html/CREC-2004-03-10-pt1-PgH946-2 .htm.

13. Jonathan Chiang et al., "Geographic and Longitudinal Trends in Media Framing of Obesity in the United States," *Obesity* 28, no. 7 (2020): 1351–57.

14. Myroslava Protsiv et al., "Decreasing Human Body Temperature in the United States since the Industrial Revolution," *eLife* 9 (2020): e49555.

15. Yann C. Klimentidis et al., "Canaries in the Coal Mine: A Cross-Species Analysis of the Plurality of Obesity Epidemics," *Proceedings of Biological Sciences* 278, no. 1712 (2011): 1626–32.

16. Juhee Kim et al., "Trends in Overweight from 1980 through 2001 among Preschool-Aged Children Enrolled in a Health Maintenance Organization," *Obesity* 14, no. 7 (2006): 1164–71.

17. Barbara King, "Can Babies Be Obese?," *13.7: Cosmos and Culture* (blog),

NPR, Jan. 14, 2018, https://www.npr.org/sections/13.7/2016/01/14/463072893/can-babies-be-obese.

18. D. J. Barker, "The Fetal and Infant Origins of Adult Disease," *BMJ* 301 (1990): 1111, https://pubmed.ncbi.nlm.nih.gov/2252919/.

19. Emily Oken and Matthew W. Gillman, "Fetal Origins of Obesity," *Obesity Research* 11, no. 4 (2003): 496–506.

20. Imogen Rogers and EURO-BLCS Study Group, "The Influence of Birthweight and Intrauterine Environment on Adiposity and Fat Distribution in Later Life," *International Journal of Obesity* 27, no. 7 (2003): 755–77.

21. Robert H. Lustig, "Childhood Obesity: Behavioral Aberration or Biochemical Drive? Reinterpreting the First Law of Thermodynamics," *Nature Clinical Practice Endocrinology and Metabolism* 2, no. 8 (2006): 447–548.

22. Robert H. Lustig et al., "Hypothalamic Obesity Caused by Cranial Insult in Children: Altered Glucose and Insulin Dynamics and Reversal by a Somatostatin Agonist," *Journal of Pediatrics* 135, no. 2, pt. 1 (1999): 162–68.

23. Pedro A. Velasquez-Mieyer et al., "Suppression of Insulin Secretion Is Associated with Weight Loss and Altered Macronutrient Intake and Preference in a Subset of Obese Adults," *International Journal of Obesity Related Metabolic Disorders* 27 (2003): 219–26.

24. Kevin D. Hall et al., "Ultra-processed Diets Cause Excess Calorie Intake and Weight Gain: An Inpatient Randomized Controlled Trial of Ad Libitum Food Intake," *Cell Metabolism* 30, no. 1 (2019): 67–77.

25. Robert H. Lustig, "Fructose, Fatty Liver, and Addiction," in *Food and Addiction*, 2nd ed., ed. Ashley N. Gearhardt et al. (New York: Oxford University Press, 2024), 328–37.

26. Carlos A. Monteiro et al., "Ultra-processed Foods: What They Are and How to Identify Them," *Public Health Nutrition* 22, no. 5 (2019): 936–41.

27. Roberto De Vogli, Anne Kouvonen, and David Gimeno, "The Influence of Market Deregulation on Fast Food Consumption and Body Mass Index: A Cross-National Time Series Analysis," *Bulletin of the World Health Organization* 92, no. 2 (2014): 99–107.

28. Bernard Srour et al., "Ultra-processed Food Consumption and Risk of Type 2 Diabetes among Participants of the NutriNet-Santé Prospective Cohort," *JAMA Internal Medicine* 180, no. 2 (2020): 283–91.

29. Bernard Srour et al., "Ultra-processed Food Intake and Risk of Cardiovascular Disease: Prospective Cohort Study (NutriNet-Santé)," *BMJ* 365 (2019): l1451.

30. Thibauld Fiolet et al., "Consumption of Ultra-processed Foods and Cancer Risk: Results from NutriNet-Santé Prospective Cohort," *BMJ* 360 (2018): k322.

31. Huiping Li et al., "Association of Ultra-processed Food Consumption with Risk of Dementia: A Prospective Cohort Study," *Neurology* 99, no. 10 (2022): e1056–66.

32. Eric M. Hecht et al., "Cross-Sectional Examination of Ultra-processed Food Consumption and Adverse Mental Health Symptoms," *Public Health Nutrition* 25, no. 11 (2022): 3225–34.

33. Committee on Addictions of the Group for the Advancement of Psychiatry,

"Responsibility and Choice in Addiction," *Psychiatric Services* 53, no. 6 (2002): 707–13.

34. Markus Heilig et al., "Addiction as a Brain Disease Revised: Why It Still Matters, and the Need for Consilience," *Neuropsychopharmacology* 46, no. 10 (2021): 1715–23.

35. Eliza L. Gordon et al., "What Is the Evidence for 'Food Addiction'? A Systematic Review," *Nutrients* 10, no. 4 (2018): 477; Roland R. Griffiths and Allison L. Chausmer, "Caffeine as a Model Drug of Dependence: Recent Developments in Understanding Caffeine Withdrawal, the Caffeine Dependence Syndrome, and Caffeine Negative Reinforcement," *Nihon shinkei seishin yakurigaku zasshi* 20, no. 5 (2000): 223–31.

36. Erica M. Schulte, Nicole M. Avena, and Ashley N. Gearhardt, "Which Foods May Be Addictive? The Roles of Processing, Fat Content, and Glycemic Load," *PLOS One* 10, no. 2 (2015): e0117959.

37. Nicole M. Avena, Pedro Rada, and Bartley G. Hoebel, "Evidence for Sugar Addiction: Behavioral and Neurochemical Effects of Intermittent, Excessive Sugar Intake," *Neuroscience and Biobehavioral Reviews* 32, no. 1 (2008): 20–39.

38. Marcia L. Pelchat et al., "Images of Desire: Food-Craving Activation during fMRI," *Neuroimage* 23, no. 4 (2004): 1486–93.

39. Magalie Lenoir et al., "Intense Sweetness Surpasses Cocaine Reward," *PLOS One* 2, no. 8 (2007): e698.

40. Joan R. Ifland et al., "Refined Food Addiction: A Classic Substance Use Disorder," *Medical Hypotheses* 72, no. 5 (2009): 518–26.

41. Tatiana Andreyeva, Michael W. Long, and Kelly D. Brownell, "The Impact of Food Prices on Consumption: A Systematic Review of Research on the Price Elasticity of Demand for Food," *American Journal of Public Health* 100, no. 2 (2010): 216–22.

42. Robert H. Lustig, "Ultra-processed Food: Addictive, Toxic, and Ready for Regulation," *Nutrients* 12, no. 11 (2020): 3401.

43. Robert N. Proctor, *The Nazi War on Cancer* (Princeton, NJ: Princeton University Press, 1999).

44. Lori Dorfman et al., "Cigarettes Become a Dangerous Product: Tobacco in the Rearview Mirror, 1952–1965," *American Journal of Public Health* 104, no. 1 (2014): 37–46.

45. Pamela Mejia et al., "The Origins of Personal Responsibility Rhetoric in News Coverage of the Tobacco Industry," *American Journal of Public Health* 104, no. 6 (2014): 1048–51.

46. K. Michael Cummings et al., "Assumption of Risk and the Role of Health Warnings Labels in the United States," *Nicotine and Tobacco Research* 22, no. 6 (2020): 975–83.

47. Patricia A. McDaniel and Ruth E. Malone, "'What Is Our Story?' Philip Morris' Changing Corporate Narrative," *American Journal of Public Health* 105, no. 10 (2015): e68–75.

48. Sue Anne Pressley, "Public Caught in the Middle of National Seat Belt Debate," *Washington Post*, Mar. 11, 1985, https://www.washingtonpost.com/

archive/politics/1985/03/11/public-caught-in-the-middle-of-national-seat-belt-de
bate/b05b13ac-dd08-40ac-ad09-5ea27a1a691f/.

49. Sungwon Yoon and Tai-Hing Lam, "The Illusion of Righteousness: Corporate Social Responsibility Practices of the Alcohol Industry," *BMC Public Health*
13, no. 1 (2013): 630; Thomas F. Babor, Katherine Robaina, and Jonathan Noel,
"The Role of the Alcohol Industry in Policy Interventions for Alcohol-Impaired
Driving," in *Getting to Zero Alcohol-Impaired Driving Fatalities: A Comprehensive Approach to a Persistent Problem*, ed. Steven M. Teutsch, Amy Geller, and
Yamrot Negussie (Washington, DC: National Academies, 2018), 477–521.

50. Andrew Joseph, "'A Blizzard of Prescriptions': Documents Reveal New Details about Purdue's Marketing of OxyContin," STAT News, Jan. 15, 2019, https://
www.statnews.com/2019/01/15/massachusetts-purdue-lawsuit-new-details/.

51. Jerrold J. Heindel et al., "Obesity II: Establishing Causal Links between
Chemical Exposures and Obesity," *Biochemical Pharmacology* 199 (2022): 115015.

52. Aseem Malhotra, Grant Schofield, and Robert H. Lustig, "The Science
against Sugar, Alone, Is Insufficient in Tackling the Obesity and Type 2 Diabetes
Crises—We Must Also Overcome Opposition from Vested Interests," *Journal of
Metabolic Health* 3, no. 1 (2018): a39; Kelly D. Brownell et al., "Personal Respon-
sibility and Obesity: A Constructive Approach to a Controversial Issue," *Health
Affairs (Millwood)* 29, no. 3 (2010): 379–87.

53. Grant Ennis, *Dark PR: How Corporate Disinformation Harms Our Health
and the Environment* (Quebec: Daraja, 2023).

54. Ambika Satija et al., "Understanding Nutritional Epidemiology and Its Role
in Policy," *Advances in Nutrition* 6, no. 1 (2015): 5–18.

55. Dean Schillinger et al., "Do Sugar-Sweetened Beverages Cause Obesity and
Diabetes? Industry and the Manufacture of Scientific Controversy," *Annals of Internal Medicine* 165, no. 12 (2016): 895–97.

56. Lenard I. Lesser et al., "Relationship between Funding Source and Conclusion
among Nutrition-Related Scientific Articles," *PLOS Medicine* 4, no. 1 (2007): e5.

57. Peter Martin, "Australian Paradox Author Admits Sugar Data Might Be
Flawed," *Sydney Morning Herald*, Feb. 9, 2014, https://www.smh.com.au/health
care/australian-paradox-author-admits-sugar-data-might-be-flawed-20140209-32
9h1.html.

58. Andrew W. Barclay and Jennie Brand-Miller, "The Australian Paradox: A
Substantial Decline in Sugars Intake over the Same Timeframe That Overweight
and Obesity Have Increased," *Nutrients* 3, no. 4 (2011): 491–504.

59. Vincent J. van Buul, Luc Tappy, and Fred J. P. H. Brouns, "Misconceptions
about Fructose-Containing Sugars and Their Role in the Obesity Epidemic," *Nutrition Research Reviews* 27, no. 1 (2014): 119–30.

60. Lisa Te Morenga, Simonette Mallard, and Jim Mann, "Dietary Sugars and
Body Weight: Systematic Review and Meta-analyses of Randomised Controlled
Trials and Cohort Studies," *BMJ* 34 (2013): e7492.

61. Gaurang Deshpande, Rudo F. Mapanga, and M. Faadiel Essop, "Frequent
Sugar-Sweetened Beverage Consumption and the Onset of Cardiometabolic Diseases:
Cause for Concern?," *Journal of the Endocrine Society* 1, no. 11 (2017): 1372–85.

62. Sanjay Basu et al., "The Relationship of Sugar to Population-Level Diabetes

Prevalence: An Econometric Analysis of Repeated Cross-Sectional Data," *PLOS One* 8, no. 2 (2013): e57873.

63. Cristin E. Kearns, Laura A. Schmidt, and Stanton A. Glantz, "Sugar Industry and Coronary Heart Disease Research: A Historical Analysis of Internal Industry Documents," *JAMA Internal Medicine* 176, no. 11 (2016): 1680–85.

64. Cristin E. Kearns, Stanton A. Glantz, and Laura A. Schmidt, "Sugar Industry Influence on the Scientific Agenda of the National Institute of Dental Research's 1971 National Caries Program: A Historical Analysis of Internal Documents," *PLOS Medicine* 12, no. 3 (2015): e1001798.

65. David Stuckler, Gary Ruskin, and Martin McKee, "Complexity and Conflicts of Interest Statements: A Case-Study of Emails Exchanged between Coca-Cola and the Principal Investigators of the International Study of Childhood Obesity, Lifestyle and the Environment (ISCOLE)," *Journal of Public Health Policy* 39, no. 1 (2018): 49–56.

66. Anahad O'Connor, "Coca-Cola Funds Scientists Who Shift Blame for Obesity Away from Bad Diets," *New York Times*, Aug. 9, 2015, https://archive.nytimes.com/well.blogs.nytimes.com/2015/08/09/coca-cola-funds-scientists-who-shift-blame-for-obesity-away-from-bad-diets/.

67. Rob Waters, "Trump's Pick to Head CDC Partnered with Coke, Boosting Agency's Longstanding Ties to Soda Giant," *Forbes*, July 10, 2017, https://www.forbes.com/sites/robwaters/2017/07/10/trumps-pick-to-head-cdc-partnered-with-coke-boosting-agencys-longstanding-ties-to-soda-giant/?sh=2be5bfc57b54.

68. Paulo M. Serôdio, Martin McKee, and David Stuckler, "Coca-Cola—a Model of Transparency in Research Partnerships? A Network Analysis of Coca-Cola's Research Funding (2008–2016)," *Public Health Nutrition* 21, no. 9 (2018): 1594–1607.

69. Tera L. Fazzino et al., "US Tobacco Companies Selectively Disseminated Hyper-palatable Foods into the US Food System: Empirical Evidence and Current Implications," *Addiction* 119, no. 1 (2024): 62–71.

70. Michele Simon, "AND Now a Word from Our Sponsors: Are America's Nutritional Professionals in the Pocket of Big Food?," Eat Drink Politics, Jan. 22, 2013, https://www.eatdrinkpolitics.com/2013/01/22/and-now-a-word-from-our-sponsors-new-report-from-eat-drink-politics/.

71. Daniel G. Aaron and Michael B. Siegel, "Supporting Public Health to Deflect Coke and Pepsi Sponsorship of National Health Organizations by Two Major Soda Companies," *American Journal of Preventive Medicine* 52, no. 1 (2017): 20–30, https://pubmed.ncbi.nlm.nih.gov/27745783/.

72. Marty Barrington, "Guest Choice Network—PRIVILEGED AND CONFIDENTIAL," PR Watch, Mar. 28, 1996, https://web.archive.org/web/20070927191427/http://www.prwatch.org/documents/berman/pm300k.pdf.

73. Eric Lipton, "Hard-Nosed Advice from Veteran Lobbyist: 'Win Ugly or Lose Pretty,'" *New York Times*, Oct. 30, 2014, https://www.nytimes.com/2014/10/31/us/politics/pr-executives-western-energy-alliance-speech-taped.html.

74. Christopher Snowdon, "Decline in Physical Activity to Blame for Rise in Obesity," Institute for Economic Affairs, Aug. 18, 2014, https://iea.org.uk/in-the-media/media-coverage/decline-in-physical-activity-to-blame-for-rise-in-obesity.

75. Robert H. Lustig, Laura A. Schmidt, and Claire D. Brindis, "The Toxic Truth about Sugar," *Nature* 487, no. 7383 (2012): 27–29.

76. Kristen Cooksey-Stowers, Marlene B. Schwartz, and Kelly D. Brownell, "Food Swamps Predict Obesity Rates Better Than Food Deserts in the United States," *International Journal of Environmental Research and Public Health* 14, no. 11 (2017): 1366.

77. Nicholas R. V. Jones et al., "The Growing Price Gap between More and Less Healthy Foods: Analysis of a Novel Longitudinal UK Dataset," *PLOS One* 9, no. 10 (2014): e109343.

78. Lissy C. Friedman et al., "Tobacco Industry Use of Personal Responsibility Rhetoric in Public Relations and Litigation: Disguising Freedom to Blame as Freedom of Choice," *American Journal of Public Health* 105, no. 2 (2015): 250–60.

79. Pia Roser, Simar S. Bajaj, and Fatima C. Stanford, "International Lack of Equity in Modern Obesity Therapy: The Critical Need for Change in Health Policy," *International Journal of Obesity* 46, no. 9 (2022): 1571–72.

80. Khadejah F. Mahmoud et al., "Personal and Professional Attitudes Associated with Nurses' Motivation to Work with Patients with Opioid Use and Opioid Use–Related Problems," *Substance Abuse* 42, no. 4 (2021): 780–87.

81. Aria Bendix, "Ozempic and Wegovy Add New Layers to the Understanding of Obesity as a Chronic Health Condition," NBC News, Feb. 8, 2023, https://www.nbcnews.com/health/health-news/ozempic-wegovy-obesity-chronic-health-condition-rcna68831.

82. Zachary J. Ward et al., "Association of Body Mass Index with Health Care Expenditures in the United States by Age and Sex," *PLOS One* 16, no. 3 (2021): e0247307.

Chapter 10

This work was presented and improved upon as part of my participation at the workshop "Agnotology: The New Science of Creating and Preventing Ignorance" hosted by Stanford University in May 2023, as well as via remarks from two anonymous reviewers. I also gratefully acknowledge Viveca Morris, with whom the bulk of research into this case was completed with a grant we received from the Climate Social Science Network at Brown University, for her feedback on this manuscript.

1. Dale Jamieson, *Reason in a Dark Time: Why the Struggle against Climate Change Failed—and What It Means for Our Future* (London: Oxford University Press, 2014).

2. Robert J. Brulle, "Institutionalizing Delay: Foundation Funding and the Creation of U.S. Climate Change Counter-movement Organizations," *Climatic Change* 122, no. 4 (2014): 681–94.

3. "Grant Founding and Endowing the Leland Stanford Junior University," Department of Special Collections and University Archives, Stanford University Libraries, SC1445, 4, accessed February 23, 2025, https://stacks.stanford.edu/file/druid:bz978md4965/su_founding_grant.pdf.

4. "Harvard Shields," Harvard University, accessed February 23, 2025, https://www.harvard.edu/about/history/shields/.

5. Naomi Oreskes and Eric M. Conway, *Merchants of Doubt: How a Handful*

of Scientists Obscured the Truth on Issues from Tobacco Smoke to Global Warming (New York: Bloomsbury, 2010).

6. Jennifer Jacquet, *The Playbook: How to Deny Science, Sell Lies, and Make a Killing in the Corporate World* (New York: Pantheon, 2022).

7. Robert N. Proctor, *Golden Holocaust: Origins of the Cigarette Catastrophe and the Case for Abolition* (Berkeley: University of California Press, 2011).

8. Leonard Zahn and Associates, "Tobacco-Health Research Grants Awarded 3 Harvard Scientists," press release, Jan. 16, 1976, box 213, ID no. gqmd0216, Council for Tobacco Research Records, Master Settlement Agreement, University of California, San Francisco (hereafter UCSF), https://www.industrydocuments.ucsf.edu/docs/gqmd0216.

9. "Harvard Scientist Challenges Heart-Smoking Relationship," *Management Bulletin: R. J. Reynolds Tobacco Company*, March 25, 1971, https://www.industry documents.ucsf.edu/docs/xrxp0061.

10. Herbert Black, "Harvard Scientist Says Elderly Smokers Needn't Quit," *Globe*, Dec. 20, 1972, https://www.industrydocuments.ucsf.edu/docs/zsgg0215.

11. Union of Concerned Scientists, *Smoke, Mirrors and Hot Air—How Exxon-Mobil Uses Big Tobacco's Tactics to Manufacture Uncertainty on Climate Science* (Cambridge, MA: UCS, Jan. 2007).

12. S. Fred Singer, "New Age Fanatics and Upper Muddle Class," *Hoover Digest*, July 30, 1998, https://www.hoover.org/research/new-age-fanatics-and-upper-muddle-class.

13. S. Fred Singer, "Sure, the North Pole Is Melting: So What?," *Hoover Digest*, Oct. 30, 2000, https://www.hoover.org/research/sure-north-pole-melting-so-what.

14. Dale Jamieson, "Responsibility and Climate Change," *Global Justice: Theory Practice Rhetoric* 8, no. 2 (2015): 23–42.

15. E.g., Donald R. Blake et al., "Global Increase in Atmospheric Methane Concentrations between 1978 and 1980," *Geophysical Research Letters* 9, no. 4 (1982): 477; Donald R. Blake and F. Sherwood Rowland, "Continuing Worldwide Increase in Tropospheric Methane, 1978–1987," *Science* 239, no. 4844 (1988): 1129–31.

16. V. Ramanathan et al., "Trace Gas Trends and Their Potential Role in Climate Change," *Journal of Geophysical Research* 90, no. D3 (1985): 5547–66.

17. Michael J. Gibbs and Lisa Lewis, *Reducing Methane Emissions from Livestock: Opportunities and Issues* ([Washington, DC]: U.S. Environmental Protection Agency, Aug. 1989), https://nepis.epa.gov/Exe/ZyPURL.cgi?Dockey=91012HP8.TXT.

18. William K. Stevens, "Methane from Guts of Livestock Is New Focus in Global Warming," *New York Times*, Nov. 21, 1989.

19. Molly O'Neill, "Cows in Trouble: An Icon of the Good Life Ends Up on a Crowded Planet's Hit Lists," *New York Times*, May 6, 1990.

20. F. M. Byers, "Eating Meat Won't Harm Environment," *USA Today*, April 27, 1990.

21. Susan Stranahan, "Of Bovines and Global Warming: Few Cows Equal Less Methane Gas," *Philadelphia Inquirer*, Nov. 12, 1993.

22. Jeremy Rifkin, *Beyond Beef: The Rise and Fall of the Cattle Culture* (New York: Penguin, 1992).

23. Jane Kay, "Anti-beef Crusade Touches Off Battle," *San Francisco Examiner,* Mar. 30, 1992.

24. Kay, "Anti-beef Crusade."

25. Jeremy Rifkin, personal communication to the author, Aug. 3, 2023.

26. See Livestock, Environment, and Development Initiative, *Livestock's Long Shadow: Environmental Issues and Options* ([Rome]: LEAD and FAO, 2006), 113, table 3.12, http://www.fao.org/3/a-a0701e.pdf.

27. Steve Cornett, "The Front Gate," *Top Producer,* Dec. 2007, ProQuest.

28. For a more extensive account, see Viveca Morris and Jennifer Jacquet, "The Animal Agriculture Industry, U.S. Universities, and the Obstruction of Climate Understanding and Policy," *Climatic Change* 177, no. 3 (2024), https://doi.org/10.1007/s10584-024-03690-w.

29. Frank Mitloehner, curriculum vitae, last modified April 14, 2021, CLEAR Center, University of California, Davis, https://clear.ucdavis.edu/sites/g/files/dgvnsk7876/files/media/documents/Frank%20Mitloehner%20CV%202021.pdf.

30. Maurice E. Pitesky, Kimberly R. Stackhouse, and Frank M. Mitloehner, "Clearing the Air: Livestock's Contribution to Climate Change," *Advances in Agronomy* 103 (2009): 1–40.

31. Sylvia Wright, "Don't Blame Cows for Climate Change," University of California, Davis, press release, Dec. 7, 2009, https://www.ucdavis.edu/news/don%E2%80%99t-blame-cows-climate-change.

32. American Chemical Society, "Eating Less Meat and Dairy Products Won't Have Major Impact on Global Warming," Mar. 22, 2010, https://www.acs.org/pressroom/newsreleases/2010/march/eating-less-meat-and-dairy-products-wont-have-major-impact-on-global-warming.html.

33. Vasile Stănescu, "'Cowgate': Meat Eating and Climate Change Denial," in *Climate Change Denial and Public Relations,* ed. Núria Almiron and Jordi Xifra (Abingdon: Routledge, 2020), 178–94.

34. David Derbyshire, "Eating Less Meat 'Won't Save Planet': Scientist Blames UN Report for Green 'Myth,'" *Daily Mail,* Mar. 23, 2010.

35. "Scientific Disciplines Mix at Chemistry Meeting," *Talk of the Nation,* NPR, Mar. 26, 2010, https://www.npr.org/2010/03/26/125216841/scientific-disciplines-mix-at-chemistry-meeting.

36. "Scientific Disciplines Mix."

37. Morris and Jacquet, "Animal Agriculture Industry."

38. Zach Boren, "Revealed: How the Meat Industry Funds the 'Greenhouse Gas Guru,'" Unearthed, Oct. 10, 2022, https://unearthed.greenpeace.org/2022/10/31/frank-mitloehner-uc-davis-climate-funding/.

39. Frank Mitloehner, "It's Time to Stop Comparing Meat Emissions to Flying," CLEAR Center (blog), University of California, Davis, Nov. 13, 2019, https://clear.ucdavis.edu/blog/its-time-stop-comparing-meat-emissions-flying.

40. "Cattle Carbon Footprint," *American Cattlemen,* June 15, 2010, https://www.americancattlemen.com/articles/cattle-carbon-footprint accessible using: http://web.archive.org/web/20190226103311/http://www.americancattlemen.com/articles/cattle-carbon-footprint (site discontinued).

41. Tim Cronshaw, "Methane's Impact on Warming Misrepresented US Scientist Tells Red Meat Sector Conference," *Otago Daily Times*, Aug. 18, 2022.

42. Beef + Lamb New Zealand, "Don't miss this great opportunity to hear directly from Dr Frank Mitloehner, a world-renowned climate change communicator," Facebook, Feb. 8, 2023, https://www.facebook.com/beeflambnz/photos/a.51 4972441865795/6506768176019495/?type=3.

43. Morris and Jacquet, "Animal Agriculture Industry."

44. Alex Kasprak, "Are Ingredients in 'Impossible' and 'Beyond Meat' Burgers Indistinguishable from Dog Food?," Snopes Fact Check, Aug. 16, 2019, https:// www.snopes.com/fact-check/impossible-burgers-dogfood/.

45. Joel Newman, "UC Davis Center Proposal Update," Institute for Feed Education and Research, memorandum, May 31, 2018, https://embed.documentcloud .org/documents/23205129-memorandum-of-understanding.

46. Oliver Lazarus, Sonali McDermid, and Jennifer Jacquet, "The Climate Responsibilities of Industrial Meat and Dairy Producers," *Climatic Change* 165, no. 1–2 (2021): 30.

47. P. J. Gerber et al., *Tackling Climate Change through Livestock—a Global Assessment of Emissions and Mitigation Opportunities* (Rome: Food and Agriculture Organization of the United Nations, 2013), https://www.fao.org/4/i3437e /i3437e.pdf.

48. Food and Agriculture Organization of the United Nations, "GLEAM 3 Dashboard," last modified 2022, https://foodandagricultureorganization.shinyapps .io/GLEAMV3_Public/.

49. "Cattle Carbon Footprint."

50. Helena Bottemiller Evich, "Group Linked to Ocasio-Cortez Seeks Ag Input after Green New Deal Backlash," *Politico*, Mar. 11, 2019, https://www.politico .com/story/2019/03/11/aoc-linked-group-seeks-ag-input-after-green-new-deal -backlash-1252568.

51. Caroline Stocks, "The Methane Myth: Why Cows Aren't Responsible for Climate Change," Medium, June 6, 2019, https://carolinestocks.medium.com/de bunking-the-methane-myth-why-cows-arent-responsible-for-climate-change-2392 6c63f2c0.

52. Robert N. Proctor, " 'Everyone Knew but No One Had Proof': Tobacco Industry Use of Medical History Expertise in US Courts, 1990–2002," *Tobacco Control* 15, supp. 4 (2006): iv117–iv125.

53. Morris and Jacquet, "Animal Agriculture Industry."

54. Cronshaw, "Methane's Impact."

55. Stănescu, "Cowgate"; Boren, "Revealed"; Hiroko Tabuchi, "He's an Outspoken Defender of Meat: Industry Funds His Research, Files Show," *New York Times*, Oct. 31, 2022; Morris and Jacquet, "Animal Agriculture Industry."

56. "Frank Mitloehner," DeSmog, accessed Sept. 4, 2024, https://www.desmog .com/frank-mitloehner/.

57. Morris and Jacquet, "Animal Agriculture Industry."

58. Wright, "Don't Blame Cows."

59. Tabuchi, "Outspoken Defender of Meat."

60. Morris and Jacquet, "Animal Agriculture Industry."

61. Amy Quinton, "Cows and Climate Change: Making Cattle More Sustainable," In Focus, University of California, Davis, June 27, 2019, https://www.ucdavis.edu/food/news/making-cattle-more-sustainable.

62. Dateline Staff, "LAURELS: Mitloehner Earns Communications Award," University of California, Davis, April 30, 2019, https://unfoldpodcast.ucdavis.edu/news/laurels-mitloehner-earns-communications-award.

63. Kat Kerlin, "Experts: Agriculture in a Changing Climate," University of California, Davis, Aug. 28, 2018, https://www.ucdavis.edu/climate/news/experts-agriculture-changing-climate.

64. "Science & Climate," University of California, Davis, last modified Oct. 23, 2023, https://www.ucdavis.edu/climate.

65. Dave Jones, "Dean Emeritus Charley Hess Dies at 87," University of California, Davis, May 3, 2019, https://www.ucdavis.edu/news/dean-emeritus-charley-hess-dies-87.

66. Kay, "Anti-beef Crusade."

67. Jacquet, *Playbook*.

68. Lisa Bero, "Ten Tips for Spotting Industry Involvement in Science Policy," *Tobacco Control* 28, no. 1 (2019): 1–2.

69. Proctor, *Golden Holocaust*, 453.

70. See, e.g., the arguments for ideological motivations made in Oreskes and Conway, *Merchants of Doubt*.

71. Samuel V. Bruton and Donald F. Sacco, "What's It to Me? Self Interest and Evaluations of Financial Conflicts of Interest," *Research Ethics* 14, no. 4 (2017): 1–17.

72. Dennis F. Thomson, "Understanding Financial Conflicts of Interest," *New England Journal of Medicine* 329, no. 8 (1993): 573.

73. Lisa Bero and Quinn Grundy, "Why Having a (Nonfinancial) Interest Is Not a Conflict of Interest," *PLOS Biology* 14, no. 12 (2016): e2001221.

74. Thomson, "Financial Conflicts of Interest."

75. Morris and Jacquet, "Animal Agriculture Industry."

76. Jacquet, *Playbook*.

77. Frank Mitloehner, "Full Disclosure: I Work to Reduce the Footprint of Animal Agriculture," CLEAR Center (blog), University of California, Davis, Oct. 31, 2022, https://clear.ucdavis.edu/blog/full-disclosure-i-work-to-reduce-the-footprint-of-animal-agriculture.

78. Carey Gillam, *Whitewash: The Story of a Weed Killer, Cancer, and the Corruption of Science* (Washington, DC: Island, 2017).

79. Bella Kumar, *Accountable Allies: The Undue Influence of Fossil Fuel Money in Academia* (Data for Progress, Feb. 2023), https://www.filesforprogress.org/memos/accountable-allies-fossil-fuels.pdf.

80. Climate Social Science Network, Institute at Brown for Environment and Society, Brown University, accessed Sept. 4, 2024, https://cssn.org.

81. "Our Values," Greenpeace, accessed Aug. 31, 2023, https://www.greenpeace.org/international/about/values/.

Chapter 11

Chapter 11, "On the Burial of the Palestinian Nakba," © 2022 from *Routledge International Handbook of Ignorance Studies*, 2nd ed., edited by Matthias Gross and Linsey McGoey, is reproduced with permission from Taylor and Francis Group LLC, a division of Informa PLC.

1. The Nakba ['catastrophe'] is the term Palestinians use to describe their expulsion by Zionist forces in 1947–9, with 'Israel' replacing 'Palestine' in official recognition, and the Palestinians dispersed among multiple 'host' countries.

2. Interview, Jerusalem 2000.

3. The United Nations Relief and Works Agency [UNRWA] was established in 1951 to provide basic welfare and social services to Palestinian 'refugees' [Hanafi et al 2019].

4. NB: curricula in Palestine were formed under British control.

5. Salman Abu Sitta [2000].

6. Eg. in 2020, Jewish settlers uprooted 8,400 olive trees: *Anadolu Agency* 19 December 2020.

7. Reuters, "Israel bans 'catastrophe' term from Arab schools" 22 July 2009.

8. Personal communication 28 January 2021.

9. Middle East Eye Staff [2021] "UK school textbooks on Middle East conflict altered to favour Israel: Report" *Middle East Eye* 1 April.

10. Personal communication 11 November 2020.

11. The six books studied by el-Borhany are: Charles Smith, *Palestine and the Arab-Israeli Conflict: A History with Documents*, 2013; Ian Bickerton and Carla Klausner, *A Concise History of the Arab-Israeli Conflict*, 1998; Benny Morris, *Righteous Victims: a History of the Zionist-Arab conflict*, 2001; Mark Tessler, *A History of the Israeli-Palestinian Conflict*, 2009; Howard Sachar, *A History of Israel: From the Rise of Zionism to Our Time*, 2007; Walter Laqueur and Barry Rubin, *The Arab-Israeli Reader: A Documentary History of the Middle East Conflict*, 2008.

12. Many scholars would contest this point, first because the Nakba was not limited to 1948, and second because many massacres carried out during 1948 have not been counted into the death toll.

13. Palestinians were attacked in Jordan in 'Black September', 1970, with heavy casualties; in Lebanon during the civil war of 1975–1990, including the massacre of Tal al-Zaater [1976]; during the Israeli invasion of Lebanon in 1982, with the massacre of Sabra/Shatila; during the Battle of the Camps 1985–1988; and again in 2007 with the Lebanese Army's attack on Nahr al-Bared camp. Palestinians were evicted from Kuwait in 1990, and again in 2003; expelled from Libya in 1994–1995; evicted by landlords in Iraq in 2003. In Syria, 4,027 have been killed and 120,00 displaced so far in the current civil war. Israeli attacks against Gaza have been continuous: 2008–2009, 2012, 2014, 2018, 2019 . . . In the Occupied West Bank, attacks by armed Israeli settlers are frequent [Amnesty 2017].

14. Middle East Studies Association, Committee on Academic Freedom, "Exposing Canary Mission" 18 April 2018.

15. Personal communication 12 February 2021.

16. Personal communication 10 February 2021.

17. Personal communication 2 February 2021.

REFERENCES

Abu El Hajj, Nadia [2017] "Some Thoughts on Facts, Politics and Tenure" in William Robson ed., *We Will Not Be Silenced: The Academic Repression of Israel's Critics* Chico California, AK Press, 163–170.

Abu Sitta, Salman [2000] *The Palestinian Nakba 1948* London, Palestine Land Society.

Abu Sitta, Salman [2016] *Mapping My Return* Cairo, American University of Cairo Press.

Achcar, Gilbert [2010] *The Arabs and the Holocaust: the Arab-Israeli War of Narratives* New York, Metropolitan Books.

Allen, Lori [2019] "Academic Freedom in the United Kingdom" *AAUP,* Fall.

Amnesty International [2017] *Fifty Years of Occupation: Fifty Years of Human Rights Violations* London, Amnesty United Kingdom Section.

Anziska, Seth [2019] "The Erasure of the Nakba in Israel's Archives" *Journal of Palestine Studies* vol XLIX [1] Autumn, 64–76.

Bhambra, Gurminder et al. [2018] *Decolonizing the University* London, Pluto Press.

Borhani, Seyyed Haili [2016] "Biases and the Question of Palestine/Israel: Textbook Treatment of the Question's History in Western Universities" *Journal of Holy Land and Palestine Studies* vol 15 [2], 225–247.

Cronin, David [2017] *Balfour's Shadow: A Century of British Support for Zionism and Israel* London, Pluto Press.

Deeb, Lara and Jessica Winegar [2015] *Anthropology's Politics: Disciplining in the Middle East* Berkeley, Stanford University Press.

Farah, Randa [2018] "Taboo Narratives: Teaching Palestine and the Palestinians" *Practicing Anthropology* vol 40 [4] Fall, 12–15.

Foreman, Geramy and Alexandre Kedar [2004] "From Arab land to 'Israel Lands'; the legal dispossession of the Palestinians displaced by Israel in the wake of 1948" *Environment and Planning D: Society and Space* vol 22, 809–830.

Gendzier, Irene [2015] *Dying to Forget: Oil, Power, Palestine, and the Foundations of U.S. Policy in the Middle East* New York, Columbia University Press.

Hadawi, Sami [1967] *Bitter Harvest: Palestine between 1914–1967* New York, The New World Press.

Hanafi, Sari, Leila Hilal, Lex Takkenberg eds, [2019] *UNRWA and the Palestinian refugees: From Relief and Works to Human Development* London and New York, Routledge.

Hovsepian, Nubar [2009] *Palestinian State Formation: Education and the Construction of National Identity* Cambridge UK, Cambridge Scholars Publishing.

Hughes, Matthew [2010] "From Law and Order to Pacification: Britain's Suppression of the Arab Revolt in Palestine 1936–39" *Journal of Palestine Studies* vol 39 [2] 6–22.

Irfan, Anne [2019] "Educating Palestinian Refugees: The Origins of UNRWA's Unique Schooling System" *Journal of Refugee Studies* fez051 lmps://doi.org/10.1093/jrs/fez051.

Masalha, Nur ed. [2005] *Catastrophe Remembered: Palestine, Israel and the Internal Refugees* London and New York, Zed Books.

Masalha, Nur [2015] "Settler Colonialism: Memoricide and Indigenous Toponymic

Memory: The Appropriation of Palestinian Place Names by the Israeli State" *Journal of Holy Land and Palestine Studies* vol 14 [1] 3–57.

Mermelstein, Hannah [2011] "Overdue Books: Returning Palestine's 'Abandoned Property" *Zochrot* September.

Mills, Charles [2015] "Global White Ignorance" *Routledge International Handbook of Ignorance Studies* 217–227.

Moughrabi, Fouad [2001] "The Politics of Palestinian Textbooks" *Journal of Palestine Studies* vol 31 [1] 5–19.

Naser-Najjab, Nadia [2020] "Palestinian education and the 'logic of elimination' " *Settler Colonial Studies DOI. 10.1080/2201473X2020.1760433.*

Nassar, Issam [2003] "In their Image: Jerusalem in Nineteenth Century English Travel Narratives" *Jerusalem Quarterly* no 19 6–22.

Pappe, Ilan [2006] *The Ethnic Cleansing of Palestine* Oxford, One World.

Pappe, Ilan [2008] "Israel, the Holocaust and the Nakba" *Socialist Review* issue 325, May.

Peled-Elhanan, Nurit [2012] *Palestine in Israeli School Books: Ideology and Propaganda in Education* New York, I.B. Tauris.

Peteet, Julie [1991] *Gender in Crisis: Women in the Palestinian Resistance Movement* New York, Columbia University Press.

Piterberg, Gabriel [2001] "Erasing the Palestinians" *New Left Review* no 10, July/August 31–46.

Sharif, Regina [1985] *Non-Jewish Zionism: its roots in Western history* London, Zed Press.

Shezaf, Hagar [2019] "Burying the Nakba: How Israel Systematically Hides Evidence of 1948 Expulsion of Arabs" *Haaretz* 5 April.

Sleiman, Hana [2016] "The Paper Trail of a Liberation Movement" *Arab Studies Journal* vol 24 [1] Spring.

Sokolower, Jody [2020] "California Model Ethnic Studies appears ready to adopt Trump's racist definition of anti-semitism" *Mondoweiss* 16 November.

Swedenburg, Ted [2003] *Memories of Revolt: The 1936–1939 Rebellion and the Palestinian National Past* Fayetteville, University of Arkansas Press.

Trouillot, Michel-Rolph [1995] *Silencing the Past: Power and the Production of History* Boston MA, Beacon Press.

Chapter 12

1. Israel Cymlich and Oskar Strawczynski, *Escaping Hell in Treblinka* (Jerusalem: Yad Vashem, 2007), 167.

2. The sign read, "Zu den Zügen Nach Bialystok und Wolkowysk." Samuel Willenberg, *Surviving Treblinka*, ed. Władysław Bartoszewski (New York: Blackwell, 1989), 107. See also Simone Gigliotti, *The Train Journey: Transit, Captivity, and Witnessing in the Holocaust* (New York: Berghahn Books, 2009), 181–83.

3. Samuel Rajzman, "Sixty-Ninth Day," in *Nuremberg Trial Proceedings*, vol. 8, Feb. 27, 1946, Avalon Project, Yale Law School, New Haven, CT, 325, https://avalon.law.yale.edu/imt/02-27-46.asp.

4. Jan Sułkowski, testimony before Judge Antoni Krytowski, Dec. 20, 1945, IPN GK 196/70, Institute of National Remembrance, Warsaw, Poland.

5. Vassili Grossman, "The Treblinka Hell," *International Literature* 12, no. 6 (1945): 371–408, esp. 378.

6. Chris Webb and Michal Chocholatý, *The Treblinka Death Camp: History, Biographies, Remembrance* (New York: Columbia University Press, 2014), chap. 6. See also Mark S. Smith, *Treblinka Survivor: The Life and Death of Hershl Sperling* (Cheltenham: History Press, 2010), chap. 12; Jacob Flaws, "Spaces of Treblinka" (PhD diss., University of Colorado, 2020), 68–69.

7. For a review of the opposing functionalist and maximalist exterminationist theories of Nazi antisemitism and the Holocaust, see A. Dirk Moses, "Structure and Agency in the Holocaust: Daniel J. Goldhagen and His Critics," *History and Theory* 37, no. 2 (1998): 194–219.

8. I.e., *Gemeinnützige Stiftung für Anstaltspflege.* Gitta Sereny, *Into That Darkness: From Mercy Killing to Mass Murder* (New York: Pimlico, 1995), 48. The name T4 refers to the Tiergartenstraße 4 euthanasia program. See also Robert N. Proctor, *Racial Hygiene: Medicine under the Nazis* (Cambridge, MA: Harvard University Press, 1988), 191.

9. Proctor, *Racial Hygiene,* 132. On toxification discourse, see Rhiannon S. Neilsen, "'Toxification' as a More Precise Early Warning Sign for Genocide Than Dehumanization? An Emerging Research Agenda," *Genocide Studies and Prevention* 9, no. 1 (2015): 83–95, esp. 83.

10. Richard Breitman, "Himmler and the 'Terrible Secret' among the Executioners," *Journal of Contemporary History* 26, no. 3–4 (1991): 431–51, esp. 438.

11. Heinz Höhne, *The Order of the Death's Head: The Story of Hitler's SS,* trans. Richard Barry (1969; repr., London: Penguin Books, 2000), 361.

12. Robert Jay Lifton, *The Nazi Doctors: Medical Killing and the Psychology of Genocide* (New York: Basic Books, 2017), 103, 135, 147–48, 150–51.

13. Christopher R. Browning, Peter Hayes, and Raul Hilberg, *German Railroads, Jewish Souls: The Reichsbahn, Bureaucracy, and the Final Solution* (New York: Berghahn Books, 2019), 27. Until Heinrich Himmler's speech before the senior SS officers at Poznań on October 4, 1943, "evacuation" (*Aussiedlung*) was a euphemism concealing the extermination of the Jews in Europe, as were the terms "resettlement" (*Umsiedlung*) and "labor in the East" (*Arbeitseinsatz im Osten*).

14. Nachman Blumental, *Słowa Niewinne* (Warsaw: Centralna Żydowska Komisja Historyczna w Polsce, 1947). The title of the book translates to "innocent words."

15. Victor Klemperer, *The Language of the Third Reich* (London: Bloomsbury, 2013), 2, 234–35.

16. Joseph Wulf, *Aus dem Lexikon der Mörder: "Sonderbehandlung" und verwandte Worte in nationalsozialistischen Dokumenten* (Gütersloh: Sigbert Mohn, 1963), 23.

17. Hannah Arendt, *Eichmann in Jerusalem: A Report on the Banality of Evil* (New York: Penguin Books, 2006), 84–86, 108.

18. Klaus Theweleit, *Male Fantasies,* vol. 1, *Women, Floods, Bodies, History* (Minneapolis: University of Minnesota Press, 1987), 70, 88, 215.

19. Primo Levi, *The Drowned and the Saved, trans. Raymond Rosenthal* (New

York: Simon and Schuster, 2017), 85. *Lager* was short for *Arbeitslager,* German for "labor camp." In 1966, the novelist Jean-François Steiner wrote of a "Sibylline language of Treblinka," but his analysis was ahistorical and inaccurate. Jean-François Steiner, *Treblinka* (New York: Signet Books, 1979), 155. On the controversy surrounding Steiner's work, see Samuel Moyn, *A Holocaust Controversy: The Treblinka Affair in Postwar France* (Lebanon, NH: University Press of New England, 2005).

20. Browning, Hayes, and Hilberg, *German Railroads, Jewish Souls,* 13.

21. Joseph Wulf, *Die SS,* 269, quoted in Höhne, *Death's Head,* 364.

22. Simon Wiesenthal, "The Murderers among Us," *Saturday Evening Post,* Mar. 11, 1967, 47.

23. Michael Lynch, *Art and Artifact in Laboratory Science: A Study of Shop Work and Shop Talk in a Research Laboratory* (London: Routledge, 2017), 158.

24. For "normative control," see Gideon Kunda, *Engineering Culture: Control and Commitment in a High-Tech Corporation* (Philadelphia, PA: Temple University Press, 1992), 11.

25. George Orwell, "The Principles of Newspeak," in *Nineteen Eighty-Four* (1949; repr., Oxford: Clarendon, 1984), 417, appendix. See also Raymond Williams, *Orwell* (London: Fontana Books, 1971), 75.

26. Karl E. Weick, *Sensemaking in Organizations,* vol. 3 (Thousand Oaks: Sage, 1995).

27. Carol Cohn, "Slick 'Ems, Glick 'Ems, Christmas Trees, and Cookie Cutters: Nuclear Language and How We Learned to Pat the Bomb," *Bulletin of the Atomic Scientists* 43, no. 5 (1987): 17, 20, 22.

28. Dwight Bolinger, *Language: The Loaded Weapon; The Use and Abuse of Language Today* (London: Routledge, 2021), 118; emphasis original.

29. For photographs taken inside the U.S. military's fictional country, Atropia, see Debi Cornwall, *Necessary Fictions* (Santa Fe, NM: Radius Books, 2020).

30. On "bad apples," see Allen Feldman, *Archives of the Insensible: Of War, Photopolitics, and Dead Memory* (Chicago: University of Chicago Press, 2015), 254.

31. Since 2013, Israel Defense Forces officers have referred to the Israeli strategy in Gaza as כיסוח דשא, or "mowing the grass." Efraim Inbar and Eitan Shamir, "'Mowing the Grass': Israel's Strategy for Protracted Intractable Conflict," *Journal of Strategic Studies* 37, no. 1 (2014): 67.

32. Daniel Akselrad and Robert N. Proctor, "Why Did Philip Morris Stop Making Cigarettes at Auschwitz? An Essay on the Geometry and Kinetics of Atrocity," *Public Culture* 36, no. 1 (2024): 48.

33. Robert N. Proctor, *Golden Holocaust: Origins of the Cigarette Catastrophe and the Case for Abolition* (Berkeley: University of California Press, 2011), 118–33, 334.

34. On power-knowledge apparatuses, see Michel Foucault, *Power/Knowledge: Selected Interviews and Other Writings, 1972–1977* (New York: Vintage Books, 1980).

35. Lorraine Daston and Peter Galison, *Objectivity* (Princeton, NJ: Princeton University Press, 2021), 32.

36. Nancy Tuana, "Coming to Understand: Orgasm and the Epistemology of Ignorance," in *Agnotology: The Making and Unmaking of Ignorance*, ed. Robert N. Proctor and Londa Schiebinger (Stanford, CA: Stanford University Press, 2008), 108; emphasis original.

37. Erving Goffman, *Asylums: Essays on the Social Situation of Mental Patients and Other Inmates* (Harmondsworth: Penguin Books, 1968), 11.

38. Rajzman, "Sixty-Ninth Day," 325.

39. Yankiel Wiernik, *A Year in Treblinka* (New York: American Representation of the General Jewish Workers' Union of Poland, 1945), 13. See also Yitzhak Arad, *Belzec, Sobibor, Treblinka: The Operation Reinhard Death Camps* (Bloomington: Indiana University Press, 1987), 122.

40. Robert N. Proctor, *The Nazi War on Cancer* (Princeton, NJ: Princeton University Press, 1999), 90.

41. Robert N. Proctor, "Nazi Medical Ethics: Ordinary Doctors?," in *Military Medical Ethics*, vol. 2, ed. Thomas E. Beam et al. (Washington, DC: Office of the U.S. Surgeon General, 2003), 418. Soviets, too, had their own pseudomedical lexicon, such as "alimentary dystrophy," a euphemism "for the wasting away of intestinal muscle due to starvation" in the gulags. See Katherine R. Jolluck, introduction to *Journey into the Land of the Zeks and Back: A Memoir of the Gulag*, by Julius Margolin, trans. Stefani Hoffman (New York: Oxford University Press, 2020), 35.

42. Smith, *Treblinka Survivor*, 85. See also Cymlich and Strawczynski, *Escaping Hell*, 130. At Bełżec, gas chambers bore a similar sign reading *"Bade und Inhalationsräume,"* or "Baths and Inhalation Rooms." Henryk Poswolski, testimony before Judge Z. Łukaszkiewicz, Oct. 9, 1945, IPN GK 196/69, Institute of National Remembrance, Warsaw, Poland.

43. Lisa Parks, " 'Stuff You Can Kick': Toward a Theory of Media Infrastructures," in *Between Humanities and the Digital*, ed. Patrik Svensson and David Theo Goldberg (Cambridge, MA: MIT Press, 2015), 355–74.

44. On ideological work, see Bennett M. Berger, *The Survival of a Counterculture: Ideological Work and Everyday Life among Rural Communards* (Piscataway, NJ: Transaction, 2003).

45. Stanley Cohen, *States of Denial: Knowing about Atrocities and Suffering* (New York: John Wiley and Sons, 2013), 81.

46. Arendt, *Eichmann in Jerusalem*, 108.

47. Sereny, *Into That Darkness*, 200.

48. On deliberate ignorance, see Ralph Hertwig and Christoph Engel, "Homo Ignorans: Deliberately Choosing Not to Know," *Perspectives on Psychological Science* 11, no. 3 (2016): 360. On "evasive thinking," see Václav Havel, *Open Letters: Selected Writings, 1965–1990* (New York: Vintage Books, 1992), 15.

49. On meso-level escalation of violence at the periphery, see Rachel Jacobs and Scott Straus, "Meso-Level Dynamics of Atrocities," in *The Oxford Handbook of Atrocity Crimes*, ed. Barbara Holá, Hollie Nyseth Nzitatira, and Maartje Weerdesteijn (New York: Oxford University Press, 2022), 235–54.

50. For *laufendes Band*, see Friedrich Entress, affidavit no. 2368, April 14, 1947, Office of the Chief Counsel for War Crimes, HLSL no. 4446, Nuremberg Trials Proj-

ect, Harvard Law School Library, Cambridge, MA, https://nuremberg.law.harvard
.edu/documents/4446-affidavit-concerning-the-euthanasia?q=evidence:no%2A.

51. Franz Suchomel, interview by Claude Lanzmann, trans. Uta Allers, transcript, Summer 2012, U.S. Holocaust Memorial Museum, Washington, DC, 5, 58,
https://collections.ushmm.org/film_findingaids/RG-60.5046_01_trl_en.pdf. The
Goldjuden were tasked with collecting and sorting gold, jewelry, bank notes, and
other valuables stolen from the arriving deportees.

52. Isadore Helfing, interview by Sandra Bradley, Mar. 9, 1992, U.S. Holocaust Memorial Museum, Washington, DC, 21, https://collections.ushmm.org/oh
_findingaids/RG-50.042.0014_trs_en.pdf.

53. Cymlich and Strawczynski, *Escaping Hell*, 42.

54. Wolfgang Sofsky, *The Order of Terror* (Princeton, NJ: Princeton University
Press, 2013), 265. See also Alfred C. Mierzejewski, *The Most Valuable Asset of the
Reich: A History of the German National Railway*, vol. 1, *1920–1932* (Chapel Hill:
University of North Carolina Press, 2013), 114. Trains to Treblinka typically con-
sisted of "two third-class passenger cars for the guards and fifty-eight boxcars for the
victims."

55. Samuel Rajzman, testimony before Judge Z. Łukaszkiewicz, Dec. 3, 1945,
IPN GK 196/69, Institute of National Remembrance, Warsaw, Poland. Rajzman
testified that Suchomel was a jeweler by trade.

56. Cymlich and Strawczynski, *Escaping Hell*, 166.

57. Edi Weinstein, *Quenched Steel: The Story of an Escape from Treblinka* (Jerusalem: Yad Vashem, 2002), 51.

58. Willenberg, *Surviving Treblinka*, 106–7. See also Stanisław Borowy, testimony before Judge Z. Łukaszkiewicz, Nov. 21, 1945, IPN GK 196/69, Institute of
National Remembrance, Warsaw, Poland.

59. Cymlich and Strawczynski, *Escaping Hell*, 167.

60. Webb and Chocholatý, *Treblinka Death Camp*, 75.

61. Arad, *Belzec, Sobibor, Treblinka*, 122.

62. Cymlich and Strawczynski, *Escaping Hell*, 167.

63. Wiesenthal, "Murderers among Us," 47.

64. Wiernik, *Year in Treblinka*, 37.

65. Sereny, *Into That Darkness*, 200.

66. Cymlich and Strawczynski, *Escaping Hell*, 167.

67. Alfred Spiess, interview by Claude Lanzmann, trans. Uta Allers, transcript,
Oct. 2013, U.S. Holocaust Memorial Museum, Washington, DC, 14, https://
collections.ushmm.org/film_findingaids/RG-60.5063_01_trl_en.pdf.

68. Steiner, *Treblinka*, 154.

69. Webb and Chocholatý, *Treblinka Death Camp*, 75.

70. Steiner, *Treblinka*, 154; Elana Gomel, *Postmodern Science Fiction and Tem-
poral Imagination* (London: Continuum, 2010), 81.

71. Maurice Rossel, interview by Claude Lanzmann, trans. Lotti Eichorn, transcript, Jan.–Mar. 2009, U.S. Holocaust Memorial Museum, Washington, DC, 33,
https://collections.ushmm.org/film_findingaids/RG-60.5019_01_trl_en.pdf.

72. Brad Prager, "Interpreting the Visible Traces of Theresienstadt," *Journal*

of Modern Jewish Studies 7, no. 2 (2008): 175–94, esp. 179. The film was later renamed *Der Führer schenkt den Juden eine Stadt,* or "The Führer Gives the Jews a City."

73. "Les pouponnieres, les hômes d'enfants, pavilions et jardins d'enfants, sont très bien installes. Une cuisines spécialisée prépare les aliments pour les petits." Maurice Rossel, "Rossel Report," in *Documents du Comite International de la Croix-Rouge concernant le ghetto de Theresienstadt* (International Committee of the Red Cross Library, June 1990), record no. 400.3/152, 85.

74. "Toutes les professions que l'on peut imaginer dans une petite ville . . . les corps de police, des dêtectives, un corps de dapeurs-pompiers . . . corporation de bouchers, de boulangers." Rossel, "Rossel Report," 79.

75. Rossel, interview, 38.

76. Roland Barthes, *Elements of Semiology,* trans. Annette Lavers and Colin Smith (1964; repr., New York: Hill and Wang, 1967).

77. Smith, *Treblinka Survivor,* 89.

78. Wiernik, *Year in Treblinka,* 18.

79. Stanisław Kon, testimony before Judge Z. Łukaszkiewicz, Oct. 7, 1945, IPN GK 196/69, Institute of National Remembrance, Warsaw, Poland.

80. Arad, *Belzec, Sobibor, Treblinka,* 215–18.

81. Cymlich and Strawczynski, *Escaping Hell,* 166–67.

82. Eugen Kogon, *Der SS-Staat: Das System der Deutschen Konzentrationslager* (Vienna: Bermann-Fischer, 1947), 349. With this expression, "failed existences," Kogon inadvertently offered the SS an alibi for their cruelty.

83. Helfing, interview, 6.

84. Cymlich and Strawczynski, *Escaping Hell,* 167.

85. Flaws, "Spaces of Treblinka," 46–47.

86. Hejnoch Brener, testimony before Judge Z. Łukaszkiewicz, Oct. 9, 1945, IPN GK 196/69, Institute of National Remembrance, Warsaw, Poland.

87. Adalbert Rückerl, *NS-Vernichtungslager im Spiegel deutscher Strafprozesse: Belzec, Sobibor, Treblinka, Chelmno* (Munich: DTV, 1977), 223.

88. Eliau Rosenberg, testimony, factual report, trans. Marc Bartuschka and Chris Webb, Dec. 24, 1947, Ghetto Fighters House, Israel, Holocaust Historical Society, United Kingdom, https://www.holocausthistoricalsociety.org.uk/contents/ treblinkadeathcamp/eliaurosenbergtestimony.html.

89. Chil Rajchman, *Treblinka: A Survivor's Memory* (London: MacLehose, 2011), 66.

90. Upper and lower camps were distinguished based on their topographical terrain. For camp topography, see Muzeum Treblinka, "Treblinka I—Topography of the Camp," accessed Aug. 19, 2024, https://muzeumtreblinka.eu/en/informacje /topography-of-the-camp-1; Muzeum Treblinka, "Treblinka II—Topography of the Camp," accessed Aug. 19, 2024, https://muzeumtreblinka.eu/en/informacje/ topography-of-the-camp.

91. Ernst Klee, Willi Dressen, and Volker Riess, eds., *"The Good Old Days": The Holocaust as Seen by Its Perpetrators and Bystanders,* trans. Deborah Burnstone (1988; repr., Old Saybrook, CT: Konecky and Konecky, 1991), 248.

92. Sereny, *Into That Darkness*, 232.

93. Rajchman, *Treblinka*, 87; Levi, *Drowned and the Saved*, 78.

94. Giorgio Agamben, Homo Sacer: *Sovereign Power and Bare Life* (Stanford, CA: Stanford University Press, 1998), 103–4.

95. Stanisław Głowa, testimony before Judge Franciszek Wesely, Sept. 18, 1947, IPN GK 196/142, Institute of National Remembrance, Warsaw, Poland; Rajchman, *Treblinka*, 87.

96. Jacob Flaws, "Sensory Witnessing at Treblinka," *Journal of Holocaust Research* 35, no. 1 (2021): 41–65.

97. Edward B. Westermann, "Stone-Cold Killers or Drunk with Murder? Alcohol and Atrocity during the Holocaust," *Holocaust and Genocide Studies* 30, no. 1 (2016): 1–19.

98. For "dirty work," see Everett C. Hughes, "Good People and Dirty Work," *Social Problems* 10, no. 1 (Summer 1962): 2–11, esp. 9.

99. On the Bulgarian Commissariat for Jewish Affairs' "entrainment" centers in Europe, see Geoffrey P. Megargee, Joseph R. White, and Mel Hecker, eds., *The United States Holocaust Memorial Museum Encyclopedia of Camps and Ghettos, 1933–1945*, vol. 3, *Camps and Ghettos under European Regimes Aligned with Nazi Germany* (Bloomington: Indiana University Press, 2018), 154. The bigram "concentration camp" more accurately describes British prison camps during the Boer War and the U.S. War Relocation Authority's so-called evacuation centers for Japanese Americans (not to be confused with internment camps for "alien enemies") under Executive Order 9066 during World War II.

100. Wiernik, *Year in Treblinka*, 26, 30. On Nazi freight cars, see Simone Gigliotti, "'Cattle Car Complexes': A Correspondence with Historical Captivity and Post-Holocaust Witnesses," *Holocaust and Genocide Studies* 20, no. 2 (2006): 256–77, esp. 272.

101. Mierzejewski, *Most Valuable Asset*, 119; Smith, *Treblinka Survivor*, 98.

102. Marek Rawecki, *Auschwitz-Birkenau Zone* (Gliwice: Wydawnictwo Politechniki Śląskiej, 2003), 13, 87.

103. Sereny, *Into That Darkness*, 51. Stangl himself signed many of the T4 death certificates representing fake causes of death.

104. Matthew S. McGlone, Gary Beck, and Abigail Pfiester, "Contamination and Camouflage in Euphemisms," *Communication Monographs* 73, no. 3 (2006): 261–82, esp. 271.

105. On the tortures now known as the Tango of Death, performed at the Janowska extermination camp in Poland, see Willem de Haan, *Tango of Death: The Creation of a Holocaust Legend* (Boston, MA: Brill, 2023).

106. Audre Lorde, "The Master's Tools Will Never Dismantle the Master's House," in *Sister Outsider: Essays and Speeches by Audre Lorde* (Trumansburg: Crossing, 1984), 110–13.

107. Edward Butscher recalls when Plath told the Scottish poet George MacBeth upon reading his poetry, "I see you have a concentration camp in your mind, too." Butscher, *Sylvia Plath: Method and Madness* (New York: Pocket Books, 1977), 335.

Chapter 13

1. [Ralph Waldo Emerson or Sophia Ripley?], "Painting and Sculpture," *The Dial*, July 1841, 79.

2. Literature has a stronger claim on representing enslavement's lived experience, and Toni Morrison, Saidiya Hartman, and Clint Smith have been important for me in problematizing monumentalisms of any kind, particularly for the unspeakable (Smith). The "diabolical" agnotology referred to in Proctor's introduction to this volume is currently floated by right-wing Christian nationalists who argue that enslavement was a method by which poorly skilled people were given "job training" for an enhanced future.

3. The author chooses to capitalize "White" as a way of marking the otherwise unmarked racial category of the dominant ethnic group; it will only be lowercase when describing the color. Similarly, "Whitening" is capitalized as a procedure enhancing the unmarked criteria and excisions required by Whiteness.

4. Powers's studio was located at 103 via dei Serragli, Florence.

5. Powers, on his intended composition, states, "The [slave sculpture] is of a young girl—nude, with her hands bound and in such a position as to conceal a portion of the figure, thereby rendering the exposure of the nakedness less exceptionable to our American fastidiousness. . . . That she is a Christian will be inferred by a cross, suspended by a chain. . . . I said a young girl, but the form will express puberty." Hiram Powers in correspondence with his patron, Colonel John Preston, Jan. 7, 1841, in Donald Martin Reynolds, *Hiram Powers and His Ideal Sculpture* (New York: Garland, 1977), 137–38.

6. Where I am concerned with Black and White abolitionist reception, Charmaine A. Nelson adduces other complexities of reception:

> Analysis of the reception of the *Greek Slave* must contend with at least two audiences—the "real," dominantly white, bourgeois, Victorian audience, who flocked to see it, and the "imagined" audience of nonwhite, nonwestern men whom Powers's racialized narrative implied. . . . The imaginary audience that the nineteenth-century viewer was compelled to contemplate was a disquieting rabble of lascivious Turks, a band of brown-skinned men pushing and crowding toward the market stage, where this chaste, moral, Christian, and let us not forget white, woman was about to be auctioned as a slave to the highest bidder. . . . The threat was not just of male sexuality but of black male sexuality. The immediate dichotomy of slave/master and colonized/colonizer, which still persisted within the contemporary practices of transatlantic slavery, was inverted and firmly reentrenched.

Nelson, *The Color of Stone: Sculpting the Black Female Subject in Nineteenth-Century America* (Minneapolis: University of Minnesota Press, 2007), 82. Whether Ottomans, Turks, or North African, the historical fact that some slave traders operating in Africa were historically "of color" is, of course, a complex issue relative to the situated discourses on race in multiple languages and traditions, beginning in the early modern period and extending into the nineteenth century.

7. Elizabeth Barrett Browning was an abolitionist who published "The Runaway Slave at Pilgrim's Point" (1849) as an exercise in empathy with the misery of an

enslaved woman. Although British, she explicitly addressed America through her poem "A Curse for a Nation," published in the Boston women's abolitionist journal the *Liberty Bell* in 1856.

8. First published as Elizabeth Barrett Browning, "Hiram Powers' Greek Slave," *Household Words*, Oct. 26, 1850; italic emphasis added.

9. The collector John Grant picked up the statue from the London docks on May 3, 1845. "It cleared customs on May 7 at 1 a.m.," and by 9 p.m., it was "installed on its pedestal in Graves's rooms at Pall Mall, the celebrated print publisher, whose facilities Powers and Grant considered particularly suitable for the exhibition of the *Slave*." Reynolds, *Hiram Powers*, 148.

10. Martina Droth, "Mapping *The Greek Slave*," *Nineteenth-Century Art Worldwide* 15, no. 2 (Summer 2016), http://www.19thc-artworldwide.org/summer 16/droth-on-mapping-the-greek-slave. See also Martina Droth and Michael Hatt, "*The Greek Slave* by Hiram Powers: A Transatlantic Object," *Nineteenth-Century Art Worldwide* 15, no. 2 (Summer 2016), http://www.19thc-artworldwide.org/sum mer16/droth-hatt-intro-to-the-greek-slave-by-hiram-powers-a-transatlantic-object.

11. Powers was granted several patents on his ingenious sculptural methods (rasps, duplication machine innovations), but there is no evidence that he was granted this intellectual property "patent." Scholar and curator Karen Lemmey has identified Powers's 1849 patent application but attests that no granted patent was found in her research. Lemmey, "Nineteenth-Century American Sculpture and United States Design Patents," in *Circulation and Control: Artistic Culture and Intellectual Property in the Nineteenth Century*, ed. Marie-Stéphanie Delamaire and Will Slauter (Cambridge: Open Book, 2021), 367–400. Further, it remains unclear in the scholarship *which* reproductions were subsequently authorized by Powers (the Minton company's, dating from 1848, could not have been—at least initially) and which replicas he profited from. The physical patent application is in the Hiram Powers Papers at the Archives of American Art, for which see *Greek Slave* patent, 1849, ID no. 17307, Hiram Powers Papers, 1819–1953, Archives of American Art, Smithsonian Institution, Washington, DC, https://www.aaa.si.edu/collections/ items/detail/greek-slave-patent-17307. For Lemmey's exhibition *Measured Perfection* at the Smithsonian American Art Museum, where this document was shown in 2015, see also Menachem Wecker, "The Scandalous Story behind the Provocative 19th-Century Sculpture 'Greek Slave,'" *Smithsonian Magazine*, July 24, 2015, https://www.smithsonianmag.com/smithsonian-institution/scandalous-story -behind-provocative-sculpture-greek-slave-19th-century-audiences-180956029/.

12. Much of the scholarship expanding Black agnotology into art history has appeared in the new millennium. For a partial account of the reception of Powers's sculpture, expanded upon here, see Caroline A. Jones, *The Global Work of Art: World's Fairs, Biennials, and the Aesthetics of Experience* (Chicago: University of Chicago Press, 2016).

13. Agnotology itself can be critiqued as a "universalist" project imagining a single discernable truth—but the current volume's editors have strenuously worked to complicate that view. I do want to cite the postcolonial thinking of art historian Sarah Rifky (who received her PhD from MIT in 2024):

In the spirit of making "equal" by eradicating blindspots or correcting ignorance, the aspirations of this Agnotology can be met with common critiques of
postwar liberalism; how can the implied universalism of a single body of knowledge corrected by Agnotology not be "moral"? How can the argument evolve
to not undermine the institutional inequalities of its own operation? How can
the project more pronouncedly shift from an "assimilationism" (thinking here
about the canon) to the mottled distribution of different kinds of knowledge,
differently local? The challenge is a subtle one that I sense needs to contend with
the desire of impartiality of art history, as one part of the broader project within
which this fits.

Rifky, personal communication to the author, Aug. 12, 2019.

14. *Powers' Statue of the Greek Slave* (New York: Craighead Printers, 1847). I
consulted the copy of this rare pamphlet in the Harvard University Libraries.

15. Droth states, "Much of the scholarship in the past few decades has demate-
rialized *The Greek Slave* into its iconography. The enormous and sustained interest
in the statue revolves around its multivalence as an image, not its intrinsic sculptural
qualities. The primary context in which we encounter the statue today is the history
of slavery, abolition, and the American Civil War. . . . But the nature of this attention creates a conundrum, for it means the importance given to *The Greek Slave* is
only oblique; essentially, it is a tool, or channel, to get at other subjects." Thus, like
many art historians, even the nuanced Droth wants to pull us away from reception
within the context of slavery to reemphasize the formal and material qualities of the
sculpture. Droth, "Mapping *The Greek Slave*."

16. Charles W. Mills, *Black Rights / White Wrongs: The Critique of Racial Lib-
eralism*, online ed. (Oxford: Oxford Academic, 2017), abstract for chap. 4, "White
Ignorance," https://academic.oup.com/book/5106/chapter-abstract/147688449.
This chapter is a further development of Mills, "White Ignorance," in *Agnotol-
ogy: The Making and Unmaking of Ignorance*, ed. Robert N. Proctor and Londa
Schiebinger (Stanford, CA: Stanford University Press, 2008), 230–49.

17. For the John Tenniel, see *Punch* 20 (London: Bradley and Evans, 1851), 236.
The sarcastic question is posed earlier in the same volume, *Punch* 20 (1851), 209. See
also Lisa Marie Volpe, "On Display: American Photography at Three Nineteenth-
Century World's Fairs" (PhD diss., University of California, Santa Barbara, 2013),
9. Tenniel's drawing, titled *The Virginian Slave: Intended as a Companion to Power's* [*sic*] *"Greek Slave,"* was published in late May or June 1851. (Some sources say
June 7, others say May, and the journal bears no inscription as to the precise date.)

18. One of the first historians to attempt to theorize African subjectivities within
the fair was performance theorist Lisa Merrill. See Merrill, "Exhibiting Race 'under
the World's Huge Glass Case': William and Ellen Craft and William Wells Brown at
the Great Exhibition in Crystal Palace, London, 1851," *Slavery and Abolition* 33,
no. 2 (June 2012): 321–36.

19. Merrill, "Exhibiting Race."

20. William Wells Brown, *Three Years in Europe, or Places I Have Seen and
People I Have Met, by W. Wells Brown, a Fugitive Slave, with a Memoir of the
Author by William Farmer, Esq.* (London: Charles Gilpin, 1852), 225.

21. William Farmer, "Fugitive Slaves at the Great Exhibition," *The Liberator*, July 18, 1851, Fair Use Repository, https://fair-use.org/the-liberator/1851/07/18/the -liberator-21-29.pdf. See also Stephen Knadler, "At Home in the Crystal Palace: African American Transnationalism and the Aesthetics of Representative Democracy," *ESQ: A Journal of the American Renaissance* 56, no. 4 (2011): 328–62.

22. Farmer, "Fugitive Slaves."

23. There is some evidence that at some point in the positioning of the sculpture at the Great Exhibition, it was placed behind a protective circular railing, keeping touching hands at bay. See the illustration in *"The Greek Slave,"* in *The Illustrated Exhibitor: A Tribute to the World's Industrial Jubilee; Comprising Sketches by Pen and Pencil, of the Principal Objects in the Great Exhibition of the Industry of All Nations* (London: John Cassele, 1851), 37. Yet this feature appears in no other known illustrations of the Great Exhibition display conditions. Whether these railings were temporary or a later development is unclear; for certain, any such barriers are directly contradicted by Farmer's account of their access to the niche during the June 21 visit. Certainly, those who had seen the sculpture just six years earlier in John Grant's London display would have had the option to rotate the sculpture to their own satisfaction—an amazing level of interactivity that has yet to be integrated into the emerging performance histories around the *Slave*.

24. See Teresa Zackodnik, *The Mulatta and the Politics of Race* (Jackson: University of Mississippi Press, 2004), 49. The group was reported to have included the British abolitionists "George and Jenny Thompson, Richard and Maria Webb, and William Farmer" in period documents. Lisa Merrill includes John Bishop Estlin. Merrill, "Exhibiting Race," 335n4.

25. The engraving of Ellen Craft is a public domain digital asset of the U.S. National Park Service, made available from the Boston African American National Historic Site, accessed Feb. 9, 2025, https://npgallery.nps.gov/AssetDetail/2326a9f6 -659f-4f9e-8fbc-babdc843608c. The engraving was produced in Boston by J. Andrews and S. A. Schoff after Luther H. Hale's daguerreotype. Titled *Ellen Craft, the Fugitive Slave*, it was widely distributed among transatlantic abolitionists and was published as the frontispiece in William Craft, *Running a Thousand Miles for Freedom; or, the Escape of William and Ellen Craft from Slavery* (London: William Tweedie, 1860).

26. Volpe, "On Display," 39. See also Barbara McGaskill, "'Yours Very Truly': Ellen Craft—the Fugitive as Text and Artifact," *African American Review* 28, no. 4 (1994): 509–30.

27. Houston A. Baker Jr., *Workings of the Spirit: The Poetics of Afro-American Women's Writing* (Chicago: University of Chicago Press, 1991), 13. See also Zackodnik, *Politics of Race*, 50.

28. Zackodnik, *Politics of Race*, 49.

29. Saidiya Hartman, "Venus in Two Acts," *Small Axe* 12, no. 2 (2008): 11.

30. This German apparatus of critical theory emerged from left-wing literary studies in the 1960s but more broadly came to designate domains of cultural reception for visual art, film, music, and all manner of artistic forms. It should be sharply distinguished from the biblical study of textual-reception history under the same name. For one influential example of the modern mediatic usage, see Peter Uwe Ho-

hendahl, ed., *Sozialgeschichte und Wirkungsästhetik: Dokumente zur empirischen und marxistischen Rezeptionsforschung* (Frankfurt: Athenäum-Fischer, 1974).

31. Reproductions generally substituted metal links painted white.

32. This positionality of "not of our world" is an ideological production. In a recent research finding, Powers is revealed to have actually used body casts to determine the configuration of the sculpture's left arm—very much "of this world" and a shocking practice that would have been fully concealed from collectors in the nineteenth century as it was considered "cheating" on the sculptor's craft. For Lemmey's exhibition displaying the body cast from Powers's studio in *Measured Perfection*, see Wecker, "Scandalous Story."

33. E. D. W. M'Kee, "Excerpta—No. III. Aesthetic Education, or Moral Uses of Art," *Christian Parlor Magazine*, May 1, 1853, 167–68, quoted in Droth, "Mapping *The Greek Slave*."

34. Saidiya V. Hartman, *Scenes of Subjection: Terror, Slavery, and Self-Making in Nineteenth-Century America* (New York: Oxford University Press, 1997).

35. Linda Williams, *Playing the Race Card: Melodramas of Black and White from Uncle Tom to O. J. Simpson* (Princeton, NJ: Princeton University Press, 2001), https://doi.org/10.1515/9780691201337.

36. Per Merrill, "Exhibiting Race," 335. Although, Zackodnik's scholarship does not place him there. Zackodnik, *Politics of Race*.

37. John Bishop Estlin to Eliza Wigham, letter, May 3, 1851, p. 4, Digital Commonwealth, https://www.digitalcommonwealth.org/search/commonwealth:2z10z690b; emphasis original. Estlin continues that he has been attempting to secure a better position for Ellen in the events—an effort he reports as pleasing her. Discussed by both Merrill and Zackodnik, the Estlin letter is held by the Boston Public Library. Merrill, "Exhibiting Race"; Zackodnik, *Politics of Race*.

38. Franklin assumed the presidency of the Pennsylvania Abolition Society in 1787 and actively circulated Josiah Wedgwood's medallions of shackled penitent slaves, yet had himself owned and profited from a series of enslaved persons between 1735 and 1781. For general background, see historian David Waldstreicher, "Benjamin Franklin, Slavery, and the Founders: On the Dangers of Reading Backwards," *Common Place: The Journal of Early American Life* 4, no. 4 (July 2004), https://commonplace.online/article/benjamin-franklin-slavery/.

39. Per Winckelmann, "A state of stillness and repose both in man and beast" is most conducive to the elevated forms of beauty. Johann Joachim Winckelmann, *Geschichte der Kunst des Alterthums* (1764), quoted in Winckelmann, *The History of Ancient Art*, trans. G. Henry Lodge (Boston, MA: J. R. Osgood, 1873), 2:113. See also the recent edition Winckelmann, *History of the Art of Antiquity*, intro. Alex Potts, trans. Harry Francis Mallgrave (Los Angeles: Getty Research Institute, 2006).

40. For a translation and commentary on Winckelmann's *Art of Ancient Greece* (only one part of the larger 1764 work *History of the Art of Antiquity*), see Joseph Comyns Carr, *Essays on Art* (London: Smith, Elder, 1879).

41. Williams, *Playing the Race Card*, 30.

42. The important scholarship of Martin Bernal is apposite here. See Bernal, *Black Athena: The Afro-Asiatic Roots of Classical Civilization* (Newark: Rutgers University Press, 1987).

43. Hiram Powers to Elizabeth Barrett Browning, letter, Aug. 7, 1853, quoted in Reynolds, *Hiram Powers*, 254.

44. A place for polychromy is being made in museums via modern archaeological reconstruction. The efforts of Frankfurt-based Professor Dr. V. Brinkmann, head of the Department of Antiquity at the Liebieghaus Skulpturensammlung, and his collaborator Dr. U. Koch-Brinkmann aim to reanimate polychromy with garish acrylic pigments on casts. To my mind and eye, these contributions are even worse than agnosis since they substitute flat plastic colors for the minerals and earth pigments rubbed into stone that constituted the historical mode of decoration in Greece. See the very recent celebration of these simulations at the Metropolitan Museum of Art in New York, *Chroma: Ancient Sculpture in Color*, an exhibition mounted in 2022–2023. "Exhibition— *Chroma: Ancient Sculpture in Color*," The Met, accessed Sept. 5, 2025, https://www.metmuseum.org/exhibitions/chroma.

45. Jones, *Global Work*, 70.

46. See the documentation from the exhibition *Sculpture Victorious: Art in an Age of Invention, 1837–1901*, Yale Center for British Art, New Haven, Connecticut, 2014, curated by Martina Droth, Jason Edwards, and Michael Hatt. "Exhibition— *Sculpture Victorious: Art in an Age of Invention, 1837–1901*," Yale Center for British Art, accessed Sept. 5, 2024, https://britishart.yale.edu/exhibitions-programs /sculpture-victorious-art-age-invention-1837-1901. In one gallery, the curators staged a confrontation between American sculptor Hiram Powers's *The Greek Slave* (1847) marble sculpture (from a collection at Newark Museum, Newark) and British sculptor John Bell's *The American Slave (A Daughter of Eve)* (ca. 1862), a bronze patinated electrotype with silver and gold plating (from the Armstrong Collection, National Trust, Cragside, Rothbury, Northumberland). As photographed by Nick Mead in 2014, the confrontation was very successful.

47. This material designation occurs in public lectures by curators Sarah Cash (National Gallery of Art), exhibition statements by curator Karen Lemmey (Smithsonian American Art Museum), and in the close analysis of art historian R. Tess Korobkin, on which see Korobkin, "*The Greek Slave* and Materialities of Reproduction," *Nineteenth-Century Art Worldwide* 15, no. 2 (Summer 2016), http://www .19thc-artworldwide.org/summer16/korobkin-on-the-greek-slave-and-materialities -of-reproduction. Without actual material analysis of the Douglass statuette, we will not know for sure.

48. Korobkin, "*The Greek Slave*." Korobkin is here citing the scholarship of Nelson, *Color of Stone*, 105; and Vivien M. Green [Fryd], "Hiram Powers's *Greek Slave*: Emblem of Freedom," *American Art Journal* 14 (Autumn 1982): 31–39. Dr. Fryd revisited her 1982 article in 2016, for which see Vivien Green Fryd, "Reflections on Hiram Powers's *Greek Slave*," *Nineteenth-Century Art Worldwide* 15, no. 2 (Summer 2016), http://www.19thc-artworldwide.org/summer16/fryd-on-reflec tions-on-hiram-powers-greek-slave.

49. Robert S. Levine, *The Lives of Frederick Douglass* (Cambridge, MA: Harvard University Press, 2016), 226.

50. "Truth compels me to admit even here in the presence of the monument we have erected in his memory, Abraham Lincoln was not, in the fullest sense of the word, either our man or our model. In his interests, in his associations, in his

habits of thought, and in his prejudices, he was a white man. He was preeminently the white man's President, entirely devoted to the welfare of white men." Frederick Douglass, "The Freedmen's Monument to Abraham Lincoln: An Address Delivered in Washington, DC, on 14 April 1876," in *The Frederick Douglass Papers*, ser. 1, *Speeches, Debates, and Interviews*, vol. 4, ed. John W. Blassingame et al. (New Haven, CT: Yale University Press, 1979–1992), 431–33, quoted in Levine, *Lives of Frederick Douglass*, 228n59.

51. Frederick Douglass, letter to the editor, *National Republican*, April 19, 1876, quoted in "Emancipation Statue," National Park Service (website), last modified Feb. 14, 2021, https://www.nps.gov/cahi/learn/historyculture/emancipation-statue.htm.

52. Browning, "Hiram Powers' Greek Slave."

53. Caroline A. Jones and Joseph Leo Koerner, "Contamination | Purification," in *Contamination and Purity in Early Modern Art and Architecture*, ed. Lauren Jacobi and Daniel Zolli (Amsterdam: Amsterdam University Press 2021), 315–60.

Chapter 14

1. Quoted in Brett T. Litz et al., *Adaptive Disclosure: A New Treatment for Military Trauma, Loss, and Moral Injury* (New York: Guilford, 2016), 30.

2. Litz et al., *Adaptive Disclosure*, 29. On cultural psychiatry, see Laurence J. Kirmayer and Harry Minas, "The Future of Cultural Psychiatry: An International Perspective," *Canadian Journal of Psychiatry* 45, no. 5 (2000): 438–46.

3. Nadia Abu El-Haj, *Combat Trauma: Imaginaries of War and Citizenship in Post-9/11 America* (New York: Verso Books, 2022), from which this essay is drawn. I have chosen to use the pronoun "he" when referring to the soldier. While there are many women who serve in the military today, it remains dominated by men. More important given the military's hypermasculinity as an institution, I use the male pronoun to index that masculinity regardless of whether the individual soldier is male or female.

4. Wendy Brown, *States of Injury: Power and Freedom in Late Modernity* (Princeton, NJ: Princeton University Press, 1005), xii.

5. Christian G. Appy, *American Reckoning: The Vietnam War and Our National Identity* (New York: Viking, 2015), 27.

6. "Kelly Defends Trump's Handling of Soldier's Death and Call to Widow," *New York Times*, Oct. 19, 2017. https://www.nytimes.com/2017/10/19/us/politics/statement-kelly-gold-star.html.

7. Helen M. Kinsella, *The Image before the Weapon: A Critical History of the Distinction between Combatant and Civilian* (Ithaca: Cornell University Press, 2011), 29.

8. Didier Fassin and Richard Rechtman, *The Empire of Trauma: An Inquiry into the Condition of Victimhood*, trans. Rachel Gomme (Princeton, NJ: Princeton University Press, 2009).

9. Fassin and Rechtman, *Empire of Trauma*, 1.

10. David Kieran, *Signature Wounds: The Untold Story of the Military's Mental Health Crisis* (New York: New York University Press, 2019).

11. Fassin and Rechtman, *Empire of Trauma*; Ben Shephard, *A War of Nerves: Soldiers and Psychiatrists in the Twentieth Century* (Cambridge, MA: Harvard

University Press, 2001); Judith Lewis Herman, *Trauma and Recovery* (New York: Basic Books, 1992); Herb Kutchins and Stuart A. Kirk, *Making Us Crazy: DSM; The Psychiatric Bible and the Creation of Mental Disorders* (New York: Free Press, 1997); Ruth Leys, *Trauma: A Genealogy* (Chicago: University of Chicago Press, 2000).

12. David Finkel, *The Good Soldiers* (New York: Farrar, Straus and Giroux, 2009), 205–6.

13. Viet Thanh Nguyen, *Nothing Ever Dies: Vietnam and the Memory of War* (Cambridge, MA: Harvard University Press, 2016), 15.

14. Nancy Sherman, *Afterwar: Healing the Moral Injuries of Our Soldier* (New York: Oxford University Press, 2015), xiii–xiv.

15. The concept of trauma borne of perpetration is not new (nor, for that matter, is the name "moral injury"), although it has often been narrated as such in the post-9/11 era. The psychological and moral pain wrought by perpetration was central to what was initially named "post-Vietnam syndrome" by radical antiwar psychiatrists working with antiwar veterans of the American war in Vietnam. See Abu El-Haj, *Combat Trauma*, chaps. 1, 4.

16. David Wood, "The Grunts: Damned If They Kill, Damned If They Don't," *Huffington Post*, Mar. 18, 2014. For Wood's essay on moral injury, see Wood, " 'I'm a Good Person and Yet I've Done Bad Things': A Warrior's Moral Dilemma," *Huffington Post*, accessed Sept. 5, 2024, https://projects.huffingtonpost.com/projects/moral-injury.

17. Wood, "Grunts."

18. Tyler E. Boudreau, "The Morally Injured," in *War and Moral Injury: A Reader*, ed. Robert Emmet Meagher and Douglas A. Pryer (Eugene: Cascade Books, 2018), 56.

19. It is worth emphasizing that even though most soldiers never engage in or see combat even among those deployed to war zones, this combat-trauma imaginary relies on the figure of the "grunt"—that is, the combat soldier on the front lines.

20. Annette Wieviorka, *The Era of the Witness*, trans. Jared Stark (Ithaca: Cornell University Press, 2006), 83. Interestingly, trauma doesn't figure in Hannah Arendt's account of Eichmann in Jerusalem. Arendt, *Eichmann in Jerusalem: A Report on the Banality of Evil* (New York: Penguin, 2006).

21. Yuval Noah Harari, *The Ultimate Experience: Battlefield Revelations and the Making of Modern War Culture, 1450–2000* (New York: Palgrave Macmillan, 2008), 1, 7, 299, 20.

22. Catherine Lutz, *Homefront: A Military City and the American Twentieth Century* (Boston, MA: Beacon, 2001), 238.

23. Roy Scranton, "The Trauma Hero: From Wilfred Owen to 'Redeployment' and 'American Sniper,' " *Los Angeles Review of Books*, Jan. 15, 2015.

24. PTSD workshop (Kansas City, MO, Oct. 2015), notes by author.

25. David Wood, *What Have We Done: The Moral Injury of Our Longest Wars* (New York: Little, Brown, 2016), 264.

26. Sherman, *Afterwar*, 3.

27. Joan W. Scott, "The Evidence of Experience," *Critical Inquiry* 17, no. 4 (1991): 776–77.

28. Amanda Anderson, *The Way We Argue Now: A Study in the Cultures of Theory* (Princeton, NJ: Princeton University Press, 2005).

29. Leys, *Trauma*.

30. Zoë Wool, *After War: The Weight of Life at Walter Reed* (Durham, NC: Duke University Press, 2015), 192.

31. Walter Benn Michaels, "'You Who Never Was There': Slavery and the New Historicism, Deconstruction and the Holocaust," *Narrative* 4, no. 1 (Jan. 1996): 1–16.

32. Theater of War Productions, dir. Brian Doerries, Columbia University, New York, NY, Nov. 6–7, 2019, notes by author.

33. Theater of War Productions, notes.

34. Theater of War Productions, notes.

35. Theater of War Productions, notes.

36. Hannah Arendt, *Responsibility and Judgment* (New York: Schocken Books, 2003), 59.

37. Allan Young, "The Self-Traumatized Perpetrator as a 'Transient Mental Illness,'" *L'Evolution Psychiatrique* 67, no. 4 (2002): 630–50.

38. Nguyen, *Nothing Ever Dies*, 224.

Index

abortion, 46–48, 54–55
abolition and abolitionists, 192–202, 205
"A calorie is a calorie": myth of, 136–39, 142–43, 146
accounting: agnogenic, 150–54
ag-gag laws, 8, 10, 86, 99
agnometrics, 7, 19–20
Agnotocene, 7, 21, 47
agnotology: adjectival, 12–13; algorithmic, 13–15, 57–59; animal, 8, 149–55; in art history, 189–93, 202; of atrocity, 185–86; attribution, 231n38, 232n57; Black, 191–97, 205–208; "blind in one eye," 21; bodily or sensory, viii, 3, 8–9, 139; Bruno Latour on, 7; cartographic, 13, 163–65; coined, 1, 7; diabolical, 8, 278n2; disaster-induced, 75; food industry, 142–44, 147–59; gun lobby, 12, 114–32; internal, 6; left-wing, 20–21, 54; ostrich, 10; pyro-, 14; structural, 8, 54–55; violent, 83; virtuous, 9, 229n26; Wikipedia entry for, 7
Alito, Justice Samuel A., 127–29
American Petroleum Institute (API), 5, 24, 35
Anthropocene, 46–47
anti-epistemology, 16, 85–99
antisemitism, 4, 161, 170, 272n7

Arendt, Hannah, 173, 220
arctic: ice-free, 30
artificial intelligence (AI): fuels misinformation, viii, 57–70; generative, 15, 58, 61–65; and pornography, 14; predictive, 58–61. See also social media
atrocity: agnotology of, 18, 185–86, 230n37; euphemized, 185; language of, 186
Auschwitz, 4, 172–75, 180–84

Balfour Declaration, 162, 170
banks and banking: Adam Smith on, 105–106; defrauded by deep fakes, 65
Berger, Peter L., 4
Big Ag, 86, 155. See also livestock
Big AI, 68–69
Big Carbon, 11, 33–37
Big Government, 101
Big Meat, 17, 147–59
Big Oil, 34–40, 67
Big Pharma, 2, 141, 158
Big Sugar, corrupts research, 8, 17, 142–44
Big Tech, 67–69
Big Tobacco: allies on U.S. Supreme Court, 5, 228n10; distraction science, 4, 67, 148, 227–228n8; "Doubt is our product," 1–4, 97;

Big Tobacco (*cont.*)
  Harvard corrupted by, 2; historians
    on payroll of, 4; invents personal
    choice, 16, 141–45; playbook for Big
    Carbon, 38, 67
biodiversity, vii, 15, 20, 42–47, 50
biofuels, 40–41, 53
births, unwanted, 8, 42–45, 53, 56,
    236n2. *See also* forced birthing
Black, James, 28–30
Boal, Iain, 6
Bolinger, Dwight, 174
British Petroleum (BP, Beyond Petro-
    leum), 35–39
Brown & Williamson, 1, 7
Brown, William Wells, 193
Browning, Elizabeth Barrett, 189–191,
    198, 202, 205, 207
Burger, Chief Justice Warren, 116, 118,
    120

Cairo consensus, 49–55
Canary Mission, 168
carbon capture and geoengineering, 10,
    31–41
carbon footprint, 11, 39
catarheumatics, 11, 33, 55, 227n1
"causes of causes," 47, 230n34
Cerf, Vint, internet evangelist, 78
CERN, 73–75
Charney report, 25, 34
ChatGPT, 62–67
checkerboarding, 90–91
Chevron Corporation, 5, 35–39, 55
child marriage, 8, 42–43, 51
Christian, Jon, 13
Christianity, evangelical, 1
cigarettes, 2, 3–7, 12–13, 19–20, 115;
    and bodily agnotology, viii; delaying
    end of, 41; "filtered," 20, 41; made
    at Auschwitz, 174; peak consump-
    tion of, 3; as uncaused cause and
    unmoved mover, 5
civilians: theorized, 214–15
climate change: Big Meat obscures,
    147–55; birthing and, 8, 14, 42–

56; catarheumatics of, 33, 55; cows
    blamed for, 17; denial, vii–viii,
    18–19, 23–41; denial abandoned,
    41; Exxon's scenarios for, 28–31;
    as hoax, 60; lobbyists corrupt re-
    search on, 228n9; most important
    fact about, 23–41; pseudo-solutions
    for, 14, 37–41. *See also* denial and
    denialism
cloud storage, 75–78, 83
Coca-Cola, 17, 143–44
coercive conception, 8–9, 14–15, 55–56
Cohn, Carol, 174
complexity: weaponized, 14, 23, 40,
    228n12
conspiracy theories, 19, 59–60, 67,
    232n54
content moderation, 7
contraception: availability, 54; enlarges
    freedom, 8–9, 14–15, 42–44, 53–56;
    funding for, cut, 50; ignored by the
    IPCC, 42–46; not considered part of
    technology, 45–46
Council for Tobacco Research, 4
COVID-19, 81, 130, 140
Craft, Ellen and William, 195–202,
    208
cyber attacks, 75

deepfakes: of the body, 9–10; defined,
    61–65; in disinformation campaigns,
    65; eroding public trust, 7, 15, 65–
    66; face-swap, 63; in filmmaking,
    64–65; puppet-master, 63; weapon-
    ized by fraudsters, 65. *See also* liar's
    dividend
denial and denialism: cigarette, vii,
    2–5, 41, 141; climate change, vii–
    viii, 18–19, 23–41; gun violence, 8,
    13, 114–32; Holocaust, vii, 12, 177;
    interpretive, 177; vs. false solutions,
    19, 41, 227n8
DeSmog, 155
Dewey Decimal System, 74
diabetes, 16–17, 143–45
digital dark age, 78,

distraction research, 4, 18, 23, 41, 67,
  227–28n8. *See also* pseudo-solutions
Doerries, Brian, 218–20
"Doubt is our product," 1–2, 97
Douglass, Frederick, 195, 202–208
drapetomania, 11
Droth, Martina, 190–191, 198, 202

eavescasting, 6, 227n1
Egg Freezing Ambassador Program, 52
Edelman (PR giant), 40, 150
Edwards, Paul N., 80
Ehinger, Adolf, 82
Ehrlich, Paul R., 50–55
endarkenment, ix
Energy Engaged Millennials (EEMs), 40
Epimetheus, 11
epistemology, vii, ix, 2, 16, 85, 96–99,
  175, 210–17
Estlin, John Bishop, 199–201, 282n37
eugenics, 4, 14, 48
euphemisms: obfuscating, 6, 10–11, 17,
  171–86, 230n29
evidentiary bar raising, 227n3
exabyte era, 15, 73–75
ExxonMobil Corporation, 30–40, 55,
  148
Eysenck, Hans, 4

Facebook, 19, 57–59, 68, 170
Farmer, William, 193–95
Final Solution, 172
Fisher, Sir Ronald A., 3
forced birthing, 42–44, 50–56
forced sterilization, 48, 51
forever wars, 18, 211, 221
forgetting: caused by corporate agendas, 81; in digital societies, 71–72,
  81; as missed opportunity, vii; of the
  Nakba, vii, 17; as survival strategy,
  229n26
fossil gas, 39–40
Franz, Kurt, 175–82
Frederiksen, Mette, 81
free-market fundamentalism, 16, 54,
  99–113

gascars, 44
generative adversarial network (GAN),
  62
generative AI. *See* artificial intelligence
  (AI)
Global Climate Coalition, 37
global warming. *See* climate change
Google, 7 13, 75–78
Great Exhibition (London), 189–93,
  198–202, 281n23
Greek Slave, The (sculpture), 187–208,
  280n11
greenwashing, 37–40
gun lobby, 16, 114–32, 256n18, 258n40
guns and gun violence: denial, 8, 13,
  114–32; Dickey Amendment bans
  research on, 257n36; in Japan, 16,
  117; John Lott's defense of, 116, 124;
  NRA on, 116; Right to Carry (RTC)
  laws, 117–18, 121–29; statistics on,
  116–17; Supreme Court on, 122–24,
  132. *See also* Second Amendment

Hartman, Saidiya V., 196–98
Harvard University: co-opted by Big
  Meat, 147–48; co-opted by Big Oil,
  38; co-opted by Big Sugar, 143;
  co-opted by Big Tobacco, 2, 148,
  227n3; parochial, 1–3; smoking allowed indoors at, 3
Hayek, Friedrich von, 104, 107, 113
Heritage Foundation: Project 25, ix
Hicks, Larry, 89
Himmler, Heinrich, 173, 178
historians: corrupted, 4
Hitler, Adolf, 172, 177, 184
Holocaust: denial, vii, 12; euphemized,
  185; Jewish state as compensation
  for, 166–67; memories of and me-
  morials to, vii, 17, 51, 163; survivor
  as witness, 215–16; in trauma genre,
  167. *See also* Treblinka
"Holy Land," 162
Hoover Institution: supports climate
  deniers, 148
Huff, Darrell, 4, 115

ignorance: bodily, viii, 3; civilian, 209–22; cognitive, 134; emotional, 134; generated by education, 20; ignored, 1–2; linguistic aspect, 10; personal vs. social, 1, 19; popular, 1; psychology of, 18–19; as social product, 19; virtuous, 9; white, vii, 18, 160, 187–97, 201, 206–208; willful, 72
ignorance studies, 7, 9, 18
individuation and invisibilization, 5–6, 22. *See also* personal choice and personal responsibility
indivisibility thesis, 101–103
Intergovernmental Panel on Climate Change (IPCC): corrupted by Big Carbon, 37; ignores corporate connivance, 5; ignores human reproduction, 42, 45, 49–50; ignores religion, 20
International Institute for Applied Systems Analysis (IIASA), 25–33
International Petroleum Industry Environmental Conservation Association (IPIECA), 35
Internet Archive, 71, 74

Kahle, Brewster (Internet Archive founder), 74
Kelly, John, 210–11
Keys, Ancel, 4
Kinsella, Helen M., 211
Knisely, Steve, 28–31
Korobkin, R. Tess, 204–206

Latour, Bruno, 7
Laurmann, John A., 25–29
lead and lead poisoning, 8, 11, 133, 140
liar's dividend, 15, 66
Lincoln, President Abraham, 205, 284n50
link rot, 15, 79
livestock: methane emissions from, 17, 148–55
Lott, John R., Jr., 114–16, 122–24, 128

male birth control, 45
market penetration time concept, 25–29
Marlboro cigarettes, 12, 143, 228n10
Massachusetts Institute of Technology (MIT), 38
mass shootings, 114, 130–32
memoricide, 17, 161–65. *See also* forgetting
metrics: misleading, 19–20, 114–17
Michaels, David, 8
Mills, Charles W., 18, 100, 160, 192
mimicry, 8, 229n21
misinformation: algorithmic amplification of, viii, 15, 57–70; climate, 154–55; correspondents, 13; gun safety, 120; judicial, 130; like dirty water and air, 21; social media and, 7, 232n56.
Mitloehner, Frank, 17, 148–58
Mouse House Massacre, 6
Musk, Elon, 51, 65–66, 84

Nakba, 12, 17, 160–69, 269n1
National Cattlemen's Beef Association (NCBA), 17, 149–150
National Rifle Association (NRA), 116–29, 255n7, 258n40
Nazis and Nazism: Black, 21; collapse of, 48; euphemisms, 180; fake railway station at Treblinka, 171–86; radio, 98; shredders used to resist, 15, 82; stagecraft, 2, 18, 172–86; used as excuse to bury Nakba, 166
Neo-Malthusianism, 42, 50–54
Nixon, President Richard M., 3, 6
Nobel laureates, 4

Obama, President Barack, 62–63, 70
obesity: vii, 12, 16, 39, 136–37; adult, 137; childhood, 136, 138; fetal, 17, 138; misconceived as gluttony and sloth, 133; personal responsibility for, 133, 135, 139–140
Orwell and Orwellian, 1, 164, 169, 173–74, 230n30

Pepsi, 17, 144

personal choice and personal responsibility: Big Tobacco's invention of, 16, 140–42; used to exculpate Big Carbon, 11, 39–41; used to exonerate cigarettes, 3–6, 16; used to exonerate sugar, 16–17, 39, 133–46

Philip Morris, 4–5, 6, 143, 156, 174, 230n28

phonaesthemics, 230n28

planetary health, 22, 28, 50, 56

Plaza Hotel: cigarette conspiracy launched at, 141, 232n55

policing, lexical, 6

population collapse, 51–52, 239n46

population control, 14, 48–49

pornography, 14, 140

Powell, Lewis, 228n10

pregnancies: rape-related, 12, 14, 42–44, 47, 51; unintended, 43–46, 50, 53, 236n2

Princeton University: oily research at, 37–38

Prometheus, 11

pronatalism, 43, 50–52

pseudo-solutions, 14, 37–41, 147, 228n8

PTSD, 212–13, 216–20

Putin, Vladimir, 51, 256n18

QAnon, 59–60, 240n7

race and racism: in art history, 187–97, 201–208; in Balfour Declaration, 170; in the Middle East, 165–70; and neo-Malthusianism, 54; George J. Stigler's, 252n36; in textbooks, 165

Ratner, Jonathan, 86–89, 247n11

Reagan, President Ronald, 1, 35, 48, 101, 116, 120

reforestation, 14, 36–38

R. J. Reynolds, 6

reproductive coercion, 43–45

reproductive liberty, 8, 14, 46–54,

reproductive technology, ignored, 45–47, 55

Rosner, David, 230n32

Scalia, Justice Antonin, 120–21

Schelling, Thomas, 33–34

Second Amendment, 16, 120–31

secrecy: anti-epistemological, 86; Babylonian, 85; and censorship, 97; governmental, 81, 86; military, viii, 85–86; shredding to protect, 15, 82; at Treblinka, 173

Selye, Hans, 3–4

Shannon, Claude E., 96, 99

Shell, Royal Dutch, 31, 35, 38, 55

Shook, Hardy & Bacon, 6

shredders: paper, 15, 82

Silver Tsunami, 51

slavery: abolitionist opposition to, 199–202; African, 200–202, 279n6; and drapetomania, 11; as freedom, 230n30; industrial work like, 103; Jewish, 178–82; as "job training," 278n2; sexual, 10, 185; and slave traders, 278n6; white, 18, 189, 201. *See also* Frederick Douglass; *The Greek Slave*

Smith, Adam, 99–113

Smithson, Michael, 230n27

social media, 7, 21, 58, 71

solar geoengineering, 10, 14, 37

Stanford University: compromised by Big Oil, 37–38; doesn't have a school of public health, 10; Energy Modeling Forum, 38; gets an F, 5; Native American sites, 13; Natural Gas Initiative, 28; Precourt Center, 38; Strategic Energy Alliance, 38

Stigler, George J., 103–13, 252n36

strategic ephemerality, 72, 81, 84

stress, 3–4

StyleGAN, 62

sugar: addictive, 139–42; blood vs. dietary, 142; causes diabetes, 143; causes obesity, 15–17, 143; conspiracy, 8; corrupts research, 142–44; as "deepfake of the body," 9; euphemisms for, 10; vs. fat, 142. *See also* Big Sugar

Sugar Research Foundation, 143

Taliban, 10, 217
Theater of War, 218–20
Tobacco Institute, 7
TotalEnergies (oil company), 24, 35, 38
trauma: combat, 18, 185, 209–21,
    286n19; intergenerational, from en-
    slavement, 197; Nazi killers suffer,
    173
trauma genre: exclusion of Palestine
    from, 167
Treblinka, 17–18, 171–86
trespass laws, 16, 86–99
Trump, President Donald J., ix, 13, 23,
    62, 48, 82, 84, 124, 130–31, 255n7
Tutu, Desmond, 11
Twain, Mark, 21, 162

ultraprocessed foods, 138–40, 144
United Nations Framework Convention
    on Climate Change: corrupted by
    lobbyists, 228n9; ignores fossil fuels,
    36; ignores population, 46
United Nations International Confer-
    ence on Population and Develop-
    ment (ICPD), 49–50
United National Refugee and Relief
    Works Agency (UNRWA), 161, 165,
    269n1
University of California, Davis, co-
    opted by Big Meat, 154–55, 158

vaccine hesitancy, 7
Venus Pudica, 189
Venus of Willendorf, 136
Vitamin M, 6
Volpe, Lisa, on abolitionist photogra-
    phy, 193, 196

Whitening of art history discourse, 18,
    191–92, 202, 278n3
Williams, Linda, 199
woke algorithms, 13–14, 21
Wood, David, 214, 217

Yad Vashem, 17, 163
YouTube, 57–61, 80

Zeus, 11, 141